Warning and Disclaimer

This book is designed to provide tutorial information about Electronics Workbench Multisim 7. Every effort has been made to make this book as complete and accurate as possible, but no warranty or fitness is implied.

The information is provided on an "as is" basis. The author and Prentice Hall shall have neither liability nor responsibility to any person or entity with respect to any loss or damages arising from the information contained in this book or from the use of the disks and programs that may accompany it.

Trademarks

"Adobe" and "Acrobat" are registered trademarks of Adobe Systems Incorporated.

"Multisim" and "Electronics Workbench" are registered trademarks of Interactive Image Technologies Limited.

"Microsoft," "Win32s," "MS-DOS," and "Windows" are registered trademarks of Microsoft Corporation.

"IBM" is a registered trademark of International Business Machines Corporation.

"Sun" is a registered trademark of Sun Microsystems Incorporated.

"Open Windows" is a trademark of Sun Microsystems Incorporated.

To my wife, Corena,

my daughters, Katarina Alexis Sierra, Laina Calysta Dine'h, Darrian Iliana Francheska,

my parents, Mom and Dad,

my cats, Tipper, Kennedy, Blaise, and Beethoven

my dogs, Samantha, Sage, and Wolfee,

my larger fish, Sedona, Flounder, Jewel, and the others my daughters have yet to name, and

my horse, Hictaris Lynx (Hickey).

Schematic Capture with Multisim® 7

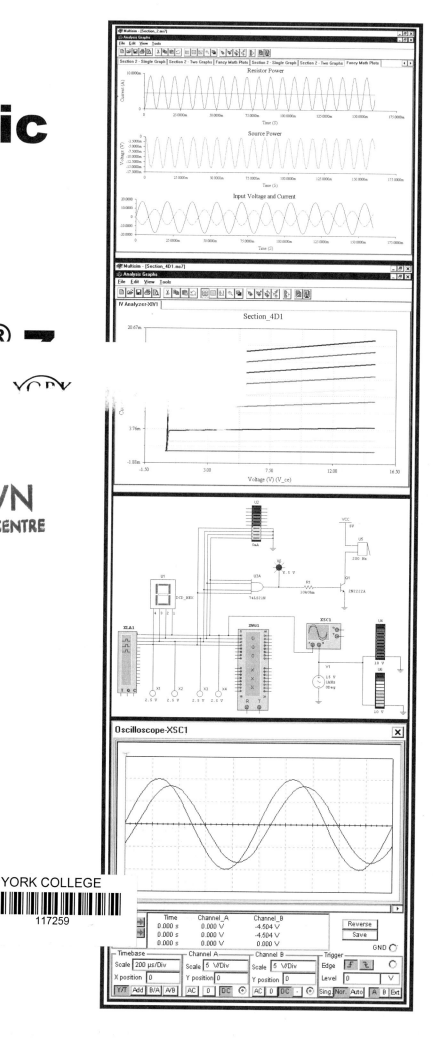

Marc E. Herniter
Associate Professor
ECE Department
Rose-Hulman Institute of Technology

PEARSON
Prentice Hall

Upper Saddle River, New Jersey
Columbus, Ohio

Senior Acquisitions Editor: Dennis Williams
Development Editor: Kate Linsner
Production Editor: Rex Davidson
Design Coordinator: Diane Ernsberger
Cover Designer: Brian Huber
Cover art: Digital Vision
Production Manager: Pat Tonneman
Marketing Manager: Ben Leonard

This book was printed and bound by Courier Kendallville, Inc. The cover was printed by Coral Graphic Services, Inc.

Copyright © 2005 by Pearson Education, Inc., Upper Saddle River, New Jersey 07458. Pearson Prentice Hall. All rights reserved. Printed in the United States of America. This publication is protected by Copyright and permission should be obtained from the publisher prior to any prohibited reproduction, storage in a retrieval system, or transmission in any form or by any means, electronic, mechanical, photocopying, recording, or likewise. For information regarding permission(s), write to: Rights and Permissions Department.

Pearson Prentice Hall™ is a trademark of Pearson Education, Inc.
Pearson® is a registered trademark of Pearson plc
Prentice Hall® is a registered trademark of Pearson Education, Inc.

Pearson Education Ltd.
Pearson Education Singapore Pte. Ltd.
Pearson Education Canada, Ltd.
Pearson Education—Japan

Pearson Education Australia Pty. Limited
Pearson Education North Asia Ltd.
Pearson Educación de Mexico, S.A. de C.V.
Pearson Education Malaysia Pte. Ltd.

10 9 8 7 6 5 4 3 2

ISBN: 0-13-118755-4

Preface

This manual is designed to show students how to use the Multisim 7 circuit simulation program from Electronics Workbench. It is a collection of examples that show students how to create a circuit, how to run the different analyses, and how to obtain the results from those analyses. This manual does not attempt to teach students circuit theory or electronics; that task is left for the main text. Instead, the manual takes the approach of showing students how to simulate many circuits found throughout the curriculum. An example could be the DC circuit shown below.

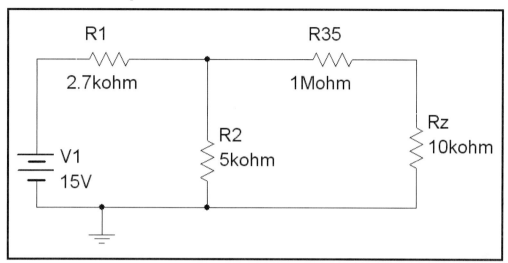

It is assumed that the student has been given enough information to analyze the circuit. This manual assumes that the student wishes to check his or her answers (or intuition) with the Multisim program. The student would construct the circuit as shown in Part 1 and then run either some of the DC simulations in section 3.A or the DC Sweep in section 4.A.

This manual was designed to be used by students for their entire educational career and beyond. Since the parts are arranged by analysis type, they contain a range of examples from circuits covered in first-semester circuit theory courses to senior-level amplifier and switching circuits. Sections that are too advanced for beginning students may be skipped without loss of continuity. All parts contain both simple circuits and advanced circuits to illustrate the analysis types. Sections do not have to be covered sequentially. Individual examples can be identified that apply to specific courses. However, the following sequence is suggested for first-time users. All beginning users should follow Parts 1 and 2 completely to learn how to draw and save schematics, and how to use the Postprocessor and Grapher. Multisim instruments such as the Multimeter and Oscilloscope are covered in the sections in which they are used. All students should follow some of the examples in Parts 3, 4, and 5 that are relevant to the course and also cover a few of the examples that may apply to earlier courses (if any). The early examples in these parts have the most step-by-step detail of how to use the software. Part 6 covers time domain analysis. The first section discusses the use of the oscilloscope, so this section should be assigned before any of the following sections are covered.

This manual contains examples that apply to courses throughout the curriculum. Introductory circuits classes usually cover DC circuits, AC circuits with phasors, and transient circuits with a single capacitor (or inductor) and a switch. Examples are given to cover these types of problems. After reviewing the examples in this manual, a student should be able to simulate similar problems. A typical first electronics course may cover transistor biasing, amplifier gain, and amplifier frequency response. Examples of these analyses are also given.

Exercises are given at the end of each section. These exercises specify a circuit and give the simulation results. The students are encouraged to work these problems to see if they can obtain the same simulation results. The exercises are intended to give students practice in using the software, not to teach them circuits. My philosophy is that simulation software should be used only to verify one's own calculations or intuition. In my classes I assign problems that are worked by hand calculation, simulated with Multisim, and then tested in the lab. The students then compare the measured results to the hand calculations and Multisim simulations. Without hand calculation, it is impossible to know if the Multisim simulations are correct.

Software Included with the Manual

The CD-ROM contains all of the circuit files used as examples in this manual. If you have a problem with one of the circuits you are simulating by following the text, you can view the circuit on the CD-ROM and see how it differs from your circuit.

Comments and Suggestions

The author would appreciate any comments or suggestions on this manual. Comments and suggestions from students are especially welcome. Please feel free to contact the author using any of the methods listed below:
- **E-mail:** Marc.Herniter@ieee.org
- **Phone:** (812) 877-8512
- **Fax:** (253) 369-9536
- **Mail:** Rose-Hulman Institute of Technology, CM123, 5500 Wabash Avenue, Terre Haute, IN 47803-3999

Acknowledgments

I would like to thank my students at Rose-Hulman Institute of Technology for giving me continued inspiration to improve this manual. Without their constant curiosity, this book would not be necessary. I would like to thank Joseph Koenig of Electronics Workbench and Dennis Williams of Prentice Hall for getting together on this project and making the book possible. I would also like to thank Luis Alves and Tien Pham of Electronics Workbench for answering my barrage of questions. Finally, I would like to express my deepest appreciation to my wife and kids, who no longer let me sit in front of my computer twenty-four hours a day.

Before You Begin

General Conventions

- This manual assumes that you have a two- or three-button mouse. The words *LEFT* and *RIGHT* refer to the left and right mouse buttons.
- All text highlighted in bold refers to menu selections. Examples would be **File** and **Run.**
- All text in capital letters refers to keyboard selections. For example, press the **ENTER** key.
- **All text in this font refers to text you will see on the computer screen. This applies to all text except menu selections**.
- `All text in this font refers to text you will type into the program`.
- The word "select" means "click the left mouse button on."

Keyboard Conventions

Throughout the manual many keyboard sequences are given as shortcuts for making menu selections. The explanations of these sequences will be given later. It is important to know the conventions used to specify the sequences.

- Many control key sequences will be specified. For example, **CTRL-R** means hold down the "Ctrl" key and press the "R" key simultaneously. **CTRL-A** means hold down the "Ctrl" key and press the "A" key simultaneously. Not all keyboards are the same; some keyboards may have a key labeled "Control" rather than "Ctrl."
- The keyboard sequence **ALT-TAB** in Microsoft Windows is used to toggle the active window. **ALT-TAB** means hold down the "Alt" key and press the "Tab" key simultaneously.

Nomenclature Used in This Manual

This manual uses many terms associated with Windows. Some of the terms are shown here:

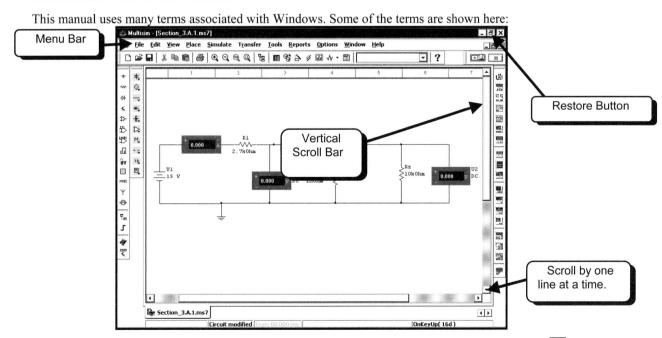

Note: The restore button shown above changes depending on the size of the window. The graphic 🗗 is the restore button and means restore the window to its previous dimensions, which were not full screen. The graphic 🗖 is the maximize button and means expand a window to occupy the entire screen.

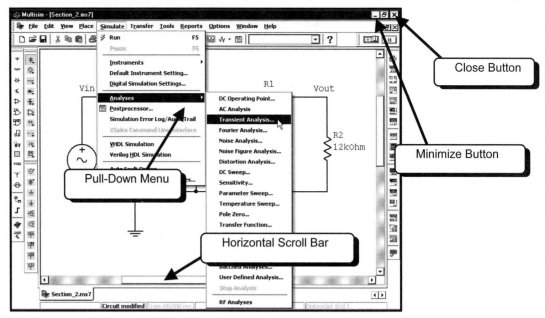
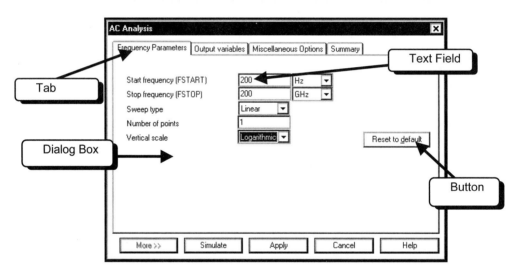

Contents

1. EDITING A BASIC SCHEMATIC 1
   1.A. STARTING MULTISIM 1
   1.B. PLACING PARTS 4
   1.C. CORRECTING MISTAKES 25
   1.D. WIRING COMPONENTS 26
   1.E. GROUNDING YOUR CIRCUIT 28
   1.F. CORRECTING WIRING MISTAKES 32
   1.G. LABELING NODES 32
   1.H. DISPLAYING AND MODIFYING THE TITLE BLOCK 34
   1.I. 3D COMPONENTS 38
   1.J. PROBLEMS 42

2. THE POSTPROCESSOR AND THE GRAPHER 47
   2.A. CREATING A PLOT WITH A SINGLE GRAPH 50
   2.B. CREATING A PLOT WITH TWO GRAPHS 61
   2.C. CREATING A PLOT WITH THREE GRAPHS 66
   2.D. SAVING AND LOADING PAGES 76
   2.E. DELETING ITEMS IN THE POSTPROCESSOR 80
      2.E.1. DELETING A TRACE FROM A GRAPH 80
      2.E.2. DELETING A GRAPH FROM A PAGE 83
      2.E.3. DELETING A PAGE 86
   2.F. MODIFYING THE PLOT PROPERTIES IN THE GRAPHER 87
      2.F.1. ADDING A GRID AND A LEGEND 88
      2.F.2. TRACE WIDTH, COLOR, AND LABEL 90
      2.F.3. CHANGING TITLES AND AXIS LABELS 95
      2.F.4. ADDING A SECOND Y-AXIS 98
   2.G. USING THE CURSORS 103
   2.H. ZOOMING IN AND OUT 110
   2.I. SAVING AND OPENING PAGES IN THE GRAPHER 118
   2.J. PROBLEMS 121

3. DC MEASUREMENTS 123
   3.A. RESISTIVE CIRCUITS 123
      3.A.1. MEASUREMENTS WITH INDICATORS 123
      3.A.2. MEASUREMENTS WITH THE MULTIMETER 133
      3.A.3. USING THE POWER METER 140
      3.A.4. DC OPERATING POINT ANALYSIS 143
   3.B. NODAL ANALYSIS WITH DEPENDENT SOURCES 147
   3.C. DIODE DC CURRENT AND VOLTAGE 150
      3.C.1. CHANGING THE TEMPERATURE OF THE SIMULATION 154
   3.D. FINDING THE THEVENIN AND NORTON EQUIVALENTS OF A CIRCUIT 156
   3.E. TRANSISTOR DC OPERATING POINT 161
   3.F. PROBLEMS 171

4. DC SWEEP 176
   4.A. BASIC DC ANALYSIS 176
   4.B. DIODE I-V CHARACTERISTIC 197
      4.B.1. DIODE I-V CHARACTERISTIC USING THE IV PLOTTER 201
   4.C. DC TRANSFER CURVES 204

4.C.1. Zener Clipping Circuit	204
4.C.2. NMOS Inverter Transfer Curve	207
4.D. Nested DC Sweep — BJT Characteristic Curves	**211**
4.D.1. BJT I-V Characteristic Using the IV Plotter	214
4.E. DC Current Gain of a BJT	**218**
4.E.1. H_{FE} Versus Emitter Current	218
4.E.2. H_{FE} Versus I_C for Different Values of V_{CE}	224
4.F. Problems	**229**

5. MAGNITUDE AND PHASE SIMULATIONS — 233

5.A. Magnitude and Phase Measurements at a Single Frequency	**234**
5.A.1. Magnitude Measurements with Instruments	234
5.A.2. Magnitude and Phase Measurements – AC Analysis	240
5.B. Bode Plots	**252**
5.B.1. Bode Plots Using the Bode Plotter Instrument	252
5.B.2. Bode Plots Using the AC Analysis	257
5.C. Amplifier Gain Analysis	**264**
5.D. Operational Amplifier Gain	**269**
5.E. Parameter Sweep — OPAMP Gain Bandwidth	**281**
5.F. AC Power and Power Factor Correction	**287**
5.F.1. Power Factor Correction	290
5.G. Measuring Impedance	**292**
5.G.1. Resistive Measurement with the Multimeter	292
5.G.2. Impedance Measurement of a Passive Circuit Using SPICE	294
5.G.3. Impedance Measurement of an Active Circuit Using SPICE	297
5.H. Problems	**301**

6. TIME DOMAIN ANALYSES — 315

6.A. Using the Oscilloscope Instrument	**315**
6.A.1. Timebase	319
6.A.2. Channel A and Channel B Volts per Division Settings	321
6.A.3. Trigger Settings	326
6.A.4. Using the Cursors	330
6.B. Phase Measurements of a Capacitive Circuit	**335**
6.C. Phase Measurements of an Inductive Circuit	**339**
6.D. Series LCR Resonant Circuit	**344**
6.E. Regulated DC Power Supply	**352**
6.E.1. Regulated DC Power Supply – Virtual Lab Simulation	356
6.E.2. Regulated DC Power Supply – SPICE Transient Simulation	365
6.F. Zener Clipping Circuit – SPICE Transient Analysis	**378**
6.G. Zener Clipping Circuit – Virtual Lab Simulation	**385**
6.G.1. Plotting Transfer Curves	389
6.H. Amplifier Voltage Swing	**392**
6.H.1. Fourier Analysis with Multisim	396
6.I. Ideal Operational Amplifier Integrator	**404**
6.J. Operational Amplifier Schmitt Trigger	**410**
6.K. Parameter Sweep — Inverter Switching Speed	**415**
6.L. Temperature Sweep — Linear Regulator	**421**
6.M. Rated Components	**427**
6.N. Problems	**432**

7. DIGITAL SIMULATIONS — 439

7.A. Digital Indicators, Signal Generators, and Instruments	**439**
7.A.1. Word Generator and the Logic Analyzer	439

7.A.2. DIGITAL PROBE AND UNDECODED BAR GRAPH LIGHTS	454
7.A.3. MIXED SIGNAL INDICATORS	459
7.B. MIXED ANALOG AND DIGITAL SIMULATIONS	**464**
7.C. PURE DIGITAL SIMULATIONS	**469**
7.D. STARTUP CLEAR CIRCUIT	**472**
7.E. REAL AND IDEAL MODE DIGITAL SIMULATIONS AND GATE DELAYS	**475**
7.E.1. IDEAL MODE DIGITAL SIMULATIONS	475
7.E.2. REAL MODE DIGITAL SIMULATIONS	477
7.F. PROBLEMS	**480**
8. INDEX	**481**

PART 1
Editing a Basic Schematic

In this part we will cover the steps for drawing a circuit using Multisim. This includes locating and placing parts, wiring parts together, grounding your circuit, and labeling nodes. In Part 2, we will run a simple simulation and then discuss how to create plots using Multisim's graphic postprocessor. Parts 1 and 2 of this manual are introductory chapters that cover the skills necessary to create circuits and view results. The remaining chapters discuss how to run circuit simulations. You should cover Parts 1 and 2 in sequence. You can then cover examples in the remaining parts at random. Some of these examples will reference techniques covered in Parts 1 and 2.

1.A. Starting Multisim

If Multisim was installed properly on your computer, it can be easily started from the Windows **Start** menu. However, depending on how the Windows desktop is configured, and your version of Windows, the **Start** menu may appear differently. The screen captures shown here were generated in Windows XP Professional using the classic version of the **Start** menu. Usually, the **Start** menu is displayed at the bottom of the desktop:

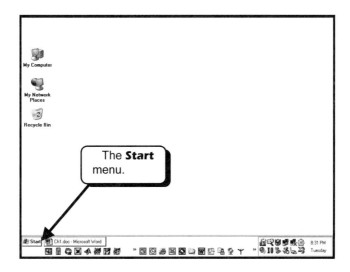

At this point, you could click on the **Start** button and continue:

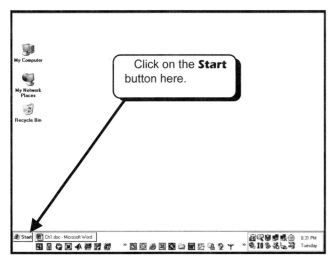

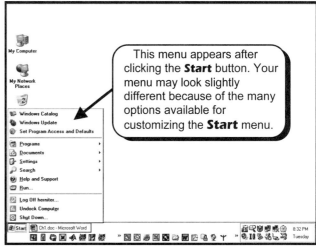

However, if the desktop appears as one of the screen captures shown below

1

the **Start** menu is hidden from view. There are two ways to make the **Start** menu appear. The first way is to bring the mouse pointer down to the bottom of the screen. After a moment, the **Start** menu should appear:

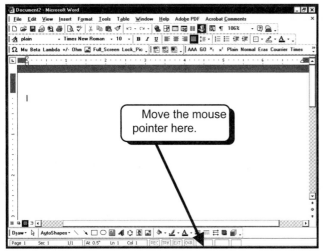 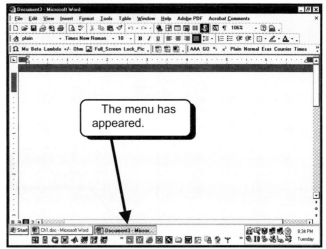

This method works if the **Start** menu is configured so that it is always on top and auto-hide mode is selected. At this point, you could click on the **Start** button ![Start] and continue:

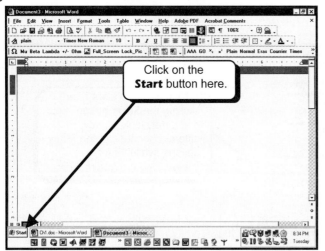 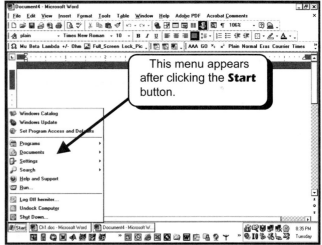

However, if you still cannot see the **Start** menu, you have three more options. (1) You can press the ![key] key on the keyboard or (2) type **CTRL-ESC**. These keys will bring the **Start** menu to the top and also select the **Start** button. (3) If you have an older keyboard and do not have the ![key] key, you can move the mouse pointer to the bottom of the screen and drag the **Start** menu up. For this example, we will press the ![key] key:

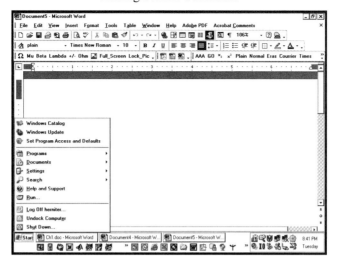

We now have the **Start** menu displayed. Select **Programs**:

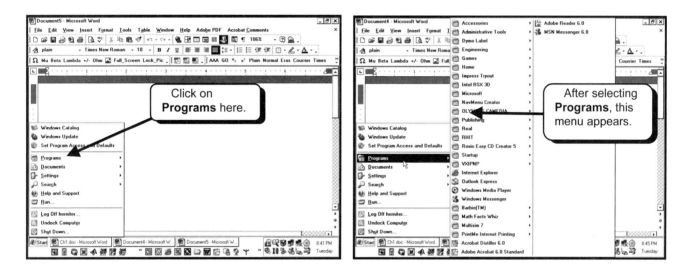

After selecting **Programs**, the programs and program groups for your computer appear. The Multisim programs are contained in the group **Multisim 7**. Click the *LEFT* mouse button on the text **Multisim 7**. This will display the programs contained in this group:

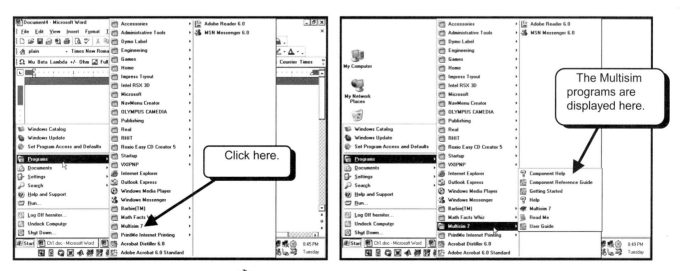

Click the *LEFT* mouse button on the item **Multisim 7** to run the Multisim circuit simulation package.

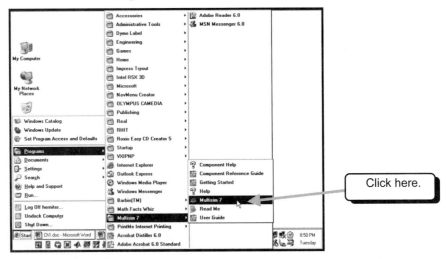

Multisim should run and display an empty schematic:

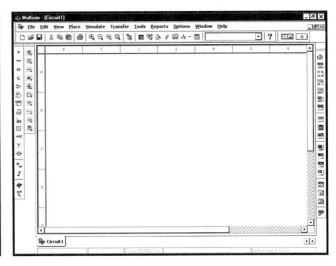

1.B. Placing Parts

Before we place any parts, we will check one of the Multisim options to be sure that we are all using the same settings. From the Multisim menus, select **Options** and then **Preferences**:

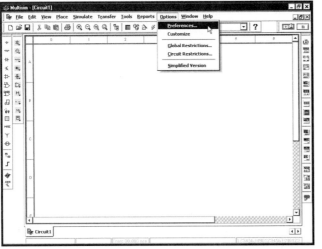
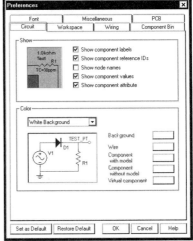

In my dialog box, the **Circuit** tab is selected. The top pane of this window allows us to specify what information will be displayed on the schematic for each component. A lot of information is displayed on the schematic that we do not need on most of our schematics, so we will turn off the display of that information here. The only information we wish to display about components is their name such as R1 or R2 for resistors, or C1 or C2, and their value, such as 1.0 kohm for a resistor. A part's name is called its component reference ID and a part's value is called its component value. We will enable the display of these two values. We will also choose to display the node names in a circuit. This is done to make it easier to use the Postprocessor. Sometimes displaying node names does clutter up the schematic, but displaying them will help us annotate the circuit and we will label some of the nodes as vin and vout. Make sure that your options match those shown below:

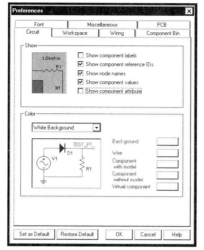

Next, select the **Component Bin** tab:

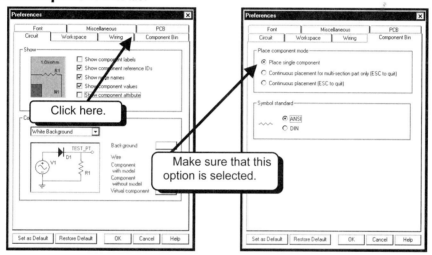

Make sure that the option **Place single component** is selected. If not, select it. With this option, we can only place a single component at a time and avoid the error of accidentally placing two identical components, one on top of the other.

Finally, select the **Miscellaneous** tab:

We will use this tab to specify a directory where we will save all of our circuit files. Click the **Browse** button and then select a directory where you would like to save your files. After you have specified a directory, the path should be listed in the dialog box:

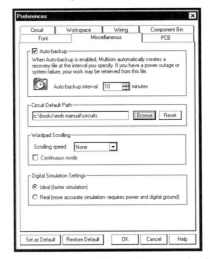

We are now done with the changes we need so that you can follow the examples in this book. You can experiment with other changes if you like. We now need to save the changes we just made. If you select the **Set as Default** button, the changes you have made to the **Preferences** will be used for all future schematics you create. If you click the **OK** button without clicking the **Set as Default** button, the changes will only apply to the current schematic. Any schematics you create in the future will use the previous option settings. Click on the **Set as Default** button and then click on the **OK** button to return to the schematic.

We will now create a schematic. The first part we will get is an AC independent voltage source. There are actually two different AC independent voltage sources available in Multisim 7. One is called the AC power source, which allows you to specify the amplitude of the source as an RMS value. The other is called the AC voltage source, which allows you to specify the amplitude of the source as the peak value. Aside from that minor difference, the two sources are the same. The only confusion possible between the two sources is that when looking at the schematic, you cannot distinguish between the two. Thus, some of the screen captures in this text may be confusing. When using this program, the confusion is easily cleared up because all you have to do is double-click on the source and it will become apparent which source you are using. We will now show how to place both sources in our schematic.

There are three methods for selecting a part: using the toolbars and component bins, using the toolbars and component dialog box, and using the Multisim menus and the component dialog box. The two AC independent voltage sources are placed with the toolbars and component bins, so we will show this method first.

On the left side of the window are two toolbars. The leftmost toolbar is the **COMPONENT** bar and gives us access to all of the parts available in Multisim 7 plus a few other features such as access to the Electronics Workbench website and EDAparts.com.

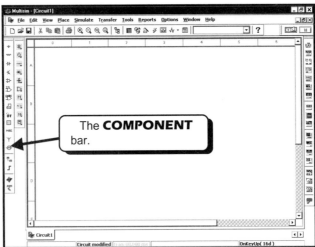

An enlarged version of the toolbar is shown below:

It takes a few more steps to use this toolbar, but it does allow us to place any part we want to use. We will discuss this toolbar in detail later when we need to use it.

The AC voltage sources can be placed quickly using the **VIRTUAL** toolbar, which is the one next to the COMPONENT toolbar:

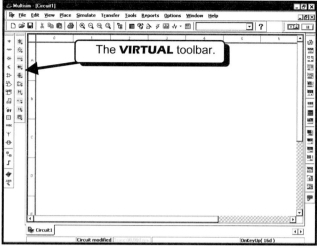

An enlarged version is shown below:

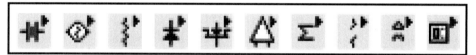

A virtual component is an idealized component like a diode, a resistor that can have a non-standard value, an ideal OPAMP, and so on. When you click on one of these buttons, another toolbar will appear, which we will call a parts bin, from which you can quickly select a part. The parts bins contain the parts you will use most. The bins are summarized below:

– Power sources – Single-phase and three-phase sources. DC power sources and ground.

– Signal sources – Signal sources: voltage and current sources, square wave, piece-wise linear, and clock sources.

– Basic parts – Basic parts including, but not limited to resistors, capacitors, and inductors.

– Diodes and Zeners.

– Transistor Components – BJTs, MOSFETs, GaAsFETs, and JFETs.

– Analog Components – Opamps and comparators.

– Miscellaneous Components – Analog switch, fuse, 7-segment displays, motor, 555 timer, etc.

– Rated Components – These are components that have physical limitations and will break down when they reach those physical limitations. Examples are resistors with wattage limitations or transistors with collector current limitations. When a device reaches its limitation in a simulation, the schematic will indicate a damaged part.

– 3D Components – Part graphics shown in three-dimensional graphic symbols such as , , and , rather than standard schematic symbols.

– Measurements Components – Voltage and current measurement devices, and digital probes.

We will place both an AC power source and an AC signal source to show the difference between the two sources. First we will place the AC power source. Click on the **Show Power Source Components Bar** button as shown. This will display the **Power Source Components** toolbar:

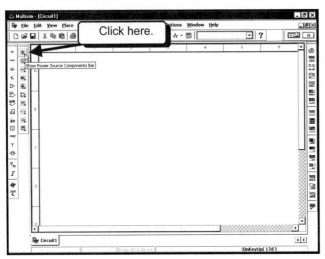
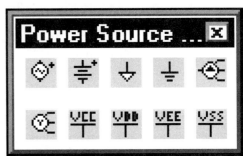

This toolbar lists common power sources that we will use in circuits such as DC voltage sources, ground connections, and three-phase sources. This bar also contains the AC source that we will use. To see the description of a button, hover the mouse pointer over a button. After a few moments, the description of the button will appear:

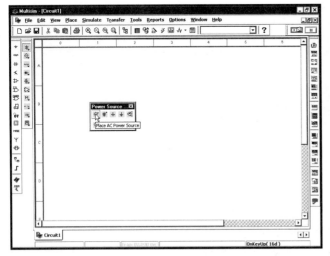
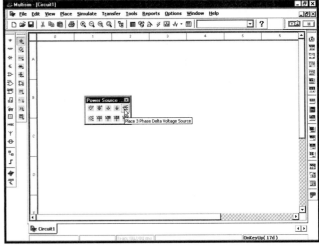

From the description and the graphic, it is usually easy to locate the part you are after. We want to place the AC power source, so click the *LEFT* mouse button on the **Place AC Power Source** button . The source will become attached to the mouse pointer and move with the mouse:

Multisim 7 — Editing a Basic Schematic — 9

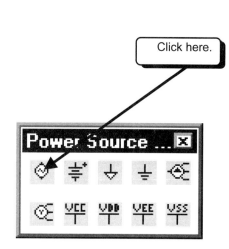
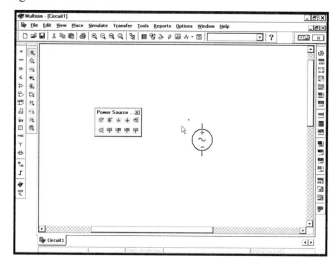

Move the source to the location where you want to place the part. Click the **LEFT** mouse button to place the part, and then move the mouse pointer.

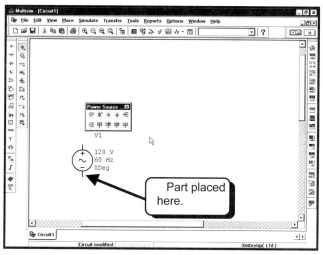

The Power Source Components toolbar is such a useful menu that we will place it next to the other menus for easy access. We can drag the toolbar to the side of the screen and place it next to the other toolbars so that they remain on our desktop, but do not clutter up the schematic. To drag the toolbar, follow the procedure given below. Click and **hold** the **LEFT** mouse button on the title as shown:

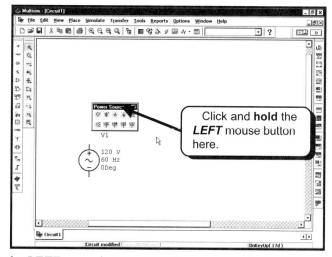

While continuing to hold down the **LEFT** mouse button, move the mouse as shown:

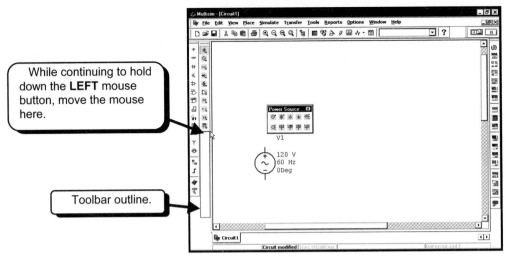

Release the mouse pointer when you see the outline of the toolbar move so that it is located where you would like the toolbar to be placed. When you release the mouse button, the toolbar will be relocated:

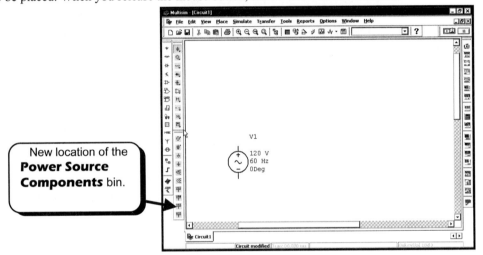

We can now easily place parts in this toolbar. The parts in this bin are used frequently, and placing the toolbar here does not reduce the area of the schematic. You can use this method to place other commonly used parts bins on the desktop as well.

We will now edit the properties of the source to make it a 12.6 V AC source at a frequency of 60 Hz. Before we do this, we will zoom in on the source. To zoom in or out, select **View** from the Multisim menus:

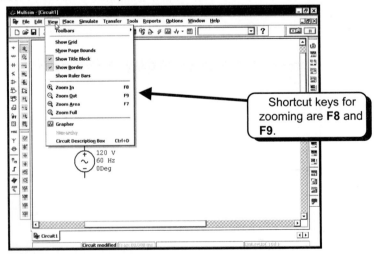

We will not use the menus to zoom in or out. However, the menus tell us which keys to press. We see that we can zoom in by pressing the **F8** key, and we can zoom out by pressing the **F9** key. The **F8** and **F9** keys have an advantage over using the menus or the toolbar buttons in that we can place the mouse pointer over the location in which we want to zoom and then press the **F8** or **F9** key. This allows us to zoom at a specific location. If you use the toolbars or the menus to zoom, it will zoom around some location, but not necessarily the area you want.

Place the mouse pointer over the AC voltage source and then press the **F8** key:

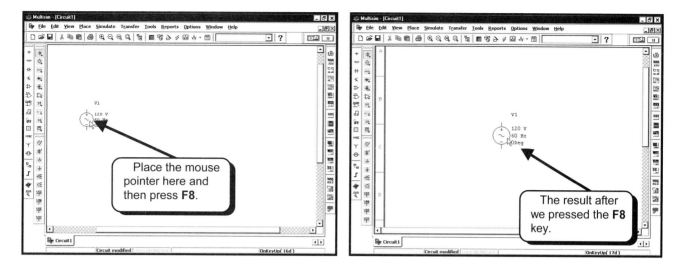

We can repeat the process several times. Place the pointer over the source and press the **F8** key. Screen captures showing the result of pressing the **F8** key a few times are shown below:

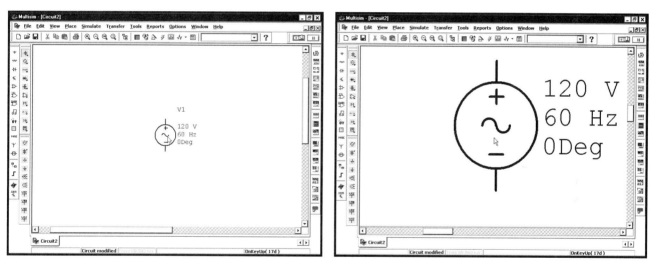

If you want to zoom out around the mouse pointer, press the **F9** key. After pressing the **F9** key several times, I have the screen capture below:

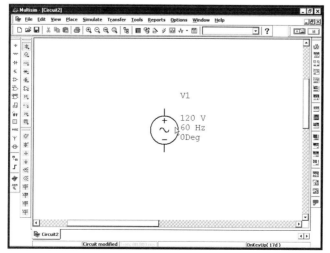

After fooling around with the zoom settings for a while, we can restore the schematic to its original size with the zoom 100% button (or by pressing the **F7** key). I recommend using the **F7** key once again because it does perform the zoom around the location of the mouse pointer. If you use the button, the zoom will be performed at a location on the screen that might not be where you would like it.

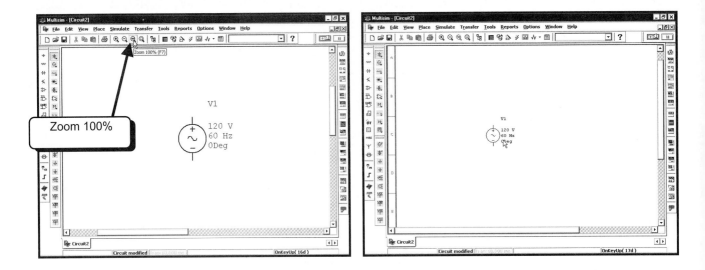

You can now choose a zoom level that you like for the current circuit you are working on.

We will now edit the properties of this AC power source. Double-click the **LEFT** mouse button on the graphic for the source. The dialog box below will appear:

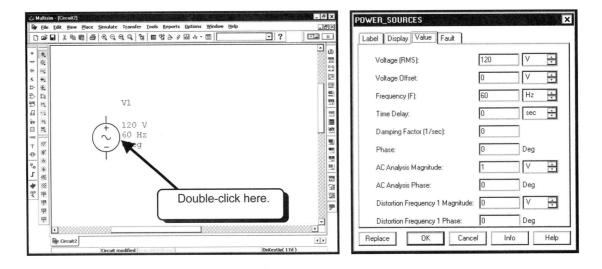

This portion of the dialog box allows us to change all of the physical parameters of the source such as the amplitude, frequency, and phase. With the AC power source we specify the amplitude of the source as an RMS value. Remember that the amplitude is the peak value of the sine wave. The relationship between the amplitude of a sine wave and the RMS value of a sine wave is:

$$RMS\ Value = \frac{Amplitude}{\sqrt{2}}$$

If you are working with RMS values in your circuit, you can specify the RMS value. If you are working with straight amplitude values, you can specify the amplitude. When you change one value, Multisim automatically changes the other.

When I said that I wanted to make the source a 12.6 V source at 60 Hz, I did not specify whether the number 12.6 was an RMS value or the peak value. We will specify 12.6 as the RMS value. Click the **LEFT** mouse button as shown to place the cursor in the **Voltage (RMS)** text field:

Multisim 7 — Editing a Basic Schematic — 13

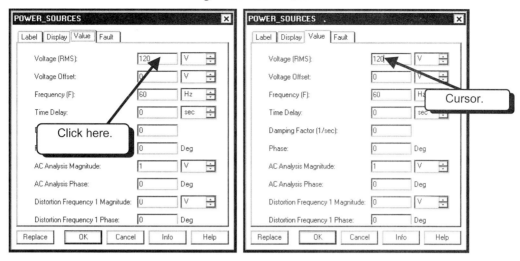

With the cursor in the text field, we can now modify its contents. Use the **BACKSPACE** key to erase the number that is already there and enter the number **12.6**:

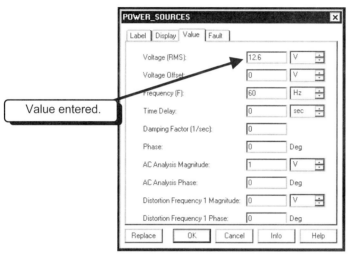

Note that we can specify the units of the amplitude. By default, the units are in volts. However, we can change the units if we wish. Click the *LEFT* mouse button on the up or down arrows as shown to change the units:

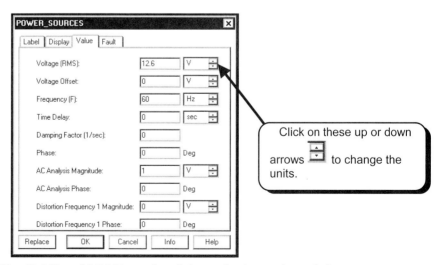

Several units are available including μV, mV, kV, and MV. Examples of changed units are shown below:

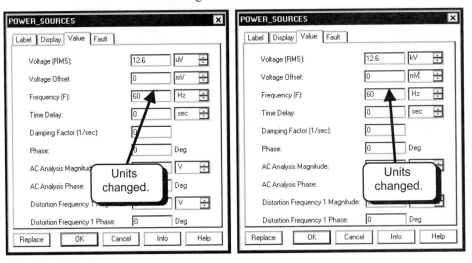

We want to create a 12.6 V RMS source, so leave the units as volts.

You can change the frequency of the source using the same techniques, but we will leave the source at 60 Hz:

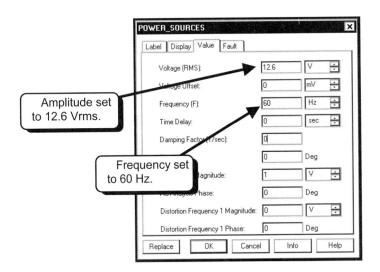

To change the name of the voltage source, we need to select the **Label** tab:

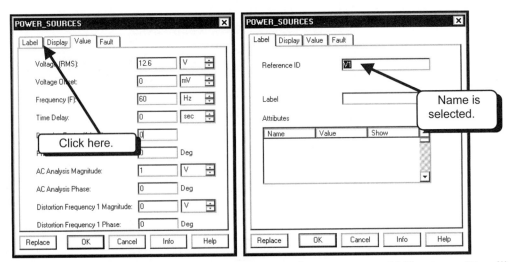

The name of the source is already selected, so all we need to do is type the name we would like to use. We will change the name to Vx, so type the text **Vx**:

Multisim 7 — Editing a Basic Schematic — 15

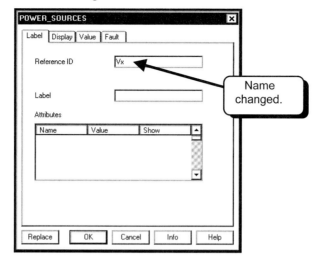

Click on the **OK** button to accept all of our changes. The changes will appear on the schematic:

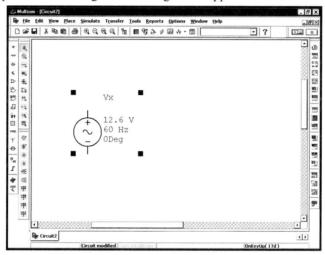

We now need to place some resistors in the circuit. To place a resistor, click on the **Show Basic Components Bar** button as shown.

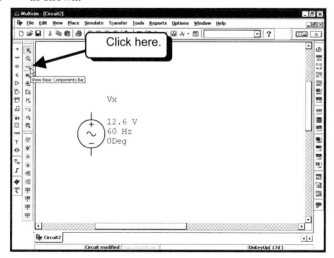

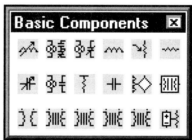

This is the toolbar that you will use to place most of the passive components such as resistors, capacitors, and inductors. The components contained in these bins are referred to as virtual components. That is, the components are not required to have standard component values. For example, a resistor can have a value of 1.395 kΩ if you so desire, and you can easily change the resistance of the resistor. There is another parts bin where you can place parts with standard value components, and it is not easy to change the value of those components.

To identify a button, hover the mouse pointer over a button. After a few moments, a description of the button will appear:

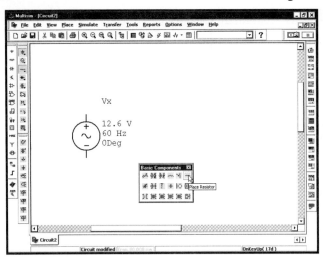 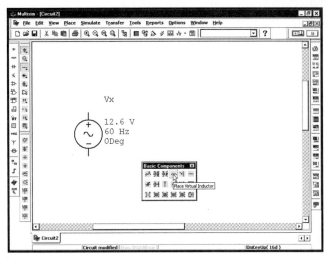

Click the **LEFT** mouse button on the **Place Resistor** button. A resistor graphic will become attached to the mouse pointer and move with the mouse:

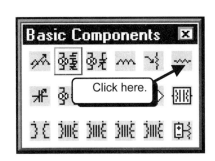 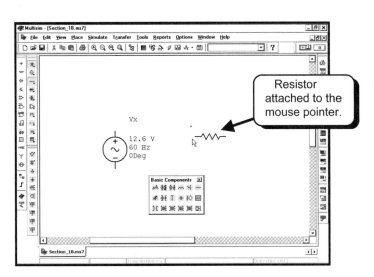

Locate the resistor as shown below and click the **LEFT** mouse button to place the part in your circuit:

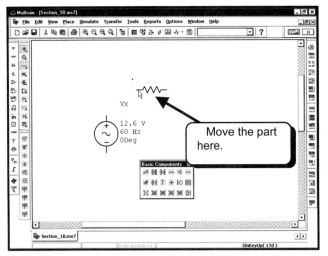 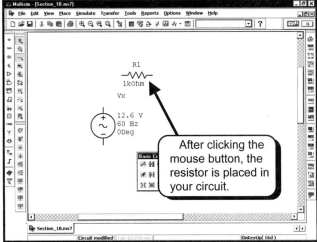

The part is now in your circuit. We would like to make this resistor a 12.5 kΩ resistor and change its name to Rx. To change the properties of the resistor, double-click the **LEFT** mouse button on the resistor graphic:

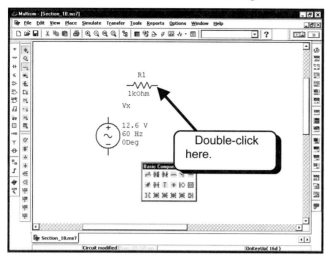
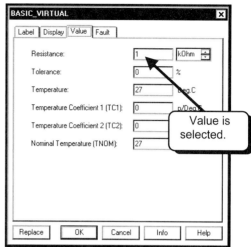

This screen lets us change the value of the resistor. We use the same procedure to change the resistance as we did to change the voltage of our source. We must first place the cursor in the text field for the resistance. Click the **LEFT** mouse button as shown below:

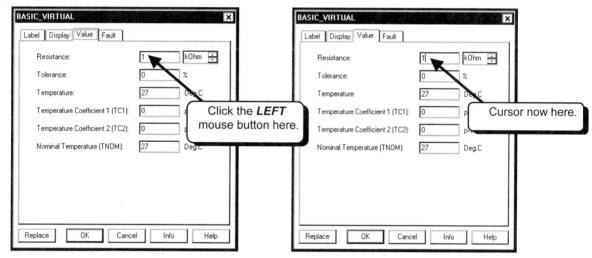

With the cursor placed as shown, type the text **12.5**. This will specify the value of the resistor as 12.5 kΩ because the units are selected as **kOhm**.

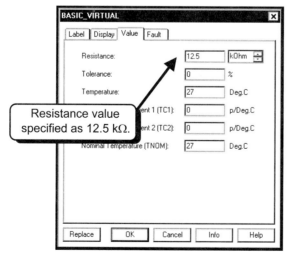

To change the name of the resistor, select the **Label** tab:

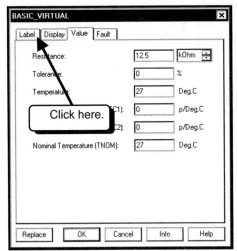 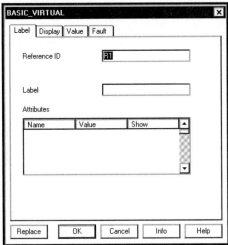

The present name of the resistor is **R1**. Since the name is highlighted, all we need to do is type the new name we want to use. We want to rename the resistor to Rx, so type the text **Rx**:

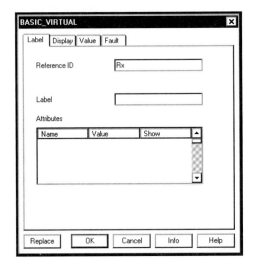

We are done changing the properties of this resistor, so click on the **OK** button. The changes will be displayed on the schematic:

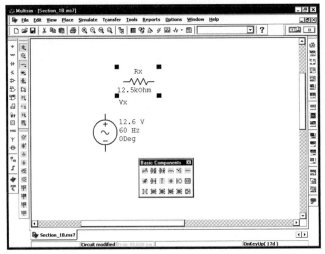

If your values are not the same as shown, edit the properties of the device again.

Next, we need to place a standard value 12 kΩ resistor. First, close the Basic Components bin by clicking on the ☒ as shown:

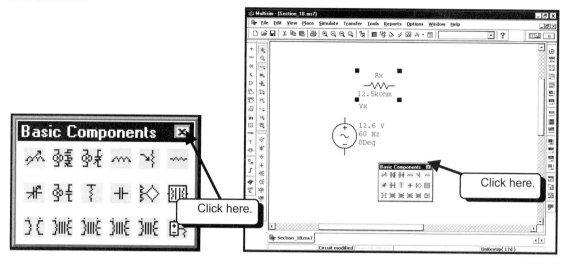

Next, click on the **Basic** button in the Components toolbar:

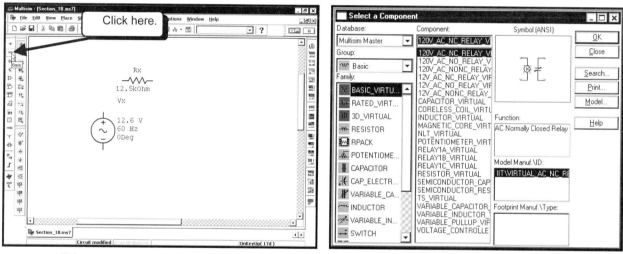

Click on the text **RESISTOR** to select the **RESISTOR Family**. This **Family** contains all of the standard 1% and 5% resistors available in Multisim 7:

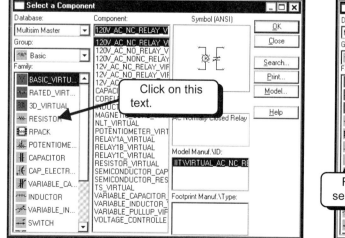
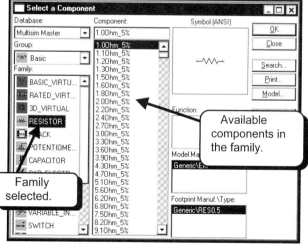

This dialog box lists many of the standard 1% and 5% resistor values you can purchase. You must scroll through this list and locate the resistor you need. Scroll down the list and find the one named **12kOhm_5%**. Click on the text to select it:

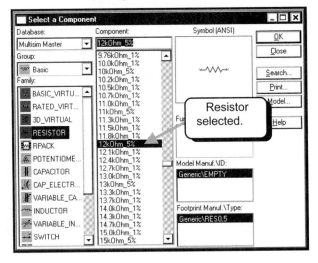

When you click on the **OK** button, the resistor will become attached to the mouse pointer and move with the mouse:

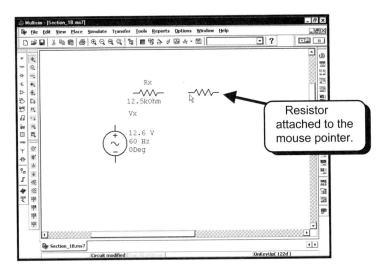

Place the resistor as shown and then click the **LEFT** mouse button to place the part in your circuit:

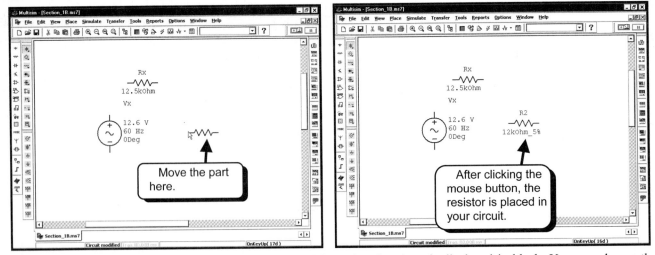

Note that a standard value resistor is displayed in blue while a virtual resistor is displayed in black. You can change the properties of this resistor using the same methods we showed for the virtual resistor. To edit this resistor's properties, double-click on the resistor graphic:

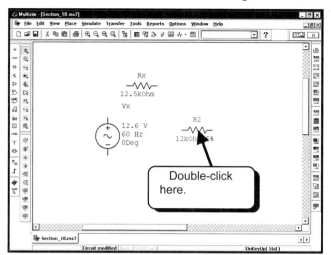 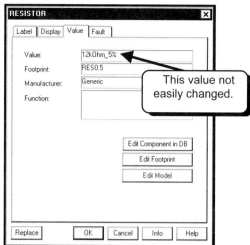

The one difference you will notice is that it is not obvious how to change the resistance value of the resistor. If you need to change the value, you should delete the resistor from your circuit and then place a new resistor with the correct value in your circuit. If you have a circuit in which you will be changing the values of components, you should use the virtual components. The one advantage of using the standard value components is that you guarantee that you are using a standard value that you can purchase.

With the above dialog box, you can still easily change the name of the resistor. Select the **Label** tab and change the name of the resistor to **Ry**:

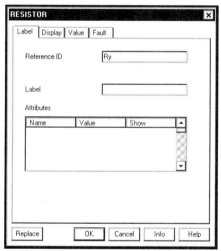

Click on the **OK** button to accept the changes. The resistor will be renamed in the schematic:

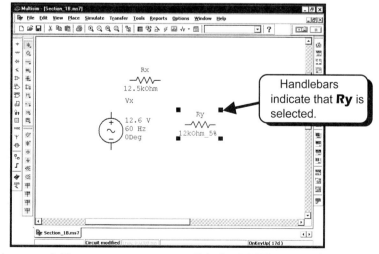

Resistor **Ry** needs to be rotated. To rotate a component, click the ***RIGHT*** mouse button on the component that has been selected. In the above screen capture, the handlebars indicate that **Ry** has been selected. If **Ry** is not selected in your schematic, click the **LEFT** mouse button on the resistor graphic to select it. Once the handlebars appear, **Ry** is selected. With Ry selected in your schematic, click the **RIGHT** mouse button on the graphic for **Ry**. A menu will appear:

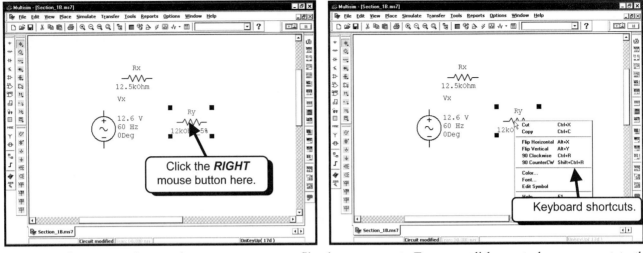

This menu shows several ways that we can rotate or flip the component. To accomplish our task, we can rotate the component either clockwise or counterclockwise. Also note that you can type **CTRL-R** or **SHIFT-CTRL-R** to rotate the part rather than use this menu. Select **90 Clockwise** to rotate the part:

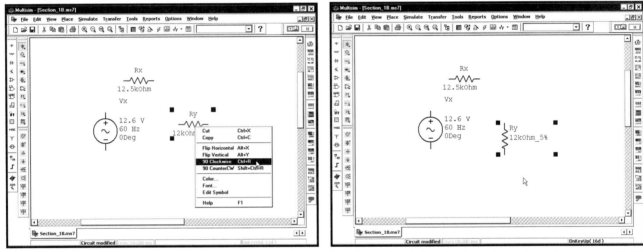

The part is now oriented the way we would like. I will reposition the resistor slightly since the resistor does move slightly when it rotates:

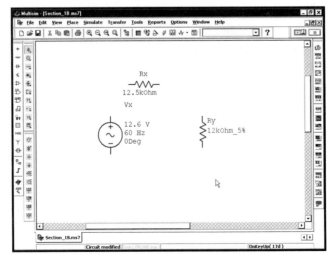

For the next resistor, we will show how to use the menus to place a part. This method is equivalent to using the component bar, but this time we will use it to place a virtual resistor rather than a standard value resistor. This takes a few more steps than using the toolbars we showed at the beginning of this section. The goal here is to show all of the different methods of placing parts, so that you can select the one that you feel most comfortable with. We will show how to place a virtual resistor, but the method shown can be used to place any component. Select **Place** and then **Component** from the menus (or type **CTRL-W**):

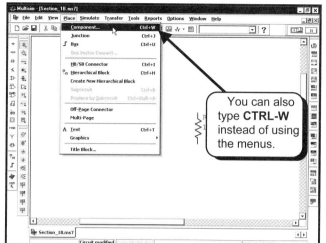
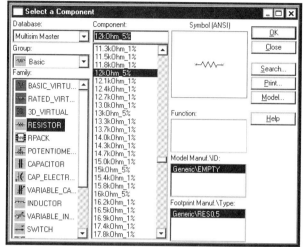

The component selector displays the last component we placed. You can browse through the parts available if you wish, and we will search through the parts available as needed. It is not always obvious where the part you need is located, and it sometimes does take a bit of searching. However, after a bit of use, common parts are easy to find.

We would like to place a virtual resistor, which is located in the **BASIC_VIRTUAL** group. Note that in my dialog box, the presently selected **Group** is **Basic**. This **Group** was selected from the last component we placed:

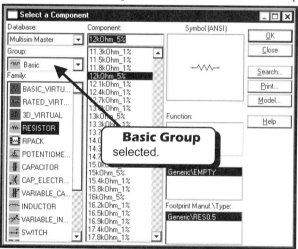

If your dialog box does not have this **Group** selected, then select the **Basic Group**. The virtual components are located in the **BASIC_VIRTUAL Family**. Click the *LEFT* mouse button on the text **BASIC_VIRTUAL** to select the **Family**:

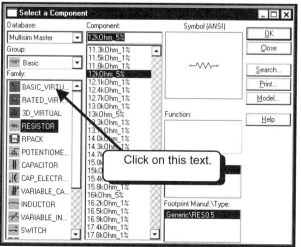
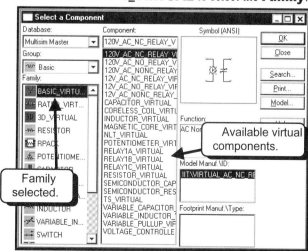

Click the **LEFT** mouse button on the text **RESISTOR_VIRTUAL** to place a resistor:

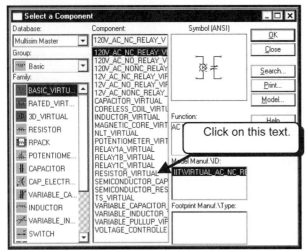 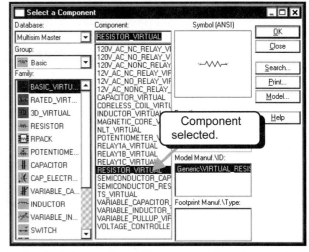

This is the same type of resistor as the first one we placed in our circuit. Click on the **OK** button to place the part in your circuit. The part will become attached to the mouse pointer and move with the mouse.

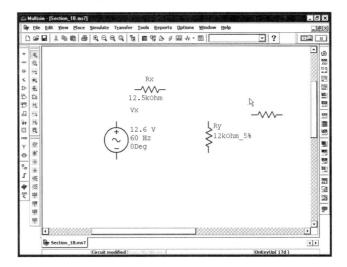

Use the same techniques to place the resistor as shown below, and change its resistance value to 12 kΩ and name the resistor R1:

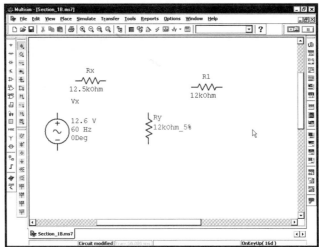

Use one of the methods shown previously to place a fourth resistor in your circuit.

Multisim 7   Editing a Basic Schematic   25

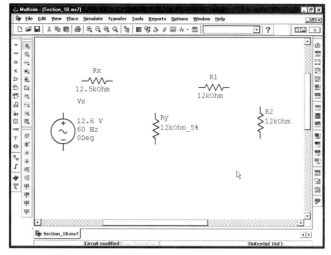

Name the resistors as shown and give the same values as shown. If you have any toolbars open, close them by clicking on the ☒ in the toolbar's upper right corner.

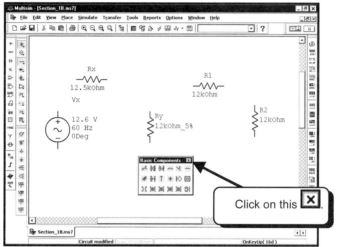

1.C. Correcting Mistakes

At this point you may have some mistakes in your schematic. To move parts follow the procedure below. For the moment we will assume that you wish to move a resistor.

1. Click the **LEFT** mouse button on the resistor graphic you wish to move. When the resistor graphic is selected, it will be surrounded by handles:

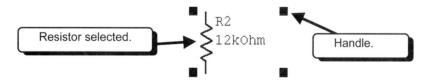

It may take several tries to highlight the resistor. Make sure that you select the resistor as shown above, and not its value or name as shown below:

2. When the resistor is highlighted, drag the resistor graphic to the desired spot in the schematic.

If you need to delete a part, follow Step 1 above. When the appropriate part is selected, press the **DELETE** key.

1.D. Wiring Components

We now must wire the components together. Place the mouse pointer over the top terminal of the voltage source. When you are near the pin the mouse pointer will be replaced by crosshairs:

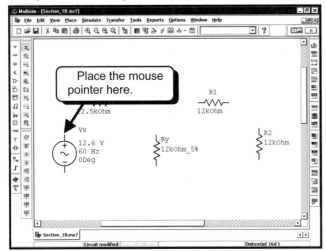
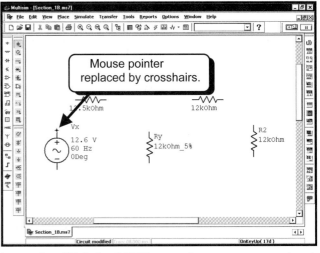

When the crosshairs are displayed, you are ready to start drawing a wire. Click the **LEFT** mouse button to start a wire. As you move the mouse away you will see a solid black line connected to the crosshairs. As you move the mouse, the line grows and shrinks with the changing position:

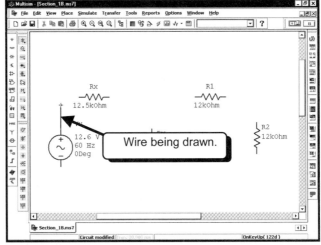

If you made a mistake and do not really want to draw a wire, press the **ESC** key, and the black line will disappear. (Don't press the **ESC** key.) Move the crosshairs to point at the left terminal of **Rx**:

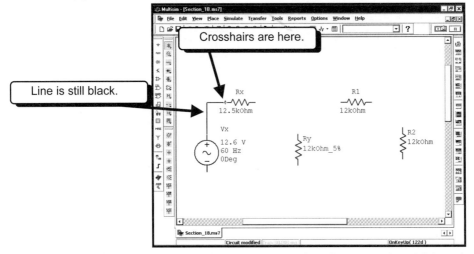

Note that the line is still black, indicating that it is not yet a wire. If we press the ESC key, the black line will disappear and a wire will not be created. With the crosshairs placed as shown above, click the **LEFT** mouse button to create the wire. Note that the black line changes to a red line, indicating that the line is now a wire connecting the two components. When you move the mouse pointer away, you will notice that the crosshairs are replaced by the mouse pointer and we are no longer drawing wires:

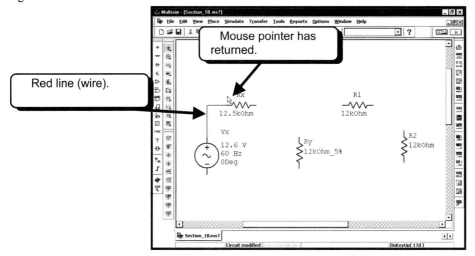

Use the same technique to connect **Rx** to **R1**:

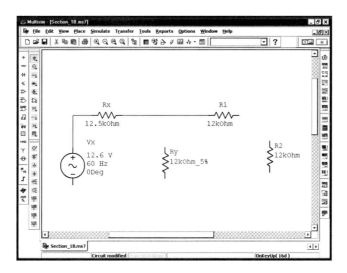

Next, we need to connect **Ry** to the node between **Rx** and **R1**. Place the mouse pointer over the top terminal of **Ry** and click the *LEFT* mouse button to start drawing a black line:

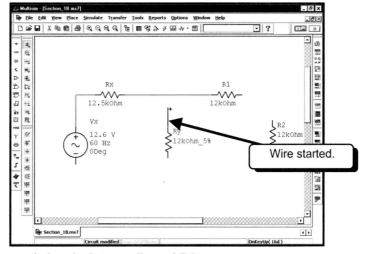

Move the crosshairs so that they touch the wire between **Rx** and **R1**:

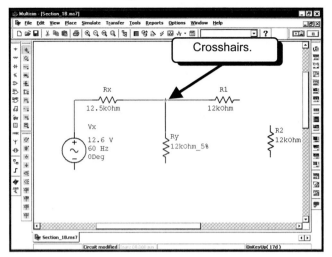

With the crosshairs as shown above, click the **LEFT** mouse button. The black line will turn red, indicating that it is a wire.

When you move the mouse away, you will notice that the mouse pointer has returned and that there is a dot ┬ at the connection:

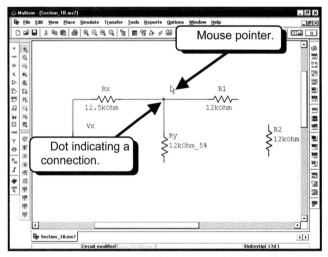

The dot indicates a connection where two wires meet. If you do not see a dot, then the wires are not connected.

Wire the remainder of the circuit as shown below:

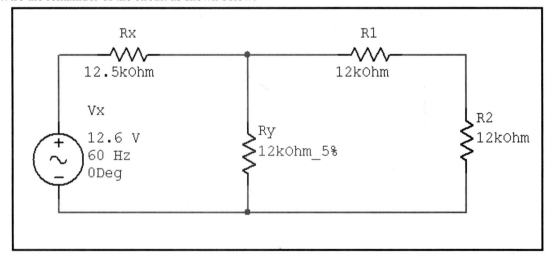

1.E. Grounding Your Circuit

To simulate a circuit on Multisim, you must have at least one ground connection in your circuit. Two ground symbols are available with Multisim, the general ground symbol (⊥) and the digital ground symbol (▽). The general ground symbol is used for all simulations except the real mode digital simulations. The general ground can be used for ideal mode digital simulations, which is the default mode for digital simulations. To see the difference between ideal and real mode digital simulations, see section 7.E. The general ground simulation works for all of the simulations covered in this text except the real mode digital simulations. When in doubt, use the general ground symbol ($\perp\!\!\!\perp$). Both ground symbols are located in the **Power Source Components** toolbar. Earlier in this chapter we placed this toolbar on the side of the screen for easy access:

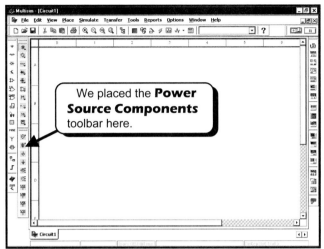

If the **Power Source Components** toolbar is not visible, then you must click on the **Show Power Source Components Bar** button as shown below to make the toolbar visible.

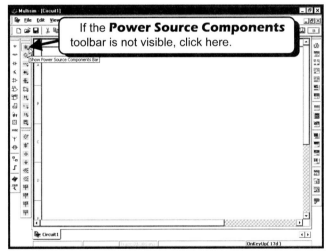

Once the toolbar is visible, all you need to do to place a ground symbol is click on the **Place Ground** button $\perp\!\!\!\perp$. The ground symbol will become attached to the mouse pointer and move with the mouse:

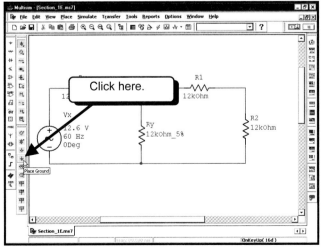 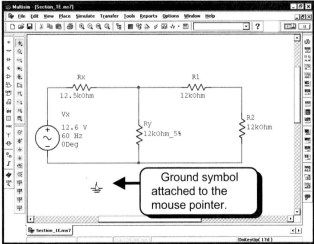

Locate the part as shown above and click the **LEFT** mouse button to place the part:

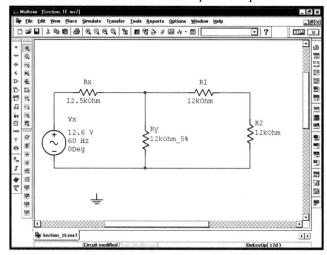

We now need to wire the ground connection to our circuit. Place the mouse pointer over the pin for the ground symbol as shown. The mouse pointer will change to crosshairs:

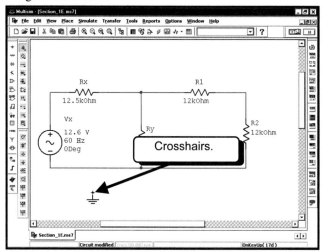

Click the **LEFT** mouse button to start drawing a wire. As you move the mouse away, a black line will be drawn:

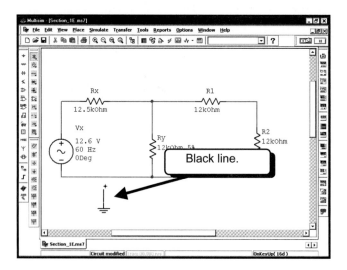

Move the mouse up to position the crosshairs on the wire as shown:

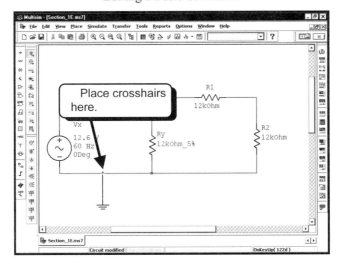

Click the **LEFT** mouse button to make the connection. The black line changes to a red line, indicating that a wire has been created:

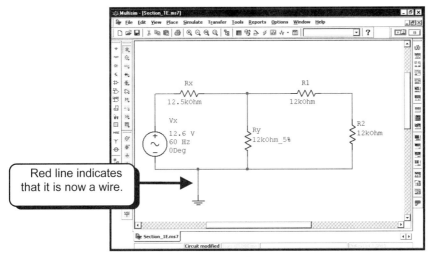

Note that using the general ground symbol (⏚) specifies a node name of 0. If you double-click on a wire connected to the ground node, you will see that the name of the node is zero (**0**):

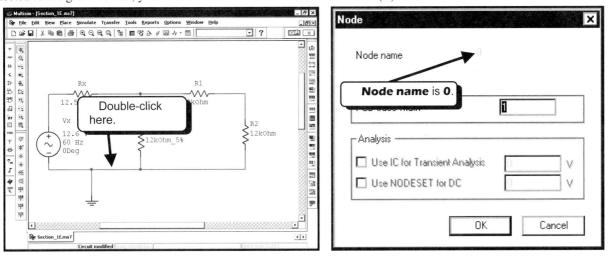

For SPICE simulations, the standard name for the ground node is zero, and all circuits must contain at least one node that is named zero (0). Note that the digital ground symbol (▽) names the node GND, not 0.

1.F. Correcting Wiring Mistakes

If you make a mistake when drawing a wire, you should use the following procedure to rework the wiring to the node to which the wire was connected. To delete a wire:

    a. Move the mouse pointer to the segment of wire that you wish to remove.
    b. Click the **LEFT** mouse button on the wire you wish to remove. This will select the wire.
    c. Press the **DELETE** key to delete the selected wire.

1.G. Labeling Nodes

Multisim automatically numbers nodes for you. With SPICE simulations, you need to know the node names in order to display traces. Before we label any traces, we will select the option to display node names on the schematic. Select **Options** and then **Preferences** from the Multisim menus:

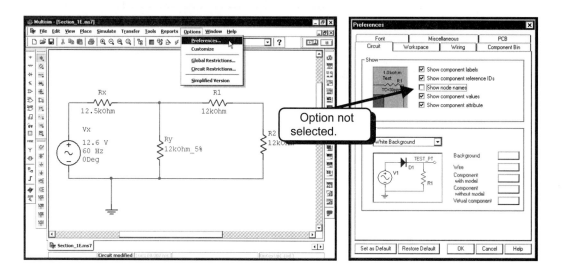

By default, node names are not displayed on the schematic. For this section, we would like to display node names, so enable the option and then click on the **OK** button:

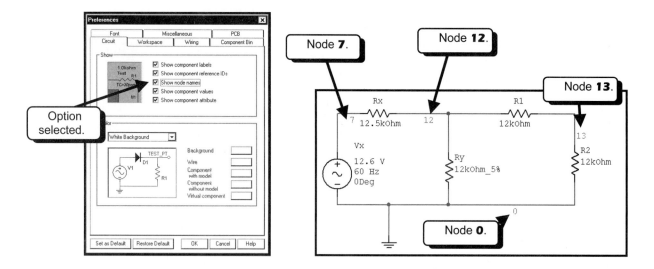

Node names are now displayed on the schematic. Note that your node names will be different from mine. Also note that if you rewire a section of your circuit, the node name of the node will change. The numbers are assigned sequentially, but you usually do not know a node name unless you give it a name.

To rename a node, double-click the **LEFT** mouse button on a wire connected to the node. For example, we will name the left node Vin. Double-click the **LEFT** mouse button as shown below:

Multisim 7   Editing a Basic Schematic   33

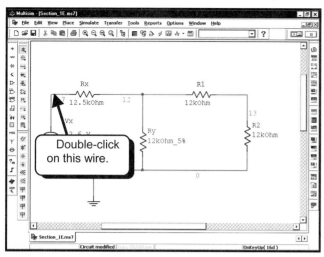 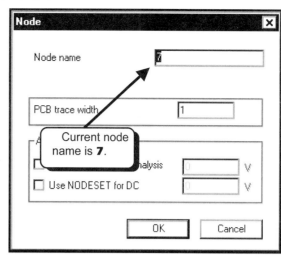

We see that the current name of the node is **7**. Since the name is highlighted, all we do to rename the node is type the new name. Type the text **Vin** and then click on the **OK** button. The new name will be displayed on the schematic:

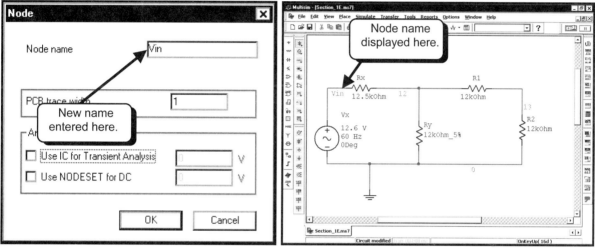

Using the same procedure, rename the center node **V_mid** and the right node **Vout**:

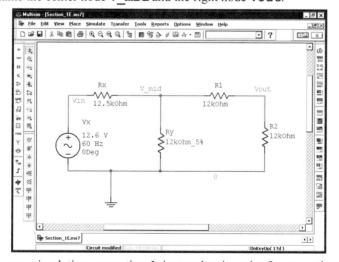

Naming nodes does not change any simulation properties. It just makes it easier for us to plot traces using the Postprocessor that is described in Part 2. **Node names should not contain any spaces.**

Now that we have finished the schematic, we need to save it. Select **File** and then **Save** from the menus, type **CTRL-S**, or click on the **Save** button in the toolbars.

1.H. Displaying and Modifying the Title Block

The title block contains information about a project such as the name, who created it, who checked it, when it was last modified, the number of pages, and so on. For our example, we will just use it to identify the project, our name, and our class. To display the title block, select **Options** and then **Preferences** from the Multisim menus:

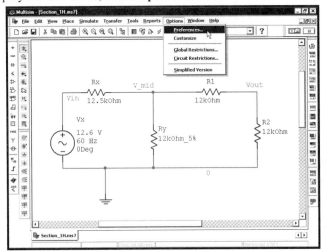
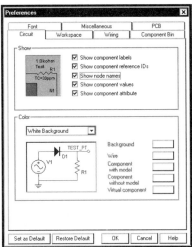

Select the **Workspace** tab:

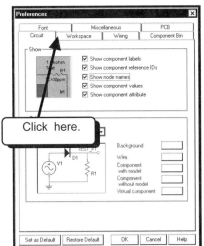
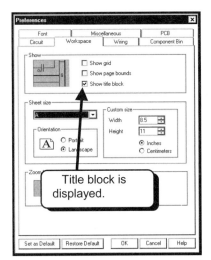

By default, the title block is displayed. If this option is not enabled, enable it and then click the **OK** button. The title block will be displayed once we place one in the schematic.

To place a title block in our schematic, we need to change the zoom level so that we can see the entire page of the schematic. Select **View** and then **Zoom Full** from the menus to display the entire schematic page:

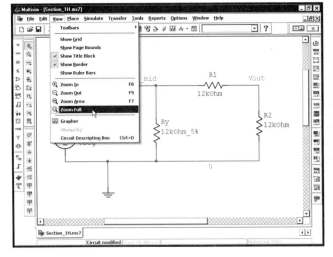
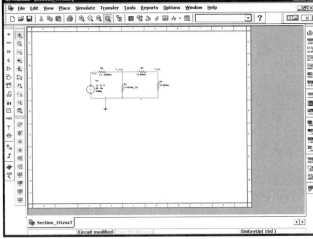

We must now place a title block in our circuit. Select **Place** and then **Title Block** from the menus:

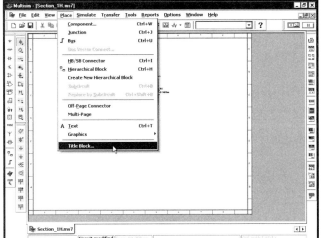
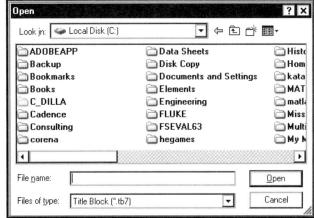

Multisim 7 title blocks are stored in files that have an extension of .tb7. If you installed Multisim 7 in the default directory, these files are stored in a directory named C:\Program Files\Multisim7\Titleblocks:

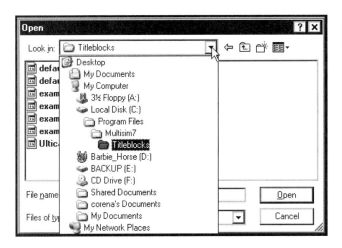
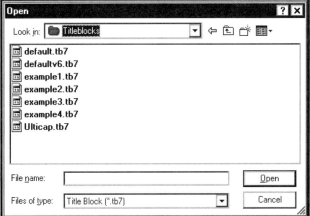

Choose a file and click the **Open** button. I will select the file named **default.tb7**. When you click the **Open** button, the title block will become attached to the mouse pointer, and you can place it in your circuit like you would any other part. Place it in the lower right corner of your schematic:

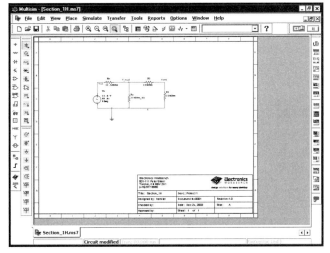

Place the mouse pointer over the title block and then press the **F8** key a few times to zoom in on the title block::

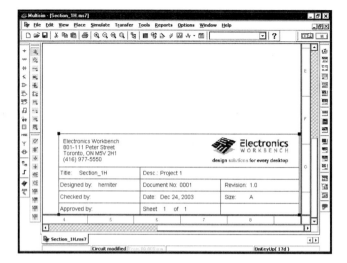

To modify the contents of the title block, select **Tools** and then **Modify Title Block Data** from the menus:

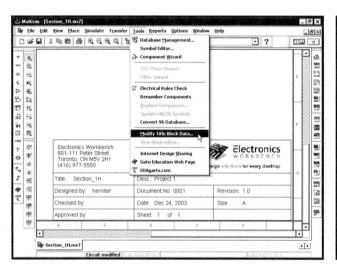
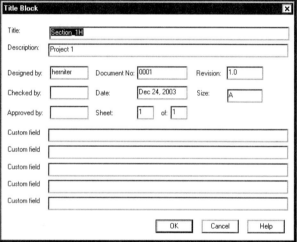

Fill in the title block fields with the information you want and click on the **OK** button. Some of the information is obvious as to where it goes, some is not. We will add some test data to see where it is placed in the title block. We can always remove it later:

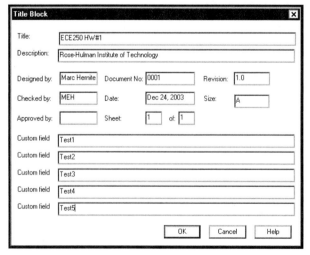

The contents of the title block are:

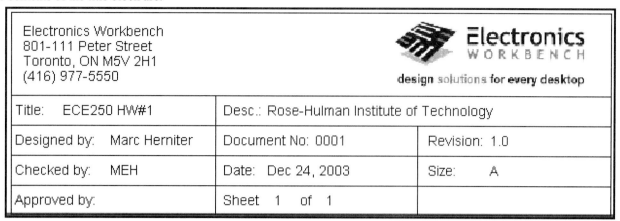

We see that the information contained the five custom fields was not used in this title block. The data is most likely used for documentation purposes rather than displayed on the title block.

Now that we have gone to the trouble of placing the title block in our schematic and adding all of the information to the title block, we can easily hide the title block if we want to. If you click the **RIGHT** mouse button on an empty space in the schematic a pull-down menu will appear:

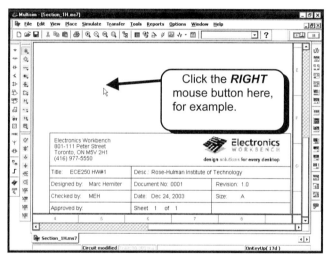
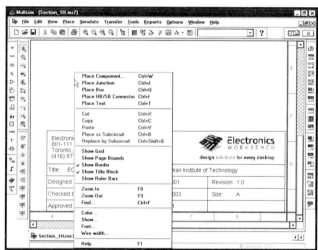

Select **Show Title Block** from the menu. This will toggle the display of the title block. In this case, it will hide the title block since it is currently being displayed:

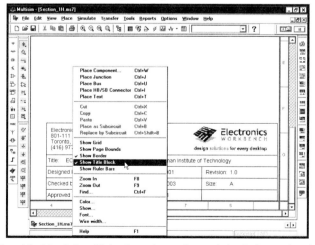
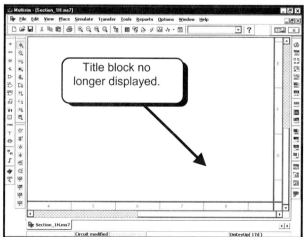

The title block is still in the schematic, and the information is still available, but it is not being displayed. To display the title block, just repeat the procedure:

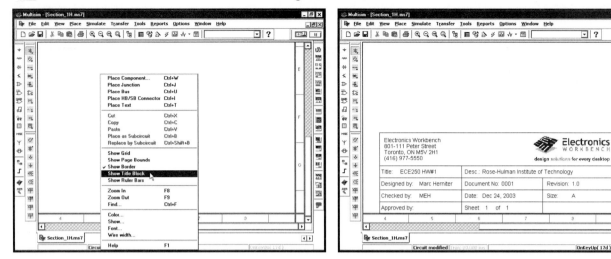

1.I. 3D Components

A new feature of Multisim 7 is the ability to place parts as three-dimensional components such as ⎯⎯ or ⎯⎯ rather than as abstract symbols like ⎯⎯/\/\/⎯⎯ and ⎯⎯▶|⎯⎯. These symbols help students visualize what a circuit would look like when they actually wire up that circuit, and help students bridge the gap between simulation and realization. We will show a small demonstration of the 3D parts here. Most of the simulations in this manual are done with the standard symbols. However, any of the 3D parts could easily be substituted if the 3D part is the correct value.

We will start with an empty schematic. To place a 3D part, we need to display the 3D Components bar. Click the **LEFT** mouse button on the Show 3D Components Bar button as shown:

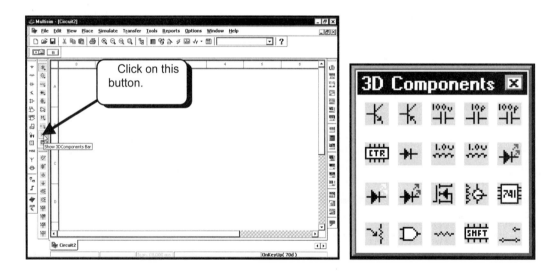

This toolbar lists most of the available 3D components. A few more components are available than are shown here, but this toolbar lists the majority of the components. We will first place a 1.0 kΩ resistor in our circuit. Click the **LEFT** mouse button on the **Place Resistor1_1.0k** button . A resistor will become attached to the mouse pointer. Place the resistor in your circuit using the techniques shown earlier:

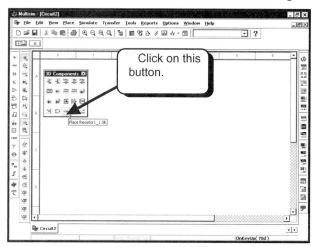
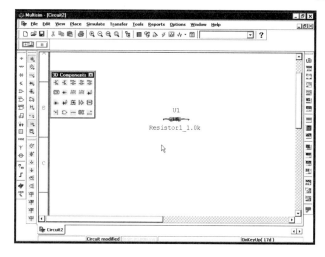

There are two things to note about the resistor. One is that it looks just like a resistor we would find in the lab, complete with color code. Second, we will find that we cannot change the value of this resistor.

Next, we will place a switch. Click the *LEFT* mouse button on the **Place Switch1** button. The component will become attached to the mouse pointer and move with the mouse. Place the switch in your circuit next to the resistor:

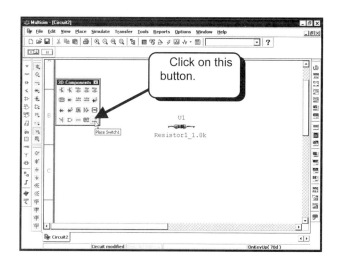

The last component we will place in our circuit is a red LED. Click the *LEFT* mouse button on the **Place Led1_Red** button and place the component in your circuit.

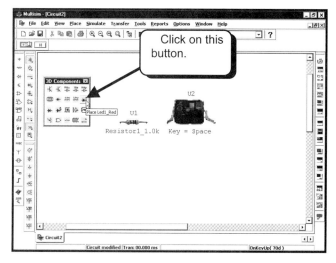
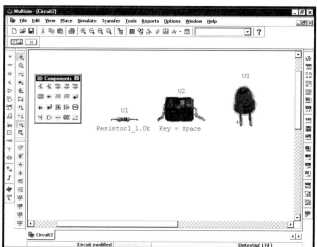

We are now done with the **3D Components** toolbar, so close it by clicking on the ⊠ in the toolbar title bar:

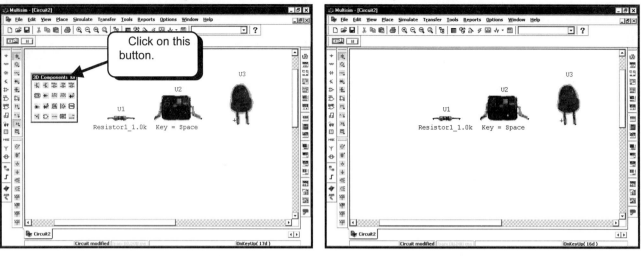

Next, add a DC voltage source as shown earlier and wire the components together as shown:

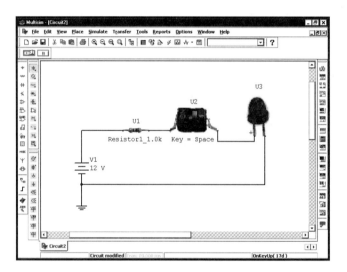

Wiring with 3D components is done the same way as with the standard components.

We now wish to run the simulation. Although simulations are covered in great detail in later chapters, because of the coolness of the 3D components, we will show a simulation here. To start the simulation, press the **F5** key.

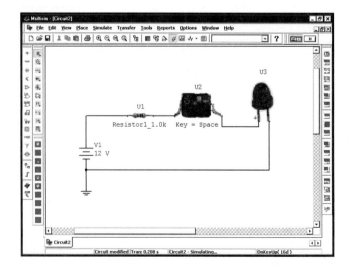

Nothing happened! This is because the switch is off and no current flows through the circuit. To turn the switch on, press the **SPACE** bar. Current will flow and the LED will illuminate:

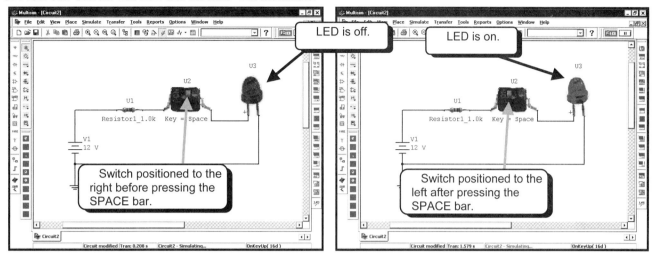

If you repeatedly press the **SPACE** bar, you will see the LED blink on and off. We will not do a more complicated simulation here, but you can use any of the components in the later simulations to get a more realistic feel for the simulations, if the components are the same as those used in later simulations. To terminate the simulation, press the **F5** key again.

There is one last thing we would like to look at before we end this section. Double-click the ***LEFT*** mouse button on the resistor graphic to edit its properties:

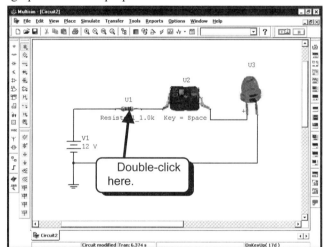 

We cannot actually change the value of the resistor using this dialog box. However, this allows us an easy way to see all of the 3D components available to us. Click on the **Replace** button:

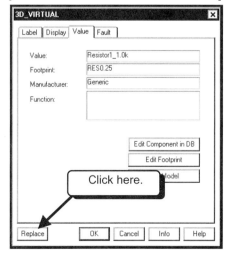 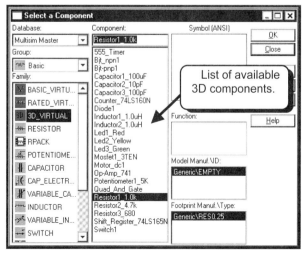

Only a handful of components are available at this time. For resistors, the only values available are 1.0 kΩ, 4.7 kΩ, and 680 Ω. If you select a component, you can see a description of the component. For example, the Bjt_npn1 is actually a National Semiconductor 2N3904 transistor:

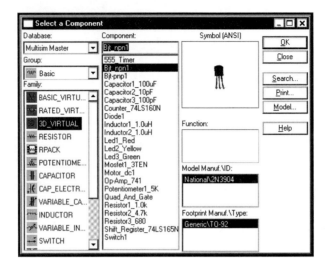

We use the 2N3904 in many of our later simulations, so you could use this 3D symbol rather than the schematic symbol if you find the 3D symbol more appealing.

We used this dialog box earlier to place components, so we know how to use it. Here, we are just showing it so that we can see what 3D components are available. After browsing through the components, click the **Close** button to return to the schematic.

1.J. Problems

For the problems below, draw the specified circuit with the same parts. Use the same part numbering and give the nodes the same names if specified. Turn in a schematic that includes a title block with your name, class, and problems specified in the title block. Most of the circuits contain components that we have not yet used.

Problem 1-1:

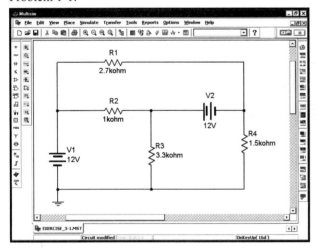

Problem 1-2:

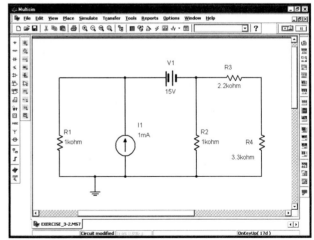

Multisim 7   Editing a Basic Schematic   43

Problem 1-3:

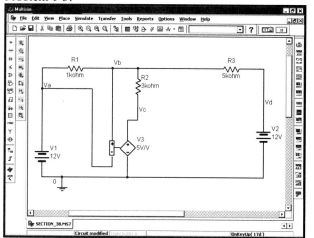

Problem 1-6:

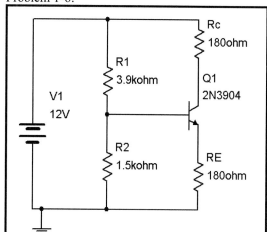

Problem 1-4:

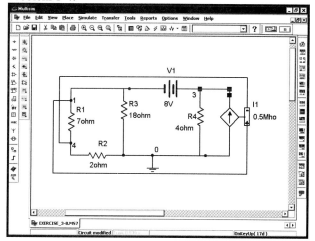

Problem 1-7:

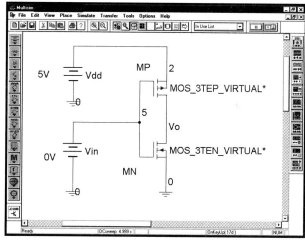

Problem 1-5:

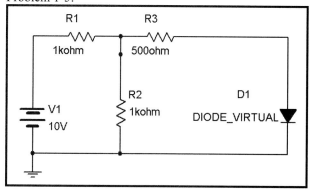

Problem 1-8:

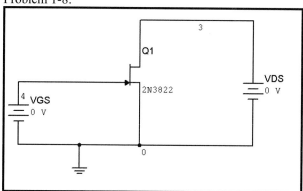

Problem 1-9:

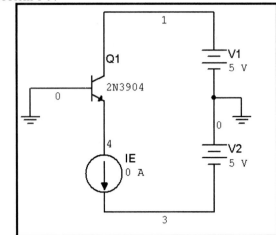

Problem 1-11:

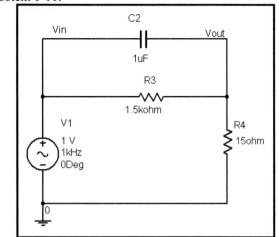

Problem 1-10:

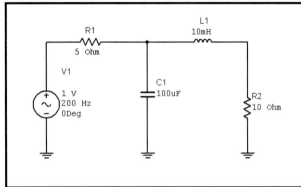

Problem 1-12:

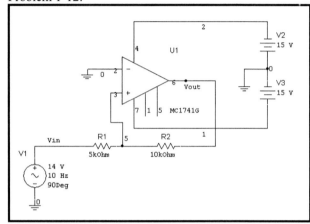

Problem 1-13:

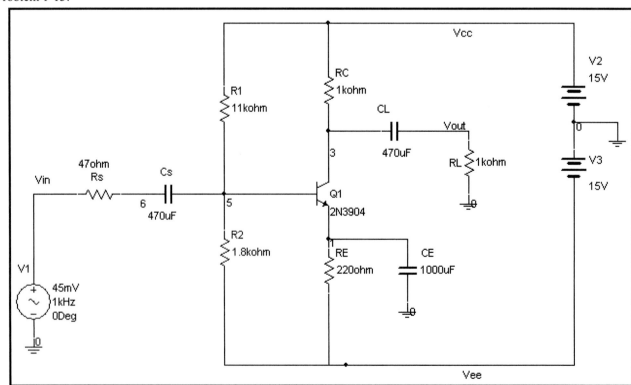

Problem 1-14:

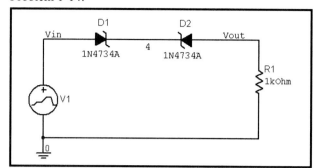

Problem 1-15:

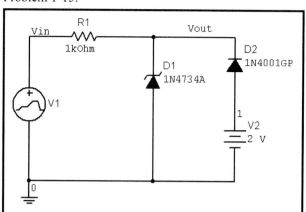

Problem 1-16:

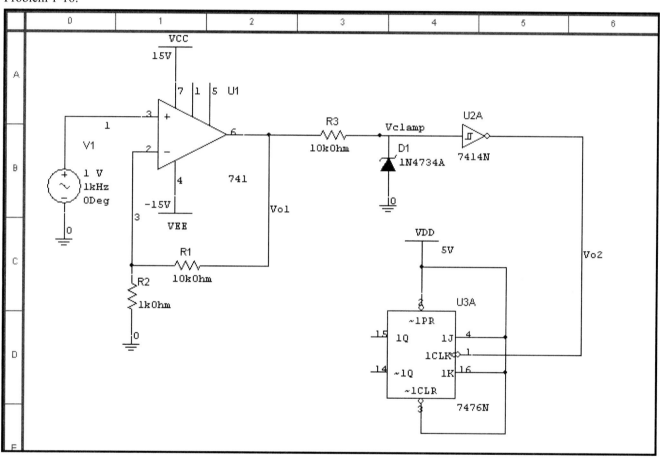

Problem 1-17:

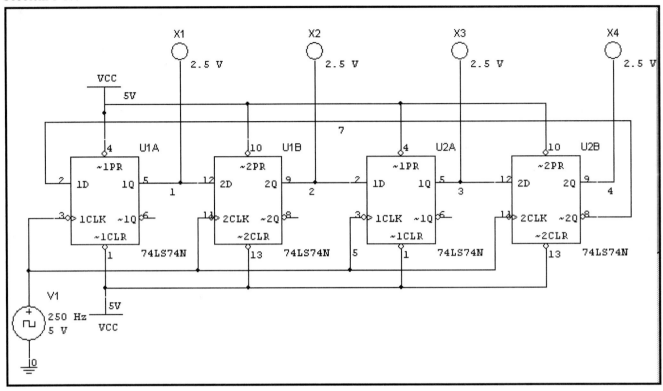

Problem 1-18:

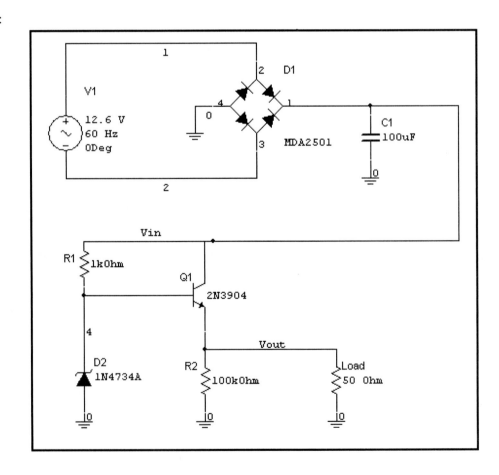

PART 2
The Postprocessor and the Grapher

The Postprocessor and the Grapher are two Multisim programs that allow you to display the results obtained from a SPICE simulation graphically. We will be using these programs extensively throughout this manual to display the results of simulations. Various aspects of the Postprocessor and Grapher are discussed in sections throughout this manual. However, if you pick specific examples you may miss those showing how to use some of the tools provided by the Postprocessor and Grapher. The result may be that a section in the manual you are currently using refers to a tool that was discussed in a section that you did not cover. To avoid this problem, we will review the most frequently used tools in the Postprocessor and the Grapher.

This entire chapter is based on the graphs we create in sections 2.A through 2.C. It is best to do this entire chapter in one sitting because the sections after section 2.C use the graphs created in the first three sections. However, if you are not able to do this, you should cover sections 2.A through 2.C, and then finish by doing section 2.D. This section shows you how to save and restore the graphs created in the first three sections. Using the methods in section 2.D, you can easily re-create your work in a few steps and then continue with the later sections.

To demonstrate the Postprocessor and Grapher, we will simulate the circuit drawn in Part 1 with a Transient Analysis:

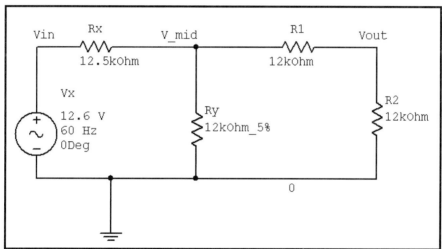

One important note about the schematic above is that we have renamed the nodes **Vin**, **V_mid**, and **Vout**. We have also selected to display the node names on the schematic. We need to know the node names in order to use the Postprocessor. See section 1.G for instructions on naming nodes and displaying node names on the schematic. Assuming that you have followed and completed Part 1, we will start by setting up the Transient Analysis. Select **Simulate**, **Analyses**, and then **Transient Analysis** from the Multisim menus:

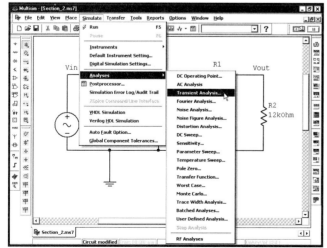
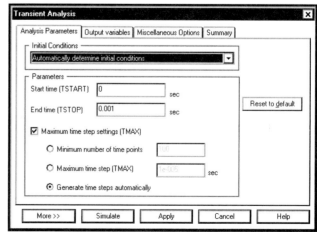

Change the parameters indicated below:

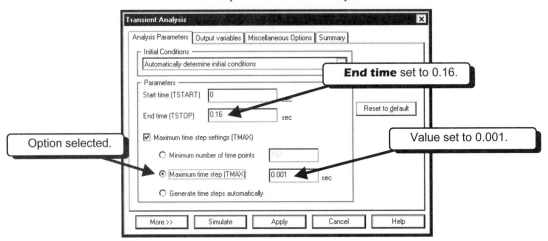

These options are discussed in detail in Part 6 and will not be discussed here.

Whenever you run a SPICE simulation, whether it is a Transient Analysis, AC Sweep, DC Sweep, or one of the others, you must specify the **Output variables** for that simulation. The **Output variables** you select are the only ones that can be plotted by the Postprocessor and the Grapher. If you have a circuit with 100 circuit elements, only the variables you specify as **Output variables** will be accessible to the Postprocessor and Grapher. Select the **Output variables** tab:

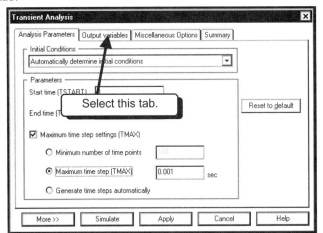
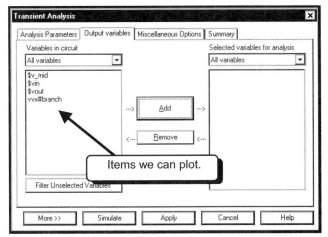

We see that the three node voltages (**$v_mid**, **$vin**, and **$vout**) are available as selections. The item vvx#branch is the current through voltage source Vx. In a typical simulation, you would only select the items that you are interested in. For this example, we will select all of the variables listed. You can add one variable at a time, or add several at one time. To add a single item, click the *LEFT* mouse button on the item to select it. For example, click the *LEFT* mouse button on the text **v_mid** to select the item:

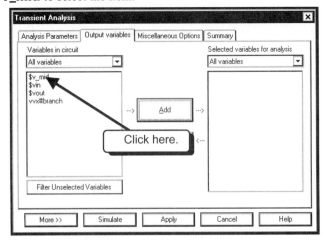
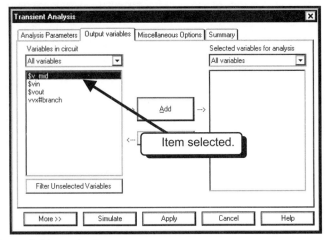

Click on the **Add** button to make the voltage data for that node available to the Postprocessor:

Multisim 7   The Postprocessor and the Grapher   49

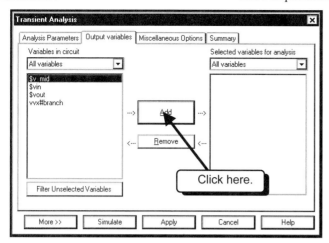 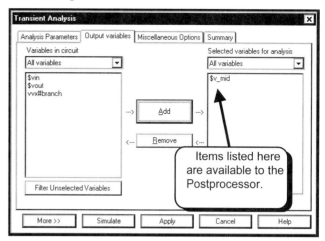

Items listed in the right pane are available to the Postprocessor and may be plotted when the simulation is complete.

We would like data for all voltages and currents to be available for our example, so we need to select more of the items in the left pane. To select multiple items, hold down the **CTRL** key and then click the *LEFT* mouse button on the items you want to select:

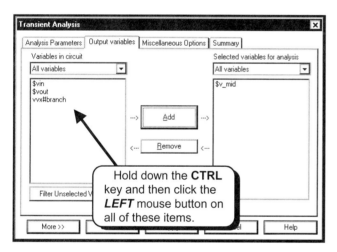 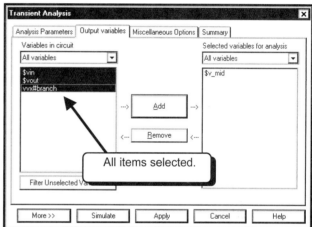

Click on the **Add** button to make the data available for the selected items:

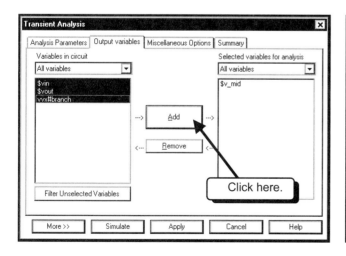 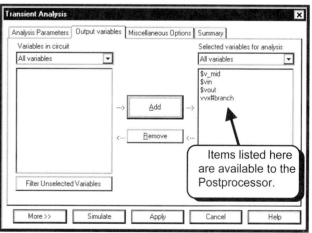

We are now ready to run the simulation, so click on the **Simulate** button. While the simulation is running, the Grapher will open and display a plot with all of the **Output variables** on the same plot:

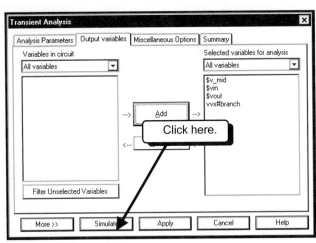 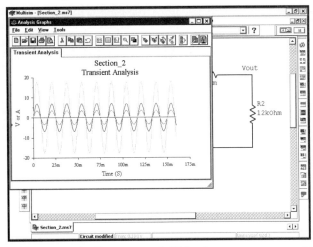

Every time you run a simulation, a plot of all the output variables is automatically generated. This plot may or may not be useful depending on how you selected the output variables. Each time you run a simulation, a new Grapher page will be generated and displayed in the Grapher, whether you want that plot or not. In a small circuit, you may only need to plot one or two items. This feature then becomes a quick way of generating a plot of the variables in which you are interested. If you are simulating a large circuit and are interested in many voltages and currents, the automatic plot will not be too useful because it plots everything on a single graph. In our graph, the current trace is useless because it is so small compared to the magnitudes of the voltages that the current looks like a straight line, which it is not. We will now use the Postprocessor to create a new plot.

We need to close the Grapher for the next section, so select **File** and then **Exit** from the Grapher menus to close the Grapher. This should leave Multisim 7 open with the circuit displayed:

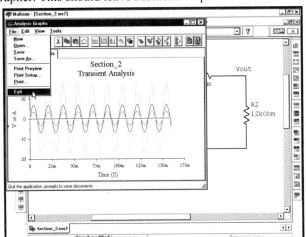 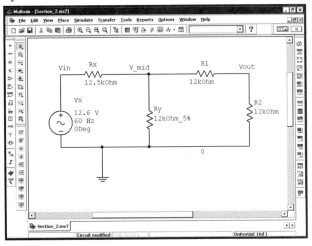

2.A. Creating a Plot with a Single Graph

We will now create a plot of the input voltage. Select **Simulate** and then **Postprocessor** to open the Postprocessor:

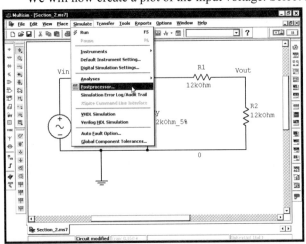 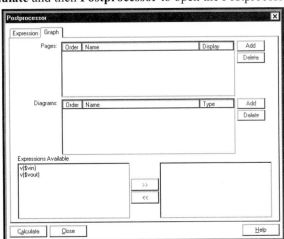

We need to create a new page within the Grapher window. Click on the **Add** button as shown:

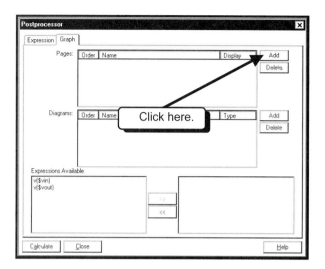 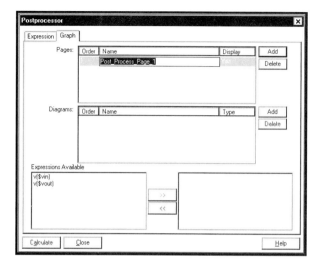

The default name of the page is **Post_Process_Page_1**. We would like to change the name to something more useful. Since the text is highlighted, we can just type in the new name, `Section 2 - Single Graph`, for example:

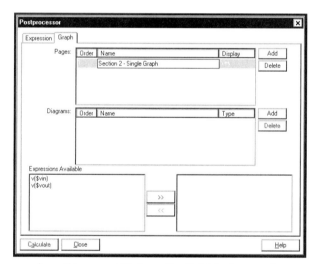

Next, we need to create a graph on this page. Click on the **Add** button as shown:

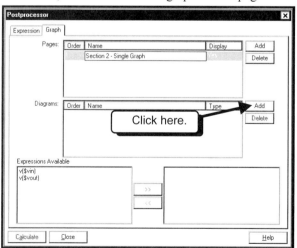 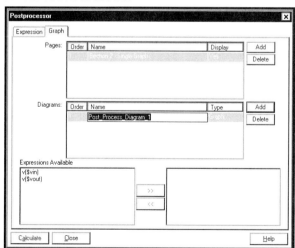

Multisim suggests a name for the graph, which will also be the title of the plot. We are going to plot a few of the circuit voltages, so name the graph `System Voltages`. Because the text is highlighted, all we have to do to rename the graph is type the new name:

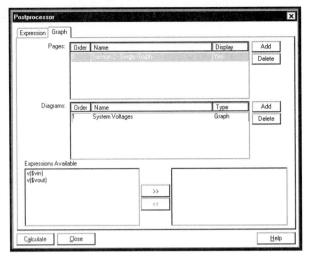

We are now ready to add a trace to the graph. We will first plot the input voltage. If you see a screen as shown below with the input and output voltage traces listed as available expressions, then you can skip this procedure and continue on page 55:

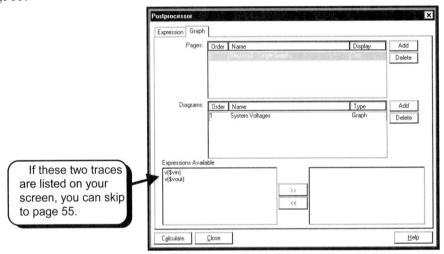

If these two traces are listed on your screen, you can skip to page 55.

However, if you do not have any available expressions as shown below, then you need to follow the procedure outlined in the next few screen captures.

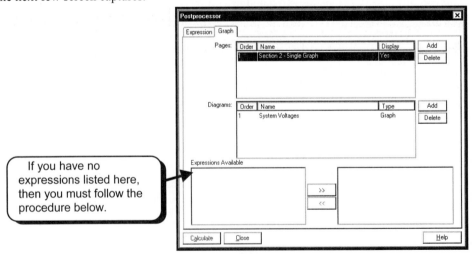

If you have no expressions listed here, then you must follow the procedure below.

We must add the trace expressions for the input and output voltages, Select the **Expression** tab:

Multisim 7   The Postprocessor and the Grapher   53

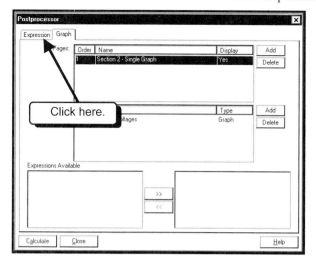
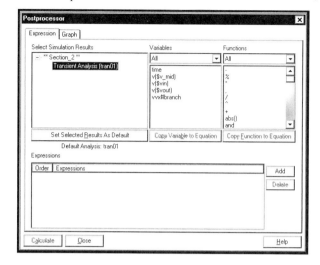

Click the **Add** button to create a new **Expression**:

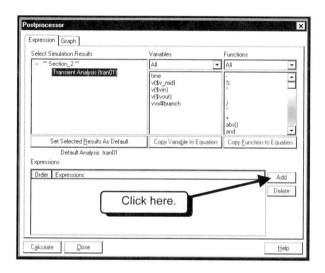
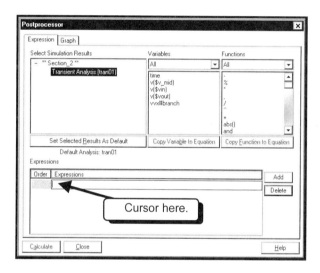

Click on the text **v($vout)** to select the text:

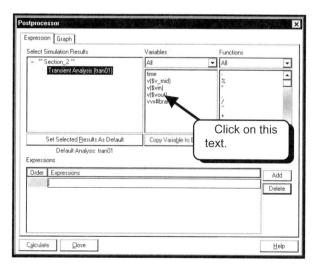
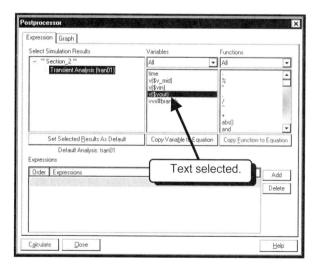

Click on the **Copy Variable to Equation** button:

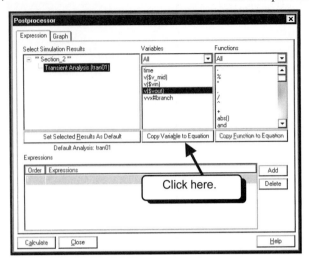 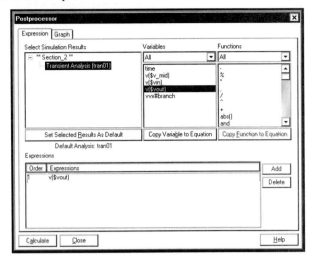

Click on the **Add** button to create a second expression and then repeat the procedure to create an expression for v($vin):

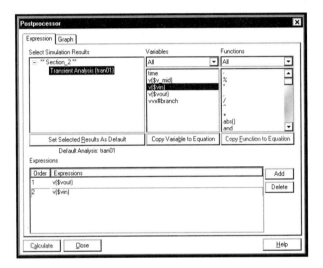

We now have the **Expressions** that we want to plot. Click on the **Graph** tab:

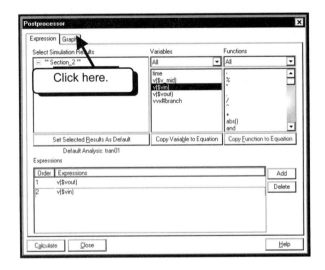 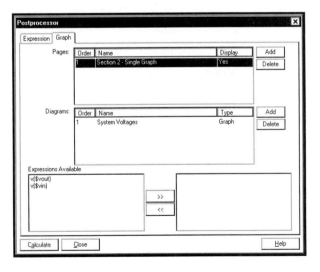

We are now at the same point we would have been at if had not had to add the traces expressions:

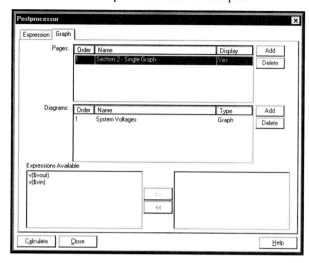

We must now select the graph we want to work with. Click on the text **1** to select the graph:

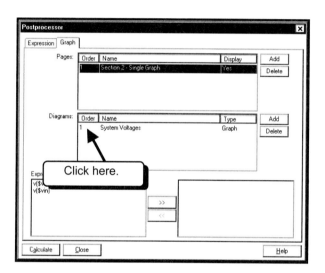
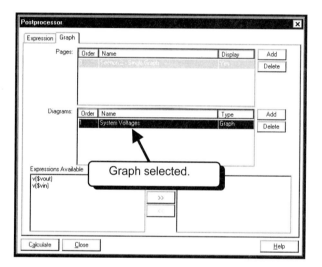

We can now add the trace expressions to the selected graph. Click on the text **v($vout)** to select the **Expression**:

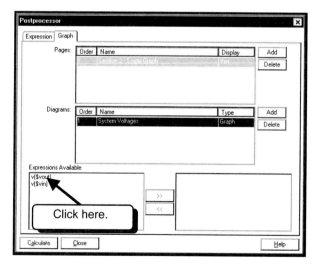
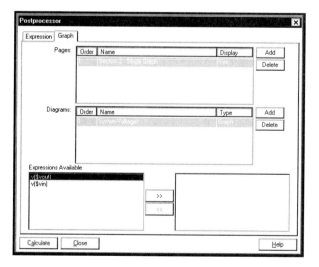

Click the >> button to add the expression to the selected graph:

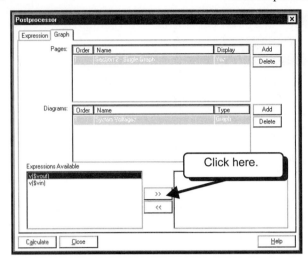 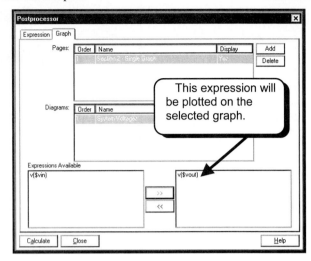

Repeat the procedure to add expression v($vin) to the right pane:

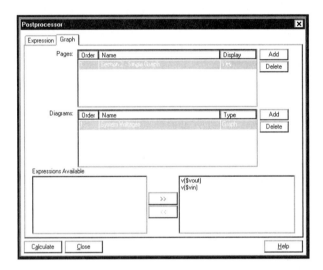

Waveforms for **Vin** and **Vout** will be plotted on the graph named **System Voltages** on the page named **Section 2 – Single Graph**. At this point we could add more traces if we wanted to. However, we will now generate the plot so that we can see what we have created thus far. To create the plot, click on the **Calculate** button:

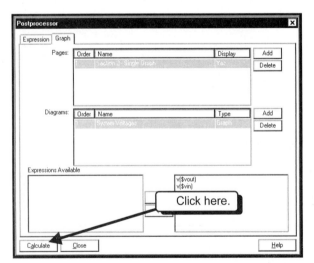 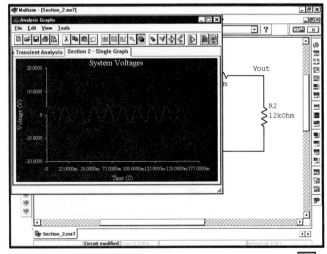

We would like to enlarge the Grapher to occupy the full window. Click the *LEFT* mouse button on the maximize icon as shown:

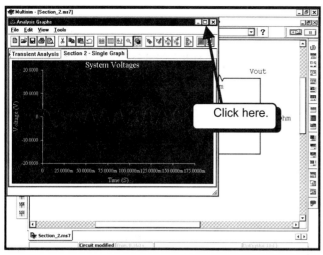 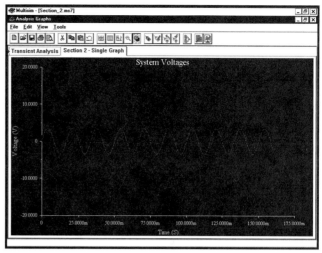

We note that all the plots have a black background. This is preferred by many people when viewing on a computer screen, but it does not work out too well when making screen captures. Thus, I will show how to reverse the colors of this plot. Select **View** and then **Reverse Colors** from the Grapher menus:

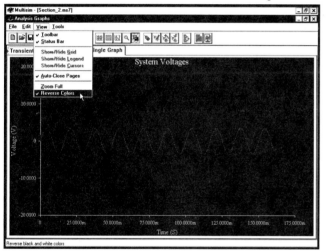 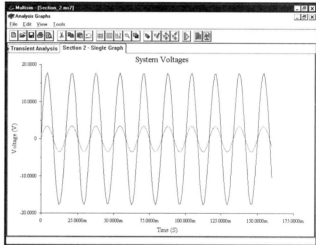

You do not have to make this change, but I will show all of my Grapher screen captures in reverse colors. This is done throughout this manual to make the screen captures more readable. Many other instruments display results graphically as well, and they also have the option of reversing the colors of the display.

Note that the Grapher has two tabs, one for each plot (or page created by the Postprocessor).

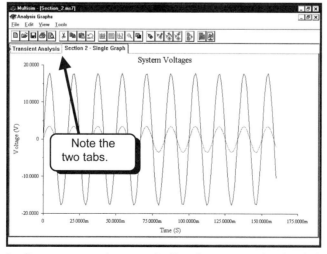

Right now we have two pages: the first was created automatically when we ran the simulation, and we generated the second using the Postprocessor. You can display either page by selecting the tab for the page:

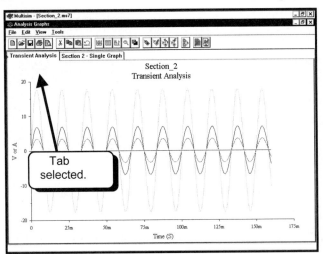
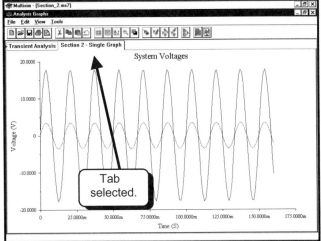

We will show how to customize graphs in later sections. For now we are just adding traces and creating pages. We will now add a third trace to this graph. We need to close the Grapher and open the Postprocessor. Select **File** and then **Exit** from the Grapher menus to close the Grapher:

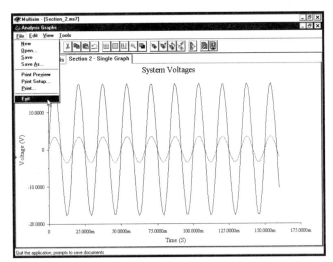
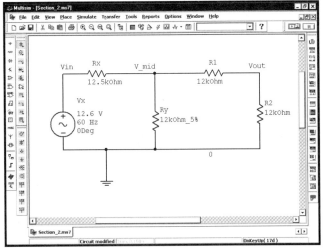

Next, select **Simulate** and then **Postprocessor** from the Multisim 7 menus to open the Postprocessor:

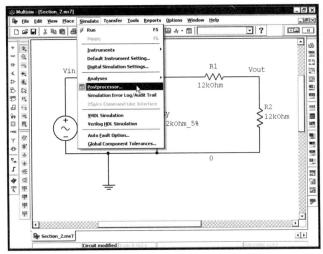
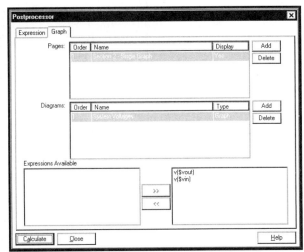

We would like to plot the voltage at note V_mid, but that trace is not listed as one of the available **Expressions**. We must add this expression, so select the **Expression** tab:

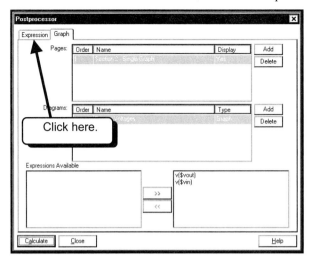 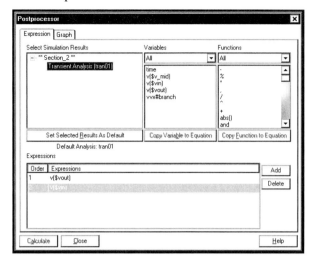

We need to create a new **Expression**, so click the **Add** button:

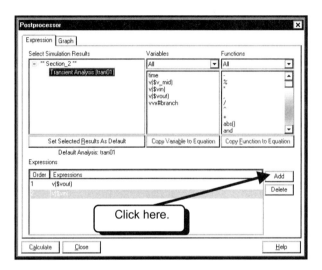 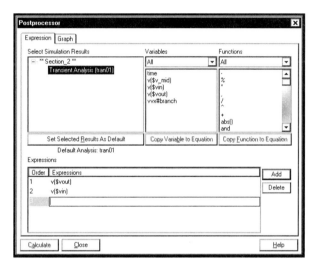

We will show later how to create complicated mathematical expressions, but for now we just want to plot the voltage at node v_mid. Type the text `v($v_mid)`

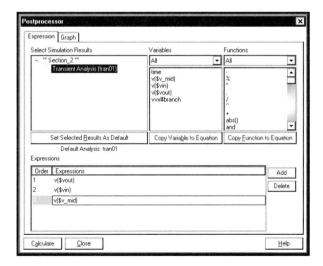

For a simple trace, this is all we need to do. Select the **Graph** tab:

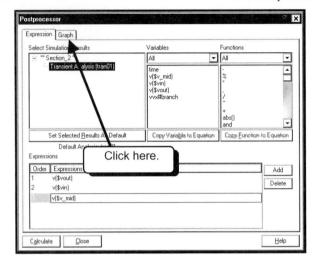 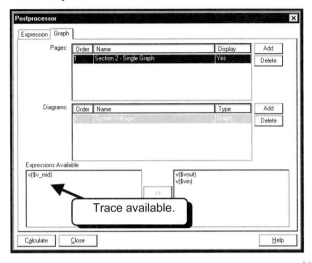

The expression is now available and we can plot it on a graph. Click on the expression to select it and then click on the >> button to add the expression to the graph:

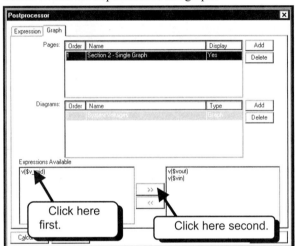 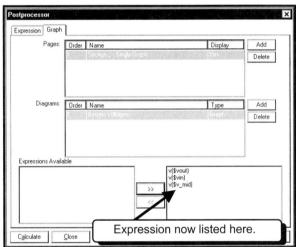

Click the **Calculate** button to create the graph:

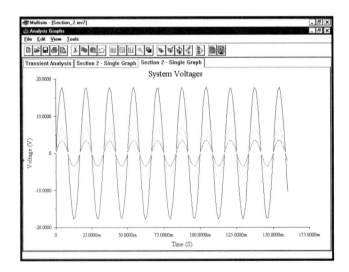

A new tab is created, and the graph that is drawn contains all of the changes that we made. We see that this plot contains three traces.

We notice that the Grapher now contains three tabs. When we click on the **Calculate** button, any earlier pages we created are not modified. Instead, a new page is created that contains all of the changes. If you select the middle tab, you will see the first page we created:

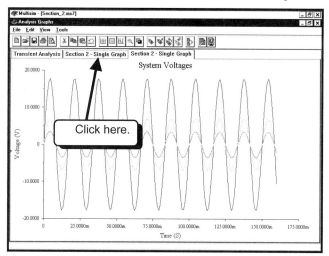
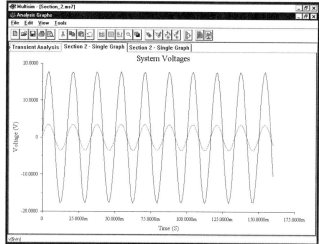

If you want to keep this page, then you don't need to do anything. If you want to delete the page, click the *LEFT* mouse button on the Cut button as shown. The page will be deleted:

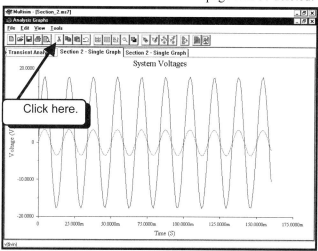
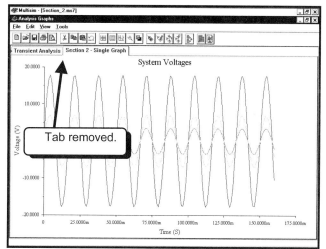

The page is deleted and only two tabs are left. The method shown can be used to delete unwanted pages in the Grapher. When you run several simulations, the Grapher can become filled with old plots that are no longer needed. Use this method to remove the unwanted plots.

2.B. Creating a Plot with Two Graphs

For our next example, we will create a new Grapher page that has two graphs on it. Assuming that we are continuing from the previous section, we must first close the Grapher. Select **File** and then **Exit** from the Grapher menus to close the Grapher:

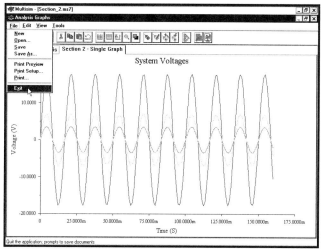
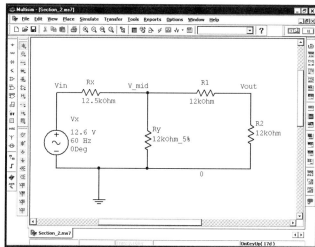

We can now open the Postprocessor by selecting **Simulate** and then **Postprocessor** from the Multisim 7 menus:

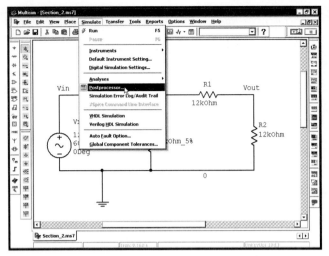 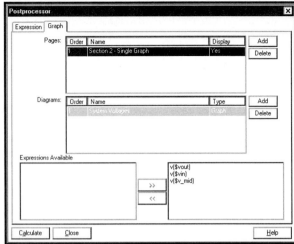

We want to keep the first page with the single graph on it and create a new page with two graphs on it, so we will create a new graph. Click on the top **Add** button as shown below to create a new page in the Grapher:

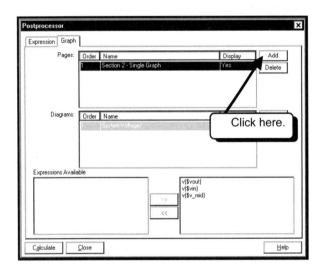 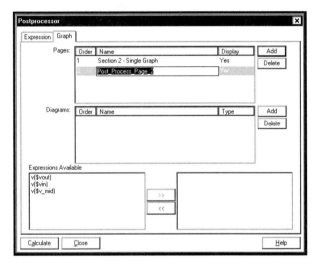

The name for the page is highlighted, so we can just type the new name to rename the page. Enter a name for the page, such as `Section 2 - Two Graphs`:

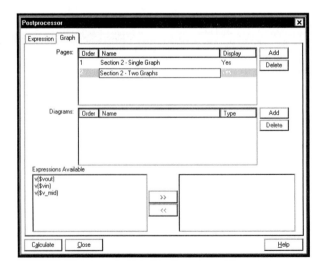

We now want to create a new graph on this page, so click on the lower **Add** button as shown below:

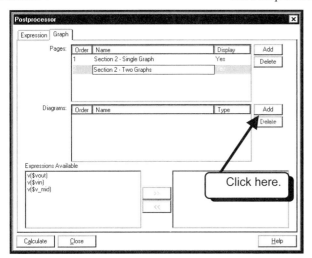 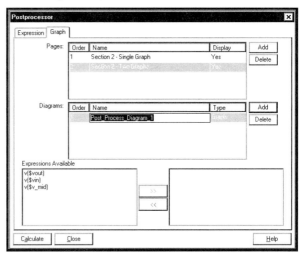

On this page I am going to plot the input voltage and the current through the source Vx. This first graph will be the current, so name the graph **Source Current**. The name for the graph is highlighted, so we can just type the new name to rename the graph.

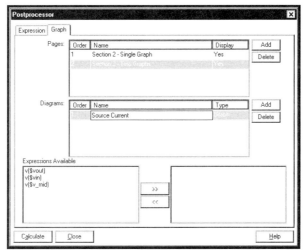

Currently, the page named **Section 2 – Two Graphs** is selected, and the graph named **Source Current** is selected. Any traces we add will be placed on the selected graph. We would now like to plot the source current on this graph. Unfortunately, the source current is not one of the expressions listed as an available expression, so click the **Expression** tab:

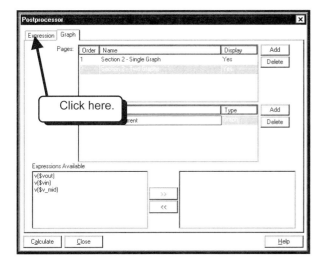 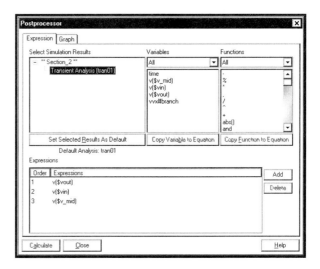

Click on the **Add** button to create a new expression and then double-click on the text **vvx#branch**:

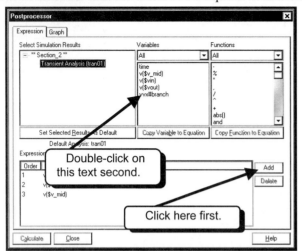 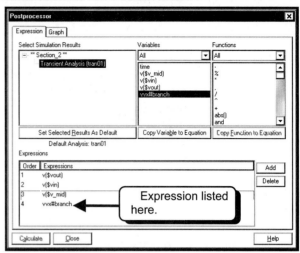

Click on the **Graph** tab:

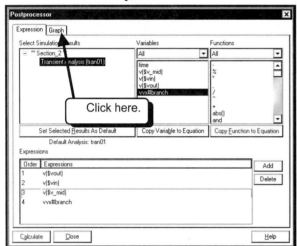 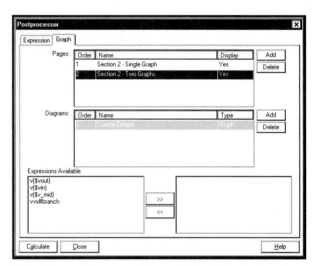

Notice that the page **Section 2 – Two Graphs** is selected and the diagram **Source Current** is selected. Any trace we add will be plotted on this page and graph. To plot the current, click the *LEFT* mouse button on the text **vvx#branch** and then click the ≫ button:

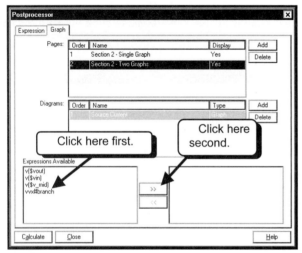 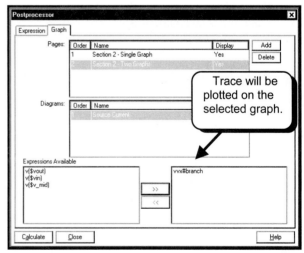

The trace is now listed in the right window pane and will be plotted on the selected graph.

We would now like to create a second graph on the selected page, which is named **Section 2 – Two Graphs**. Click on the lower **Add** button to create a second graph on the selected page and enter a name for the graph, such as `Input Voltage`:

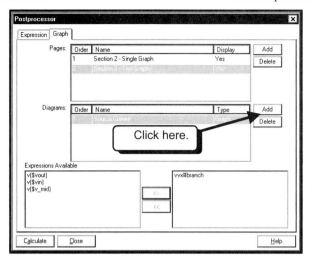 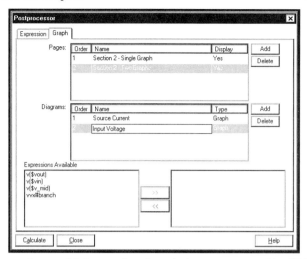

The page **Section 2 – Two Graphs** is selected and the diagram **Input Voltage** is selected. Any trace we add will be plotted on this page and graph. To plot the input voltage, click the *LEFT* mouse button on the text **v($vin)** and then click the ≫ button:

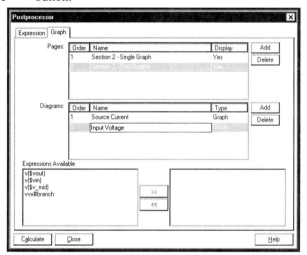 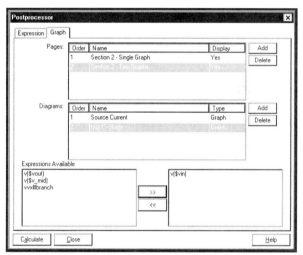

The trace is now listed in the right window pane and will be plotted on the selected graph.

We are now done with these graphs and need to draw the page. Click on the **Calculate** button to create the page in the Grapher:

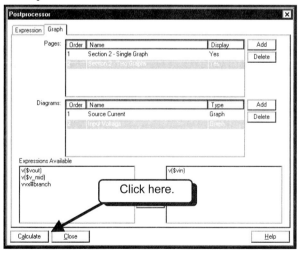 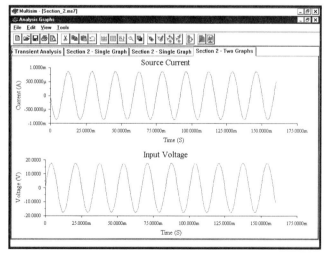

The page looks as expected. One thing we note about a branch current is that it is positive when it flows into the positive voltage terminal of the source. Because our voltage source supplies current to our circuit, the current is actually leaving the positive terminal when the voltage is positive. This is why the current is shown as negative when the voltage is positive.

Another thing we notice is that every time we click on the Calculate button, it creates a new page for every page listed in the Postprocessor. Our Postprocessor had two pages defined. When we clicked on the Calculate button, both pages were added to the Grapher window, not just the page we were working on. In our example, this had the effect of creating a second copy of the page titled **Section 2 – Single Graph**, because this page was already displayed in the Grapher from the previous example. Thus, if you define multiple pages in the Postprocessor, all of the pages will be created as new pages in the Grapher each time you click on the Calculate button. In our example, this has the effect of creating several unwanted pages in the Grapher. To fix this problem in our example, we will delete the unwanted pages. Click the ***LEFT*** mouse button on one of the tabs labeled as **Section 2 – Single Graph**:

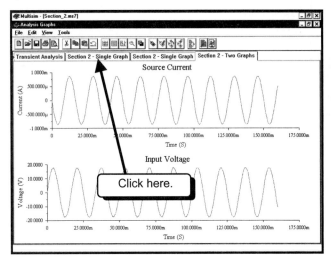 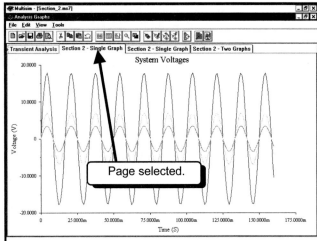

Click on the **Cut** button to delete the selected page:

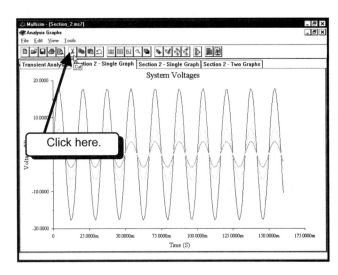 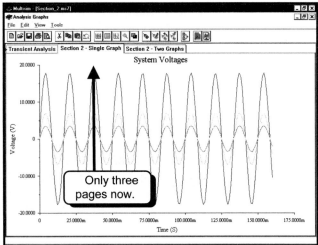

The page was deleted. You will need to use this technique to clear out many of the plots that appear in the Grapher.

2.C. Creating a Plot with Three Graphs

As a last example, we will create a new page with three graphs on it. On this page we will show some of the math functions available in the Postprocessor. Select **File** and then **Exit** to close the Grapher and then select **Simulate** and then **Postprocessor** from the Multisim 7 menus to open the Postprocessor. Next, click on the top **Add** button to create a new page and name it `Fancy Math Plots`:

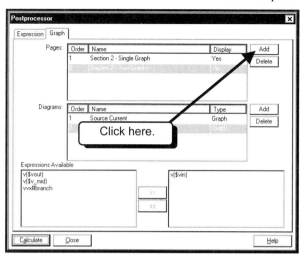 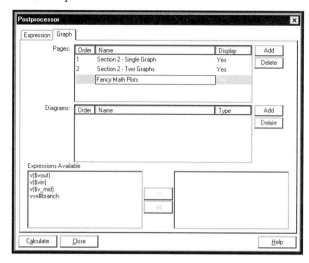

We now want to create three graphs on this page. Click on the lower **Add** button and name the graph `Resistor Power`:

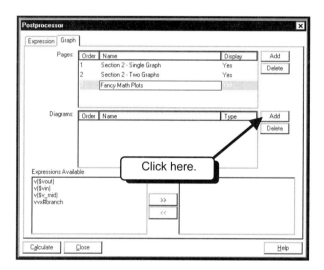 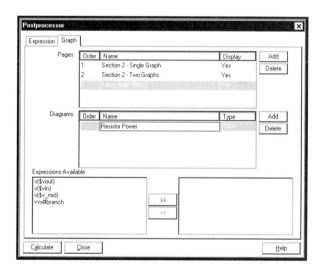

Click on the lower **Add** button and name the graph `Source Power`:

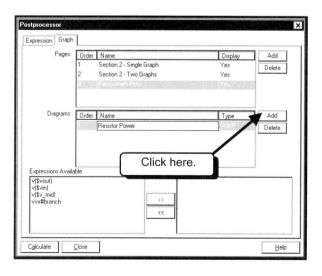 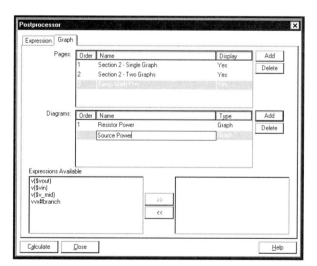

Click on the lower **Add** button and name the graph `Input Voltage and Current`:

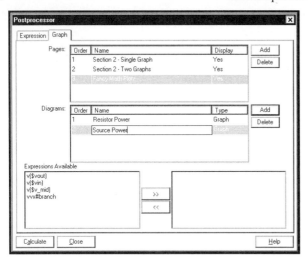 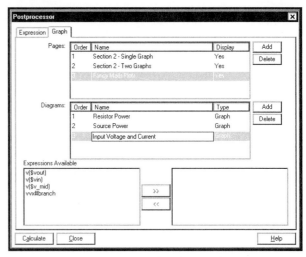

We would like to plot the source current and input voltage on the same plot. The only problem is that the current is about 10,000 times smaller than the input voltage in magnitude. In order to be able to see the current trace on the same set of axes as the voltage trace, we need to multiply the current trace by 10,000. (We can also solve this problem by using a different y-axis for the current trace. This method is covered in section 2.F.4.) We need to create a new trace expression 10000*vvx#branch, or modify the existing expression.

First, select the **Expression** tab and then click on the text **vvx#branch** to select the expression:

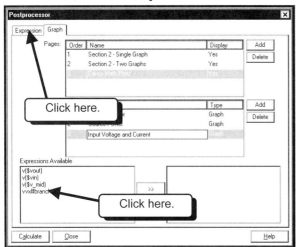 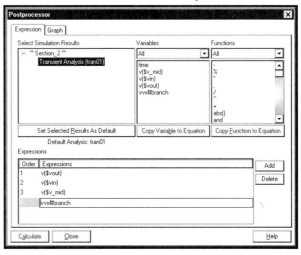

Notice that the cursor is placed to the left of the text **vvx#branch**, which is right where we need it. All we need to do is type the text **10000***

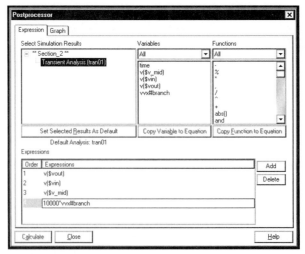

While we are in the **Expression** dialog box, we will create the expressions of the other traces we wish to plot. One trace is the power supplied by the source Vx. In terms of the variables used by Multisim 7, the equation for the power is v($vin)*vvx#branch. Click the **Add** button and then create the new expression:

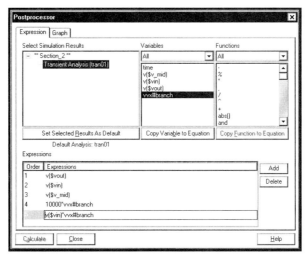

Lastly, we would like to plot the instantaneous and average power absorbed by Rx. The instantaneous power absorbed by Rx is the voltage across Rx times the current through Rx: `-vvx#branch*(v($vin)-v($v_mid))`. Click the **Add** button and then create this expression. You can type it out or use a combination of typing, double-clicking, and using the **Copy Variable to Equation** button:

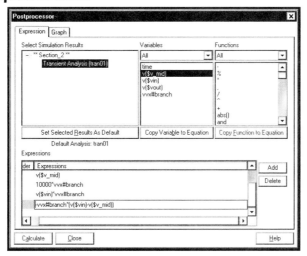

We would now like to create an expression that calculates the average power dissipated by the resistor. Multisim does not have a time average function; however, it does have a mean function. This function calculates the mean of all of the elements of a trace. The result of the mean function is a single number that is the mean of all data in the trace. This is not the same thing as a running time average, but it can be used as an approximation for calculating the time average power. Because the function returns a single value, it will be plotted as a straight line. First, click the **Add** button to create a new expression:

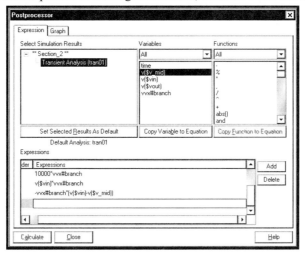

To see the available functions, click the *LEFT* mouse button on the down triangle ▼ as shown:

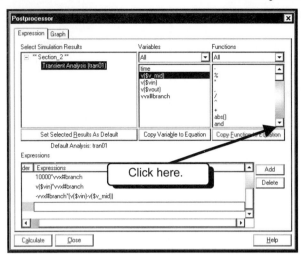 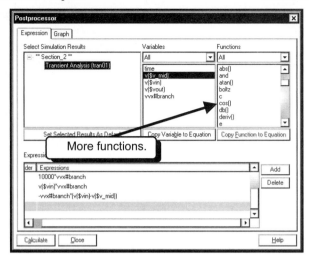

Some of the functions are obvious, others are not.* Scroll down the list until you see the text **mean()**. Double-click on this text to add it to the new expression:

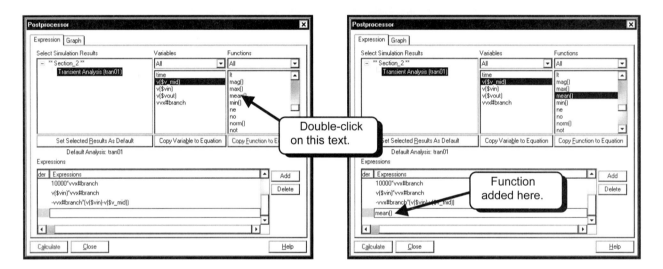

Use the mouse to place the cursor between the parentheses:

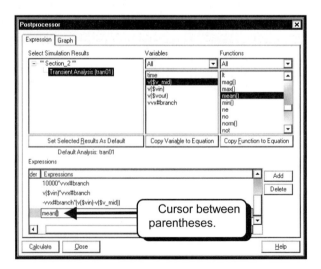

---

* To see a list of the functions and their properties, click on the **Help** button.

With the cursor located between the parentheses, you can create the remainder of the text string using the techniques covered earlier:

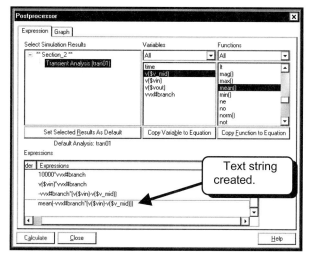

We have now created all of the trace expressions we need for all of the graphs we will be creating.

We can now add the traces to each graph. Select the **Graph** tab:

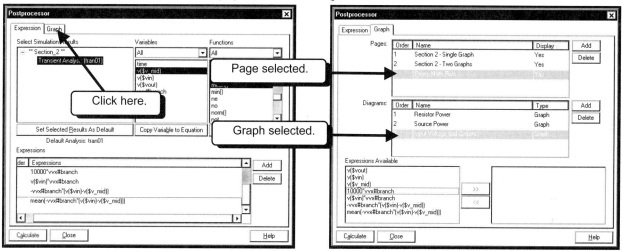

On my dialog box the page **Fancy Math Plots** is selected, and the graph **Input Voltage and Current** is selected. Any trace expression we plot will be plotted on the selected graph. On this graph, we would like to plot the expressions **v($vin)** and **10000*vvx#branch**. Click the *LEFT* mouse button on this text to select the expressions:

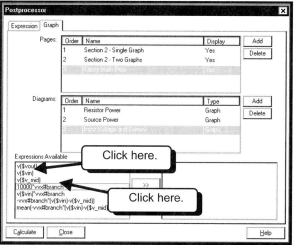 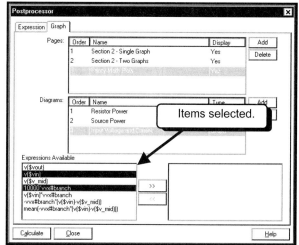

Next, click on the >> button to plot the selected expressions on the selected graph:

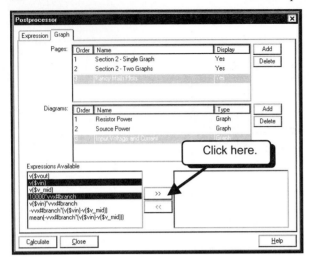
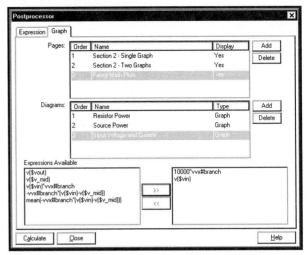

Next, we would like to work with the graph named **Source Power**. Click the *LEFT* mouse button on the text **Source Power** to select the graph:

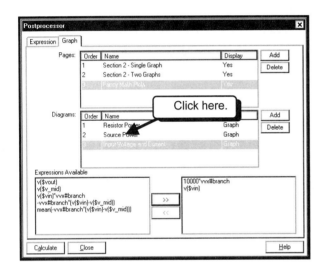
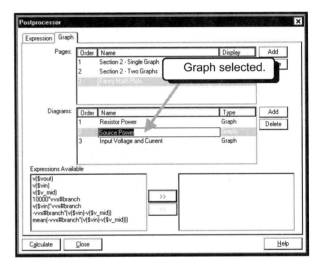

On this graph, we would like to plot the expression **v($vin)*vvx#branch**. Click the *LEFT* mouse button on this text to select the expression:

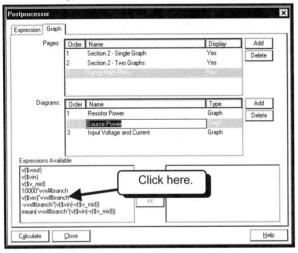
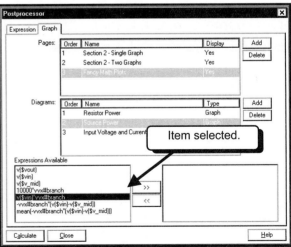

Next, click on the >> button to plot the selected expression on the selected graph:

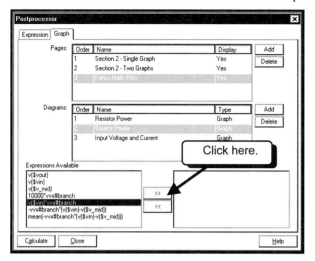 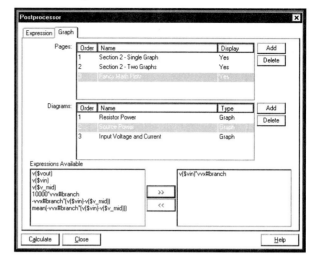

The last thing we want to do is add two traces to the Resistor Power graph. Select the graph named **Resistor Power**:

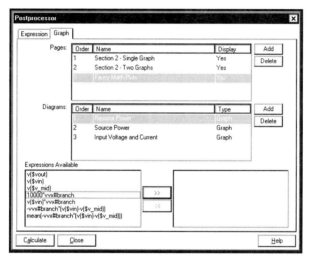

We would like to plot the average and instantaneous power absorbed by resistor Rx on this graph. Add these two expressions as shown below:

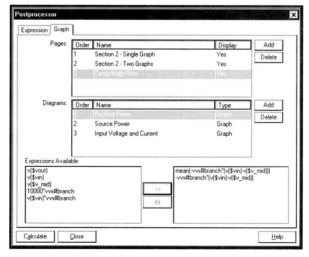

We are now done with selecting all of the graphs, so click on the **Calculate** button to create the graphs:

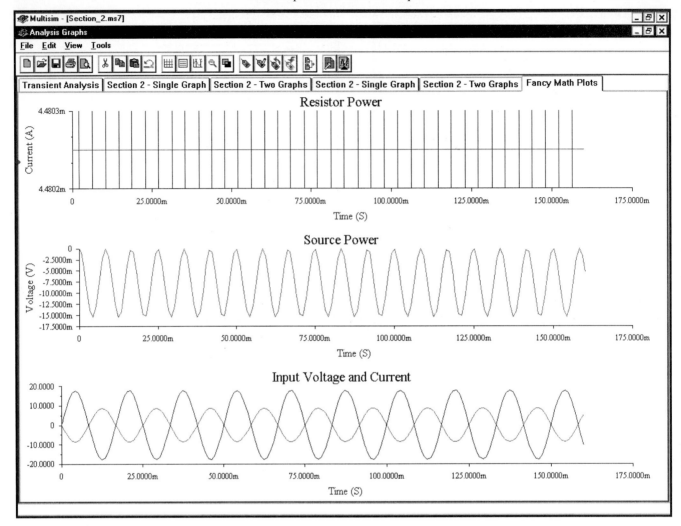

If you are using a Windows screen resolution of 800 by 600 or less, you will not see the titles **Resistor Power**, **Source Power**, and **Input Voltage and Current** on your page because there is not enough room to display the titles. Also note that when we clicked on the Calculate button, three plots were added to the Grapher window. We will need to delete some of the duplicate tabs.

Also note that the top plot does not have the correct y-axis scale, so the plot looks like a bunch of vertical lines. To fix this problem, click the *RIGHT* mouse button somewhere on the top plot and then select **Properties**:

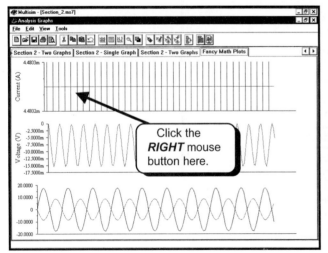
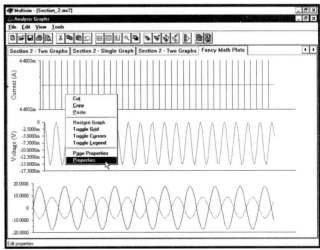

A dialog box for modifying the properties of the top plot will open:

Multisim 7 The Postprocessor and the Grapher 75

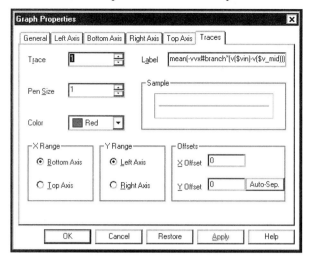

Select the **Left Axis** tab:

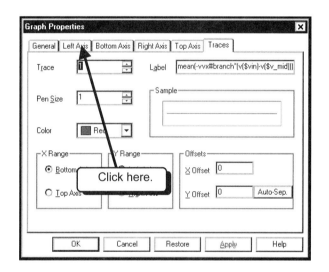 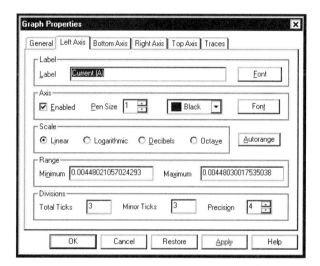

Change the minimum range to **0** and the maximum range to **0.01**. These settings specify the limits for the y-axis of the top plot:

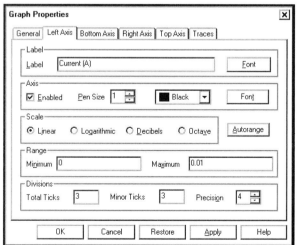

Click the **OK** button to apply the changes:

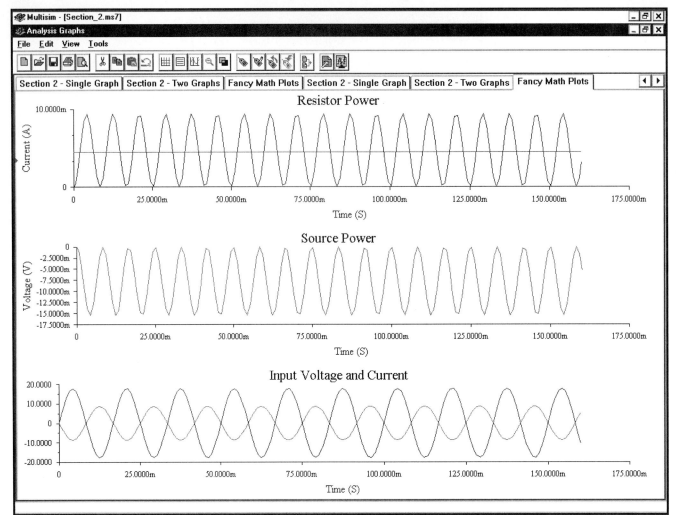

Before stopping and exiting Multisim, you may want to complete the next section because it shows you how to save and recover all the work we just did. After completing the next section, you can stop and then complete the remaining sections at a later time.

2.D. Saving and Loading Pages

We have spent a great deal of time creating and setting up pages and graphs with the Postprocessor. We would like to save this setup so that we can easily create these pages and graphs anytime we run this circuit. It turns out that the Postprocessor setup is automatically saved when we save a circuit. Thus, before we exit we must make sure that we save the circuit. First, we must exit the Grapher. Select **File** and then **Exit** from the Grapher menus to close the Grapher:

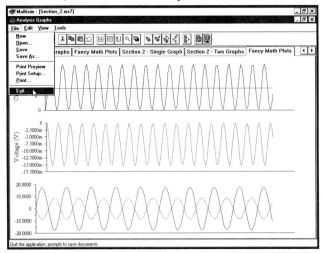
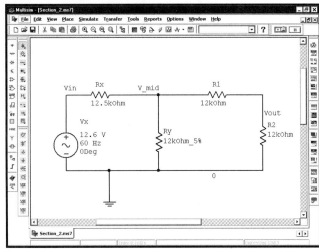

We now need to save the circuit, so select **File** and then **Save** from the Multisim 7 menus:

Multisim 7 — The Postprocessor and the Grapher 77

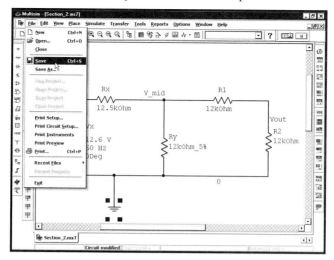

If this is the first time you have saved your circuit, you will have to specify a name for the circuit at this point.

We have now saved the Postprocessor pages and the circuit, so we can exit Multisim. At a later time, we can easily regenerate all of the simulation results with a few steps. Select **File** and then **Exit** to close down Multisim 7.

Now that we have saved the information, we will show how to restore the graphs from the previous sections. Open Multisim 7. You will start with a blank schematic:

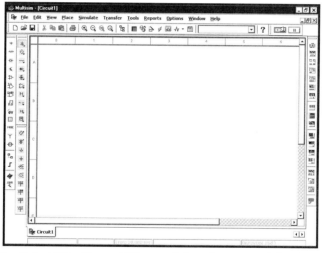

We must first open the circuit file. Select **File**, **Recent Files**, and then the name of the circuit file we were just using. In my example, I called the file **Section_2.ms7**:

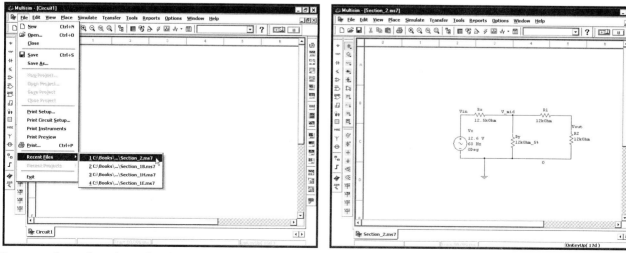

You can enlarge the schematic window and center the circuit if you wish.

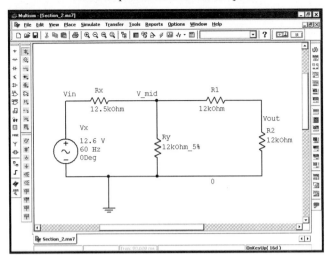

We must now rerun the simulation to generate the data for the circuit. We must repeat the simulation we ran earlier. Select **Simulate**, **Analyses**, and then **Transient Analysis** from the menus:

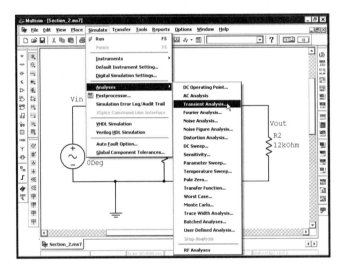

We must verify that the **Analysis Parameters** and the **Output variables** tabs of the dialog box are the same as before. They should be the same as before, because this information is saved with the circuit. However, it never hurts to check and make sure the information is correct:

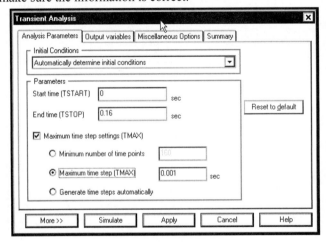
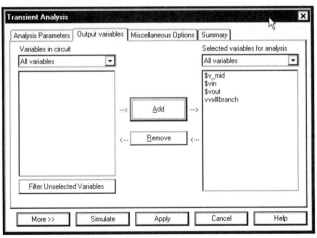

Both windows look correct, so click on the **Simulate** button to rerun the simulation:

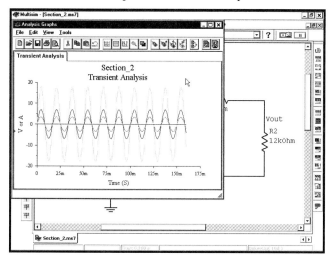

We must now open the Postprocessor to "calculate" the pages we created earlier. First, select **File** and then **Exit** to close the Grapher:

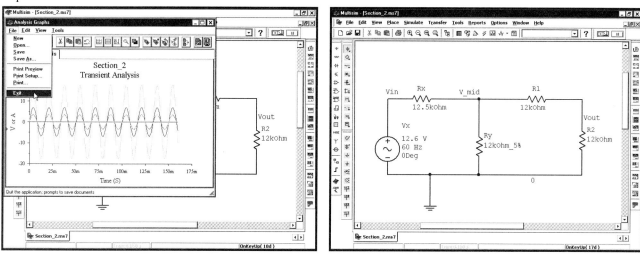

Next, select **Simulate** and then **Postprocess** from the Multisim 7 menus to open the Postprocessor:

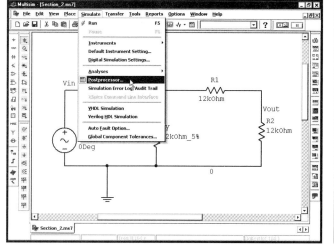
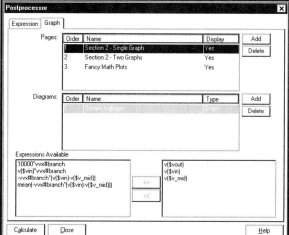

We note that all the pages and graphs we created in sections 2.A to 2.C are still listed in the Postprocessor. To display the pages and graphs, all we need to do is click on the **Calculate** button:

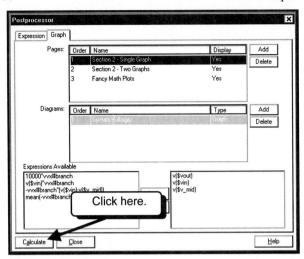
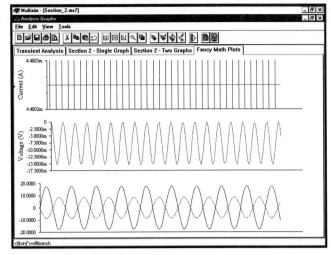

All three Postprocessor pages are created and are shown below:

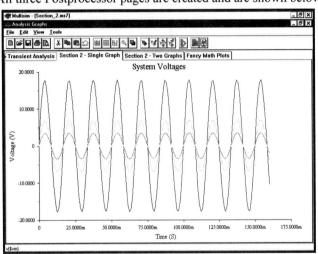
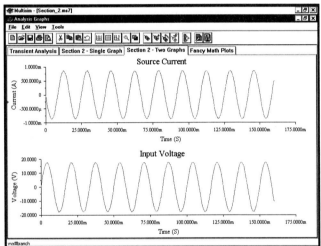

2.E. Deleting Items in the Postprocessor

Now that we have spent a great deal of time creating Postprocessor pages and graphs and traces, we will show you how to delete them.

2.E.1. Deleting a Trace from a Graph

We will first show how to delete a trace from a graph on a specific page. We will show how to delete the current trace on the graph labeled Input Voltage and Current on the page named Fancy Math Plots. Before we show how to delete a trace, we will empty the Grapher window to eliminate any confusion when new pages are drawn. If the Postprocessor is open, close it. If the Grapher window is not open, type **CTRL-G** to open the Grapher window:

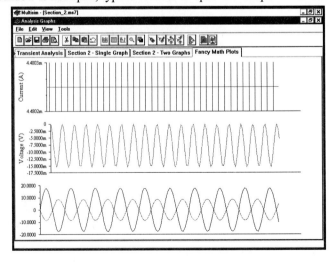

We want to clear all of these plots, so select **Edit** and then **Clear Pages** from the Grapher menus:

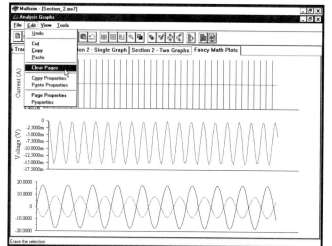
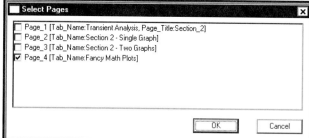

We want to delete all of the pages displayed in the Grapher, so select all of the pages and then click the **OK** button:

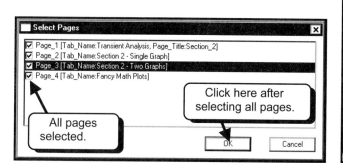
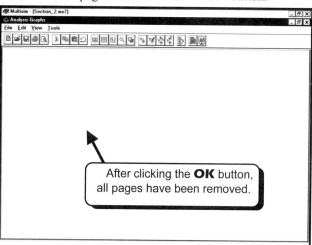

Now when we create new plots, it will be easy to identify them. Had we not cleared all the pages, we would have old plots and newly created plots, and there could be some confusion as to whether we are looking at old or new plots, and if the changes we made had any effect.

We must now open the Postprocessor and select the correct page, graph, and trace. Select **File** and then **Exit** to close the Grapher, and then select **Simulate** and then **Postprocessor** to open the Postprocessor. In my window, the correct page is not displayed:

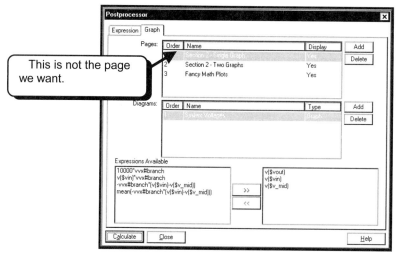

We want to work with the page named **Fancy Math Plots**, so click the *LEFT* mouse button on the text **Fancy Math Plots** to select this page:

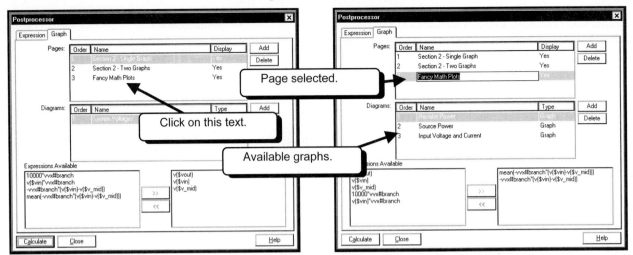

We have now selected the page named **Fancy Math Plots**. This page has three graphs on it, and we want to modify the graph named **Input Voltage and Current**. This graph is not yet selected. Click the *LEFT* mouse button on the text **Input Voltage and Current** to select the graph:

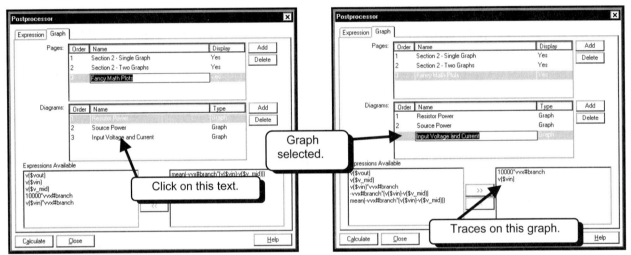

This graph has two traces on it. We want to remove the current trace, so click the *LEFT* mouse button on the text **1000*vvx#branch** to select it:

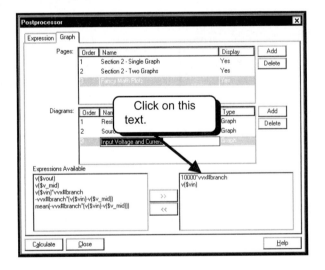
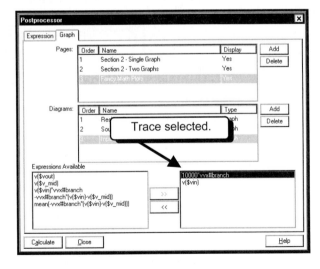

To remove the selected trace, click on the << button:

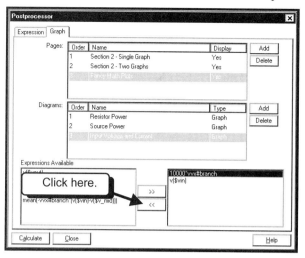 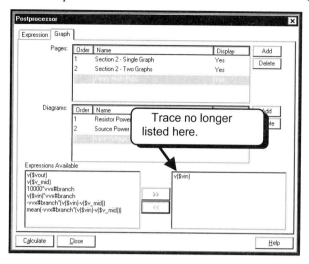

The trace has been removed from the graph. To see the effects of our change, we must click on the **Calculate** button to redraw all the Postprocessor pages. This will duplicate some earlier plots, but the newest version of the page named **Fancy Math Plots** will have the changes we just made:

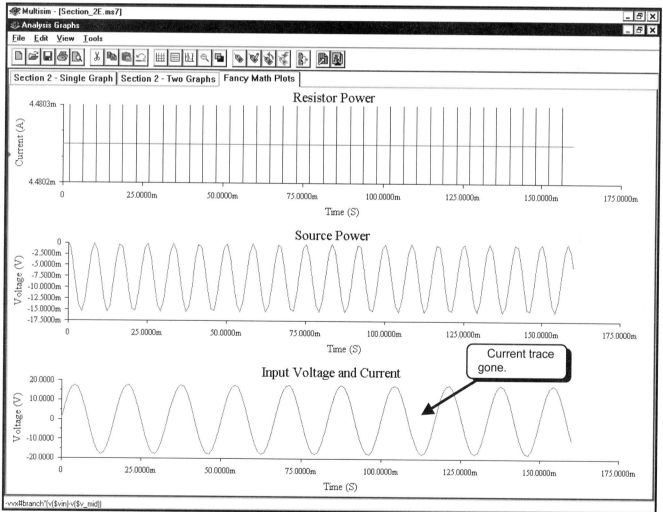

2.E.2. Deleting a Graph from a Page

We will now delete a graph from a page. Before we continue, delete all pages from the Grapher using the procedure shown at the beginning of section 2.E.1 on pages 80-81:

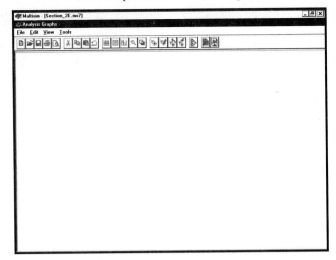

We would like to delete the graph named Input Voltage on the page named **Section 2 – Two Graphs**. Close the Grapher and open the Postprocessor:

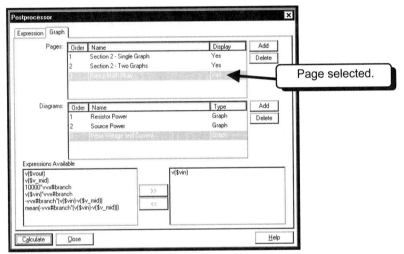

On my Postprocessor window, the page **Fancy Math Plots** is still selected from the previous example. We want to use the page named **Section 2 – Two Graphs**. To select this page, click the *LEFT* mouse button on the text **Section 2 – Two Graphs**:

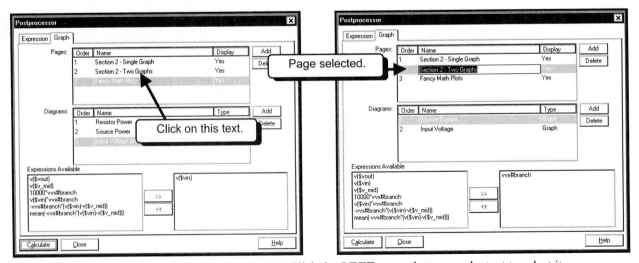

We must now select the graph named **Input Voltage**. Click the *LEFT* mouse button on the text to select it:

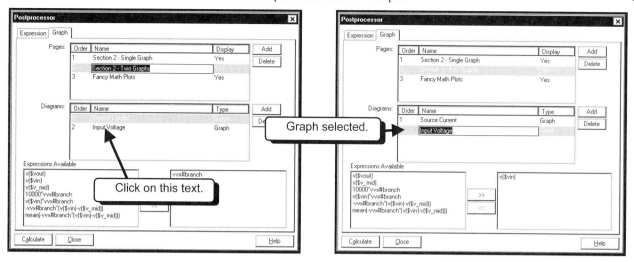

We have now selected the graph we want to delete, so click on the lower **Delete** button:

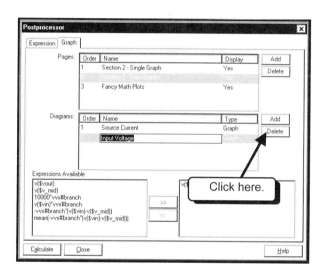
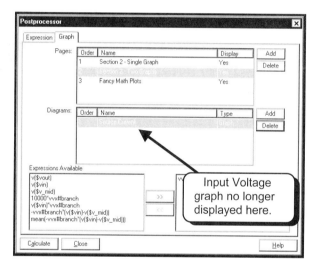

We see that the Input Voltage graph is no longer listed. This Postprocessor page only has a single graph now.

To see the results of the changes we have made, click on the **Calculate** button. Select the page named **Section 2 – Two Graphs**:

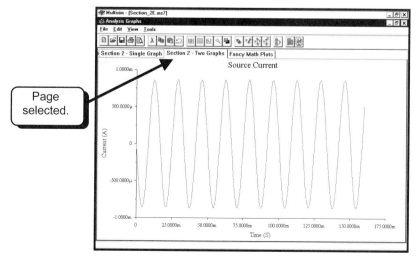

We see that this page now contains only a single graph.

2.E.3. Deleting a Page

We will now delete an entire page. Before we continue, delete all pages from the Grapher using the procedure shown at the beginning of section 2.E.1 on pages 80-81:

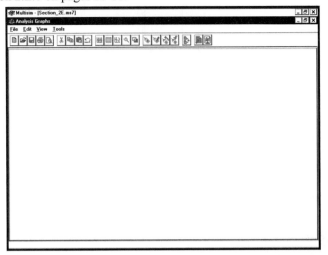

Close the Grapher and then open the Postprocessor. We would like to delete the page named **Section 2 – Single Graph**. Open the Postprocessor. In my window, the correct page is not displayed:

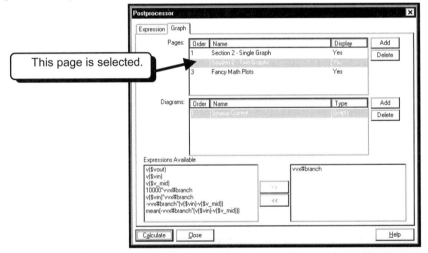

To select the page, click the **LEFT** mouse button on the text **Section 2 – Single Graph**:

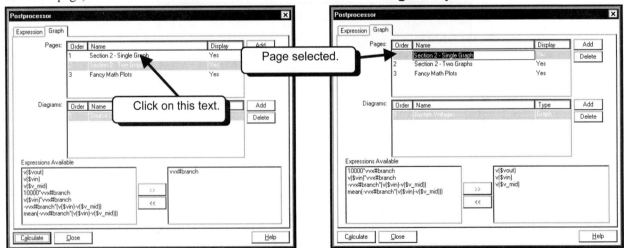

We have now selected the page we want to delete. To delete the page, click on the upper **Delete** button:

Multisim 7 — The Postprocessor and the Grapher — 87

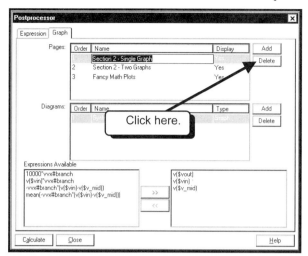 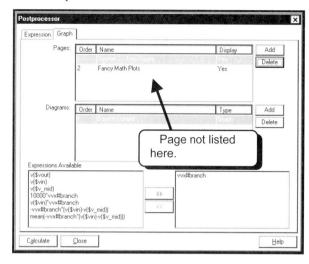

The page is no longer listed.

To see the results of the changes we have made, click on the **Calculate** button. The Grapher will now have only two tabs, indicating only two pages. The two pages are shown below:

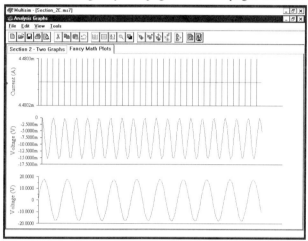 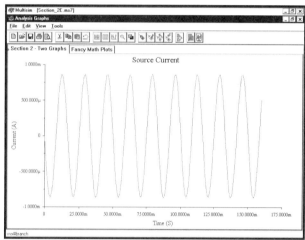

No graph titles are shown on the left plot because this is an 800 × 600 resolution screen capture. In earlier screen captures of the page with three plots, the screen captures were taken at 1024 × 768 resolution and these plots displayed the titles of each graph.

2.F. Modifying the Plot Properties in the Grapher

The Postprocessor allows us to create graphs and place traces on those graphs. The Grapher allows us to customize the graphs. This includes adding a grid and a legend, modifying trace widths, changing labels, and changing the axis scales. We will continue with the pages generated in the earlier sections. If you do not have these pages displayed, follow section 2.D to load and regenerate the plots. We will use the Grapher page with a single plot:

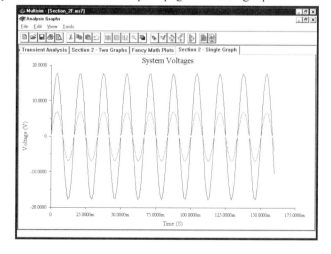

2.F.1. Adding a Grid and a Legend

There are two ways to add a legend and a grid to a graph. We will show both techniques here. On the page shown above, click the ***RIGHT*** mouse button on an empty space in the graph. A menu will appear:

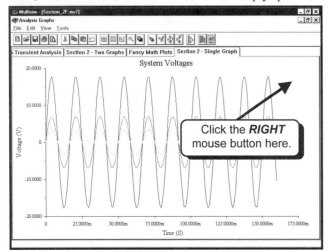
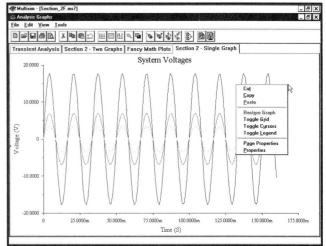

To turn on the grid, select **Toggle Grid** from the menus:

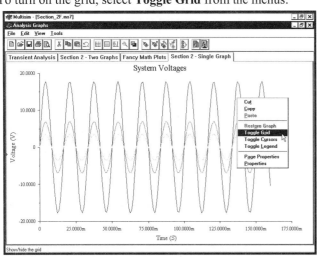
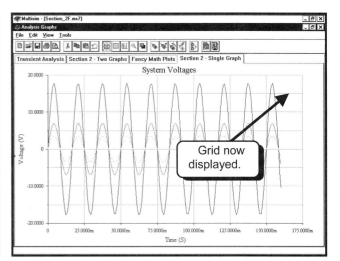

To display the legend, repeat the above procedure, but select **Toggle Legend**:

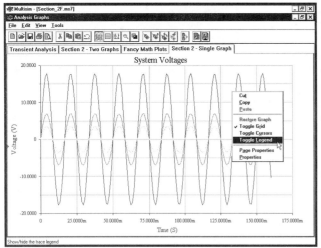
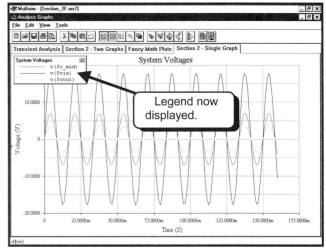

If you want to turn off the grid or the legend, just repeat the above steps.

There is a second way to turn on or off the display of the grid and the legend. For the second example, we will use the page that contains two graphs:

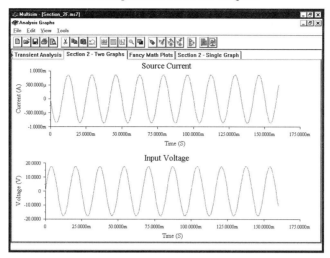

Each graph is modified separately. Thus, we will change the grid and the legend of one of the graphs. Click the **RIGHT** mouse button on an empty space in the bottom graph. A menu will appear:

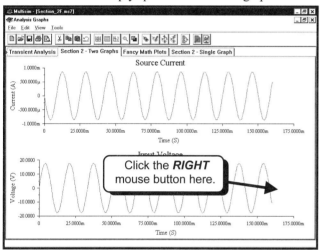
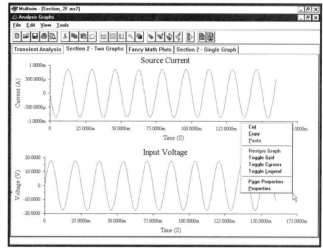

Select **Properties**:

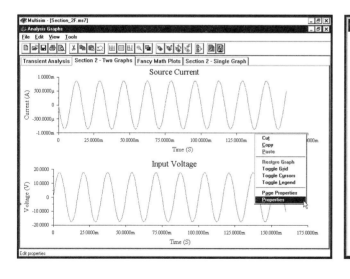
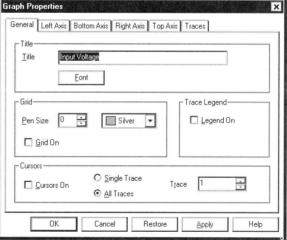

This dialog box allows us to modify all settings for a graph. For now, we will only change the display of the grid and the legend. On my dialog box the **General** tab is selected. If you do not have the **General** tab selected, then click on the tab to display it. Note that this portion of the dialog box allows you to display both the grid and the legend. By default, they are not displayed:

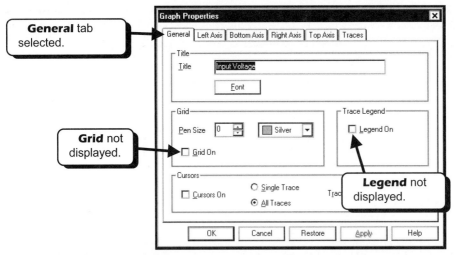

To display the **Grid** and the **Legend**, click on the square ☐ next to the options to place a checkmark in the box ☑:

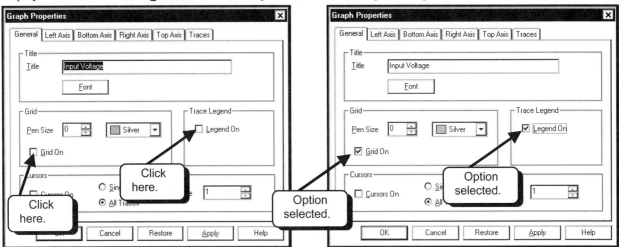

When you click on the **OK** button, the grid and the legend will be displayed on the bottom graph only:

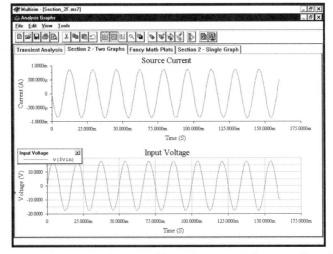

To display the grid and the legend on the top graph, you will need to repeat the procedure for the top graph. To obtain the properties dialog box of the top graph, you will need to click the **RIGHT** mouse button on an empty space in the top graph.

2.F.2. Trace Width, Color, and Label

For this example, we will work with the page that has a single plot. Click the **RIGHT** mouse button on an empty space in the graph. A menu will appear:

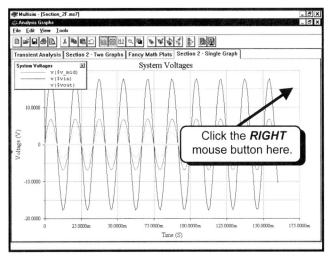 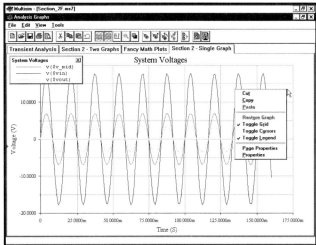

Select **Properties**:

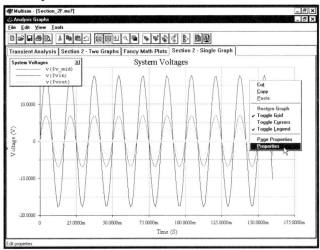 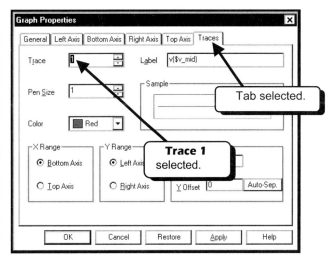

On my dialog box, the **Traces** tab is selected. This is the tab we need to use. Select the **Traces** tab if it is not selected.

By default, **Trace 1** is selected, so we will modify the properties of this trace first. We will first change the width of the trace. Click the **LEFT** mouse button as shown below to place the cursor in the **Pen Size** field:

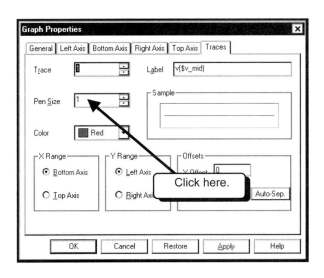 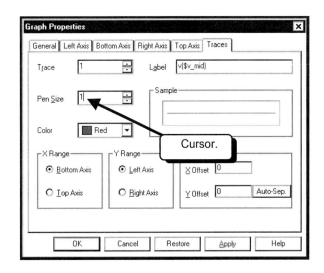

We can use the up and down arrows (⬆⬇) on the keyboard to change the value. We would like to change the thickness to 2, so press the up arrow key ⬆ once:

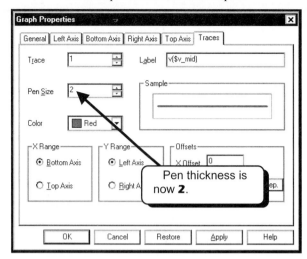

Next, we will change the color of the trace. Click the **LEFT** mouse button on the down triangle ▼ as shown:

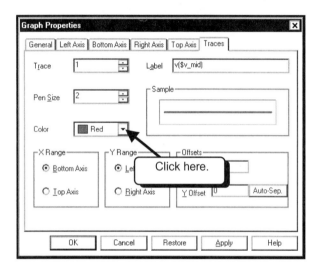
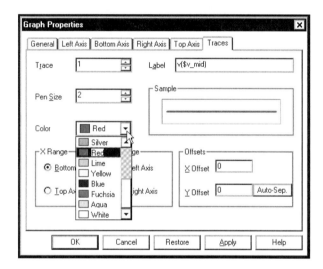

Select the color you want to use. I will select **Fuchsia**:

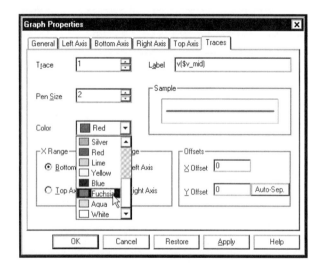
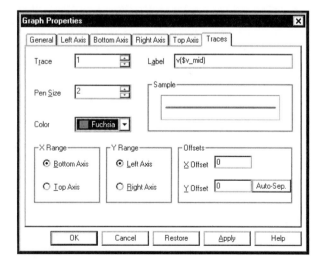

The last thing we will do is modify the label for this trace. Click the **LEFT** mouse button as shown to place the cursor in the **Label** text field:

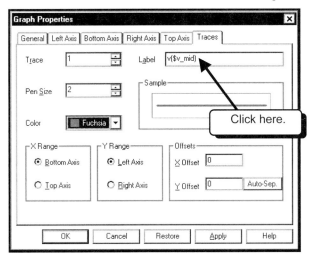 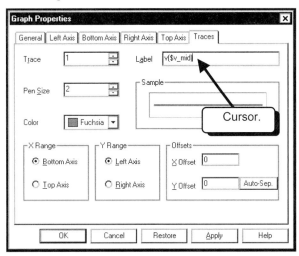

We can now modify the text to suit our needs. Use the **BACKSPACE** key to erase the text and then enter the text `Midpoint Voltage (Vin)`:

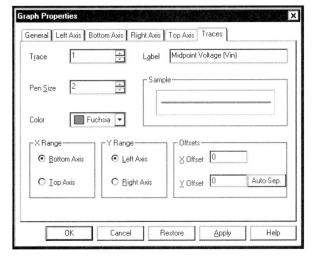

We are now done with this trace.

This graph has three traces, and we can modify the properties of each trace independently. Presently **Trace 1** is selected. To switch to another trace, click the *LEFT* mouse button as shown below to place the cursor in the **Trace** field:

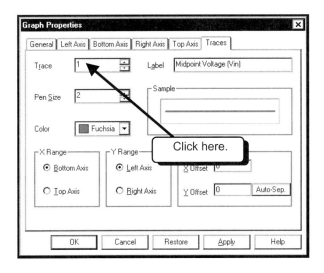 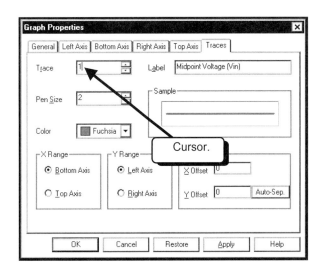

We can now use the up and down arrows (⬆⬇) on the keyboard to change the value. We would like to change to trace 2, so press the up arrow key ⬆ once:

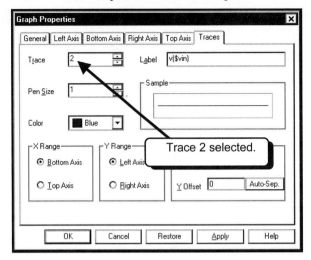

We are now working with **Trace** number **2**. We note that the properties of this trace are completely different than those for Trace 1. We can use the techniques shown earlier to modify the properties of this trace. Change the **Pen Size** to 3, the color to silver, and the **Label** to `Input Voltage (Vin)`:

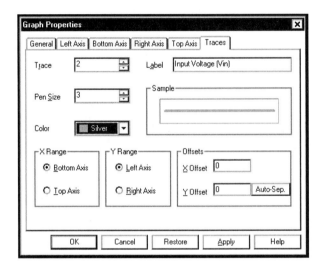

Next, modify the properties of trace 3. Change the color to Red, the thickness to 4, and the label to `Output Voltage (Vout)`:

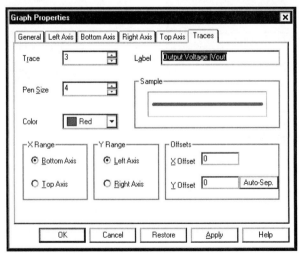

To see the changes and return to the Grapher, click on the **OK** button:

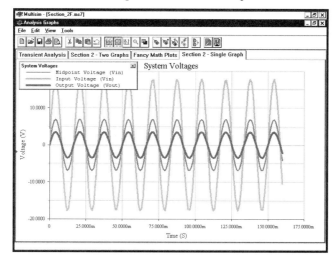

You can decide on the best trace color and thickness for your plots. For most of the screen captures in this text, we will use a Pen Size of 2. There is no way to change the default Pen Size, so you will need to make the changes for each graph you create.

2.F.3. Changing Titles and Axis Labels

Each graph on a page has a title, a y-axis label (called the left axis label), and an x-axis label (called the bottom axis label):

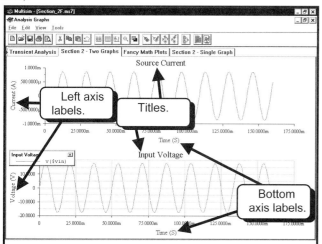
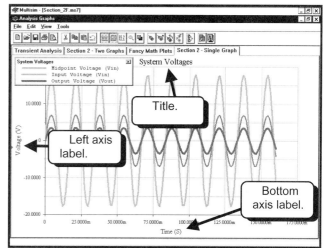

Once again, we can obtain the properties dialog box for each graph, and change the items for each graph separately. We will work with the page with a single graph. Click the **RIGHT** mouse button on an empty space on the graph. A menu will appear:

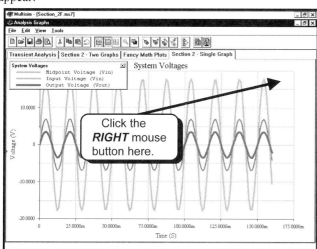
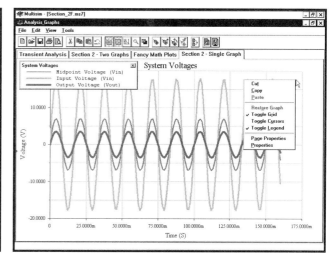

Select **Properties**:

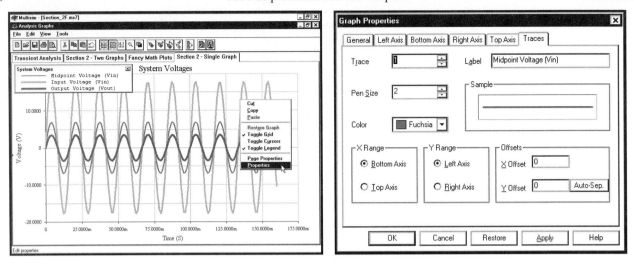

The **General** tab allows us to change the **Title** of the selected graph. If the **General** tab is not selected, click on the tab to select it:

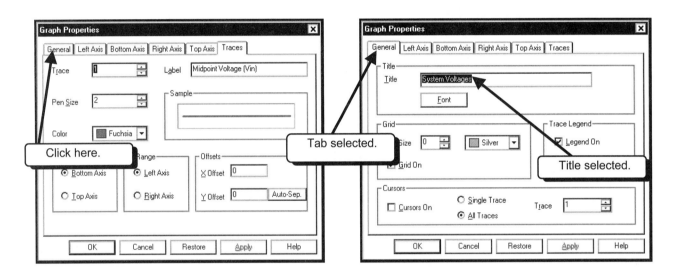

In my dialog box, the text for the **Title** is selected. To change the title, all we need to do is type the new title. Type the text
`Three Resistor Circuit - Voltages`:

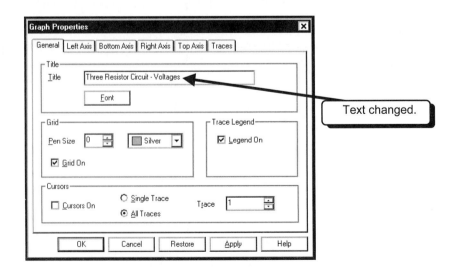

To change the y-axis settings, select the **Left Axis** tab:

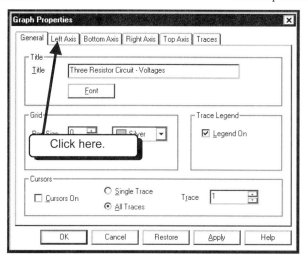 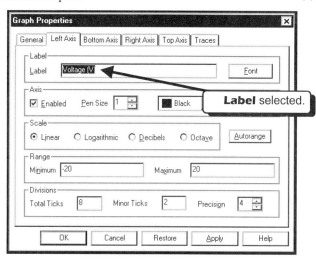

This dialog box can be used to change all of the properties for the y-axis. You should experiment with the settings to see their effects. In my dialog box, the text for the **Label** is selected. To change the **Label**, all we need to do is type the new **Label**. Type the text `Voltage (Volts)`:

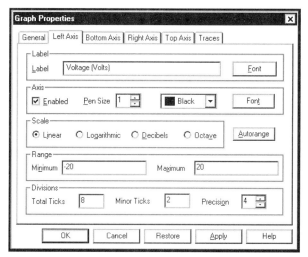

To change the x-axis settings, select the **Bottom Axis** tab:

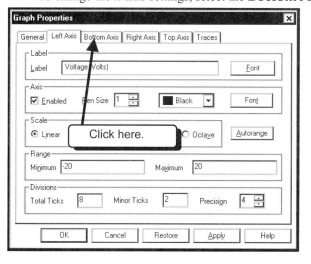 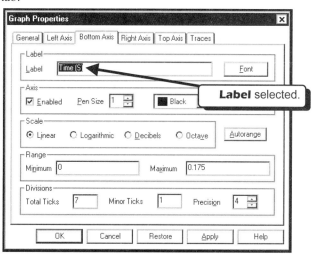

This dialog box can be used to change all of the properties for the x-axis. You should experiment with the settings to see their effects. In my dialog box, the text for the **Label** is selected. Once again, all we need to do is type the new **Label**. Type the text `Time (Seconds)`:

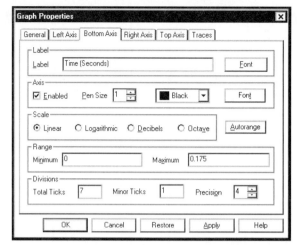

To make the changes take effect, you can click on the **Apply** button or the **OK** button. I will click on the **OK** button. This will apply the changes and close the dialog box:

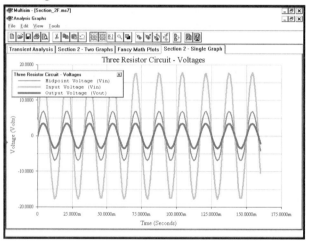

2.F.4. Adding a Second Y-Axis

The Grapher has the ability to use two y-axes on a single graph. This is a useful feature if you are trying to plot two traces with very different magnitudes on the same graph. An example would be plotting a current and a voltage. For many circuits, the voltage will be in volts while the current may be in milliamperes or microamperes. When plotted on the same y-axis scale, the current trace will look like a straight line through zero because the current is so small compared to the voltage. To fix this problem, we can use different axes for the voltage and current.

To illustrate the procedure, we will create a new Postprocessor page that plots the source voltage and source current for the circuit we simulated at the beginning of this chapter. Follow the procedures of section 2.A and use the Postprocessor to create a new page with a single graph. Plot the source voltage (vin) and the source current (vvx#branch) on the same graph:

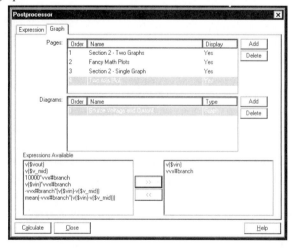
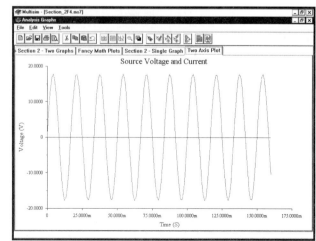

We need to obtain the Graph Properties dialog box to make the necessary changes. Click the **RIGHT** mouse button on an empty area of the graph to obtain the menu:

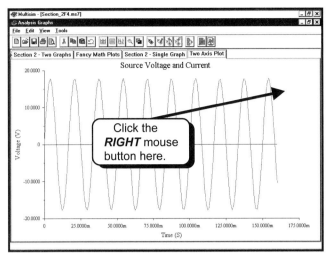
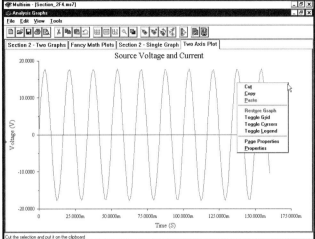

Select **Properties**:

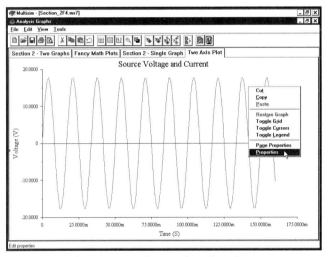
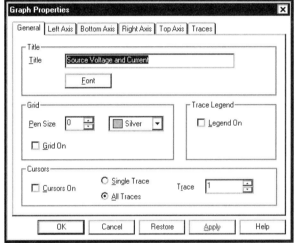

We need to use the **Traces** tab, so select the tab:

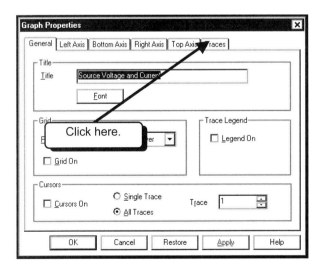
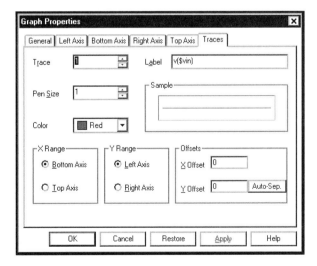

Note that **Trace 1** is the input voltage and it uses the **Left Axis** as its scale. We want this trace to use the left axis, so we will not change this setting. However, we will change the **Label** and **Pen Size**:

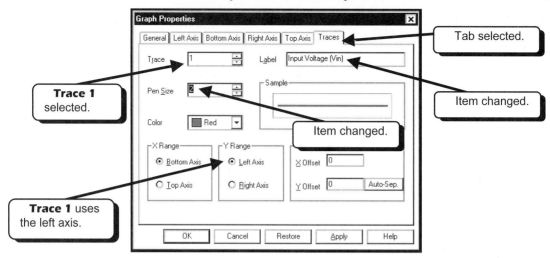

We would like trace 2 to use the right axis, which is our second y-axis. Click the **LEFT** mouse button as shown to place the cursor in the **Trace** field:

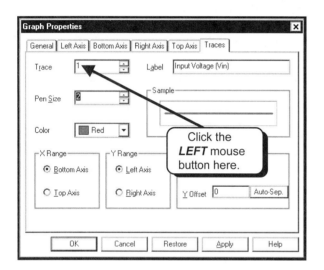
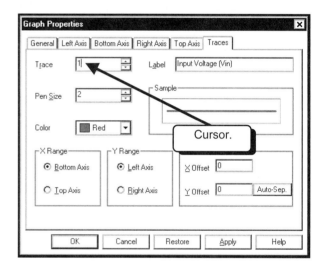

With the cursor as shown we can use the arrow keys (↑↓) on the keyboard to change the trace. Press the up arrow ↑ key on the keyboard to select **Trace 2**:

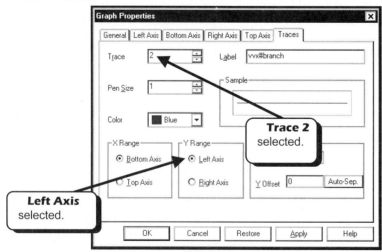

We see that Trace 2 also uses the left axis. We want this trace to use the right axis, so click the **LEFT** mouse button on the text **Right Axis** to select the option:

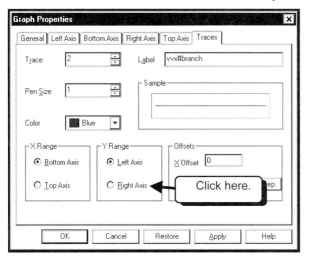 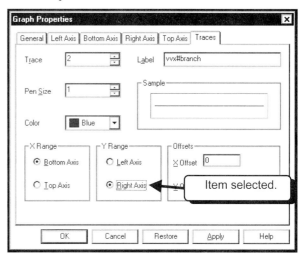

Change the **Pen Size** to **2** and the **Label** to `Source Current`:

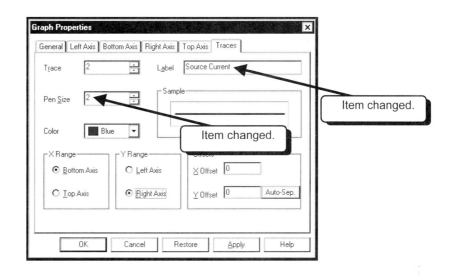

The second trace now uses the right axis. We must now look at the properties of that axis, so select the **Right Axis** tab:

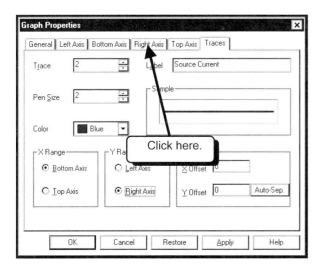 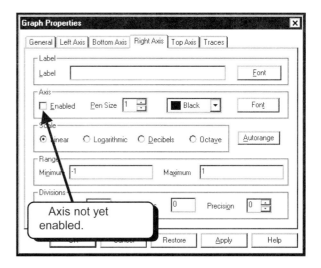

Nothing is specified for this axis. First, we must enable the axis. Click the *LEFT* mouse button on the text **Enabled** to select the option:

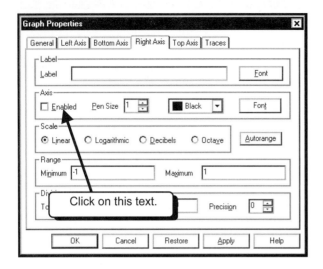 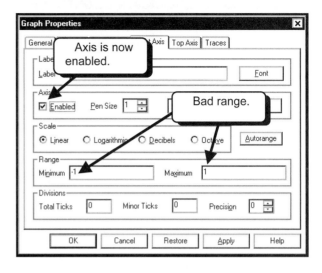

We note that the range is set for –1 to 1. This is not correct for the current we are plotting. We can either enter values for the range or have Multisim pick the range. Click on the **Autorange** button to have Multisim select the **Range** and **Divisions** settings.

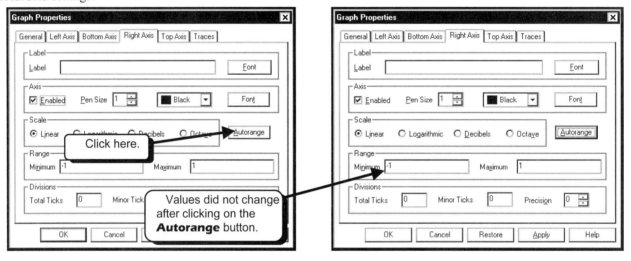

We did not see the Range change after we clicked the **Autorange** button, so we will have to enter the range manually. From previous plots, we know that a range of –1 mA to 1 mA will be sufficient, so we will enter this range manually. Set the **Total Ticks** to 20, the **Minor Ticks** to 5, and the **Precision** to 0. These settings specify the numbering of the axis and take a bit of experimentation to set up correctly:

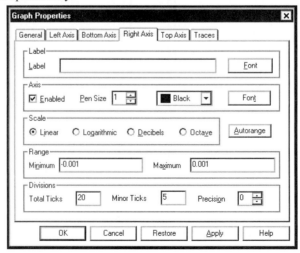

The last thing we need to do is to add a **Label** for this axis. Fill in the **Label** field as `Current (Amps)`:

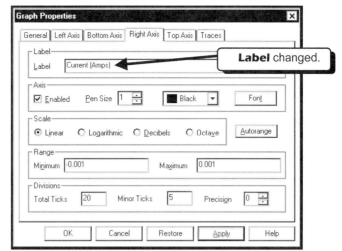

To view the results of our changes, click on the **OK** button:

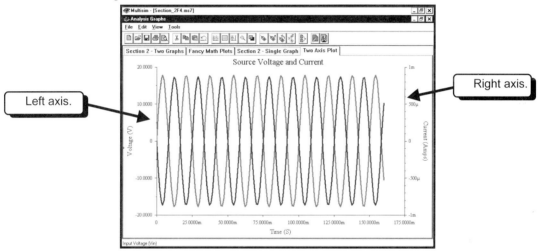

You can turn on the legend and the grid as shown in section 2.F.1:

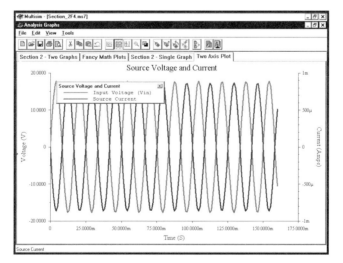

2.G. Using the Cursors

The cursors allow us to obtain numerical values from a plot. We can use the cursors on a page with a single graph with a single trace, a page with a single graph with multiple traces, or a page with multiple graphs that have multiple traces. First, we will use the graph we created earlier:

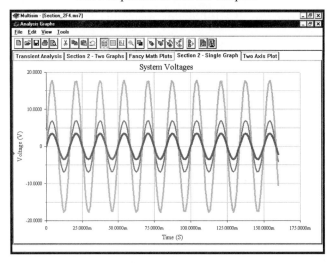

There are two ways to turn on the cursors: using the **RIGHT** mouse button or using the Graph Properties dialog box. We will show both here. We will first use the **RIGHT** mouse button. Click the **RIGHT** mouse button on an empty area of the graph:

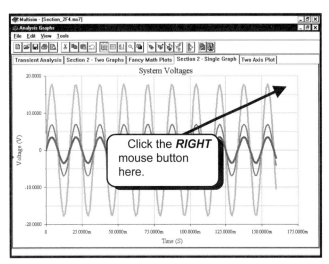
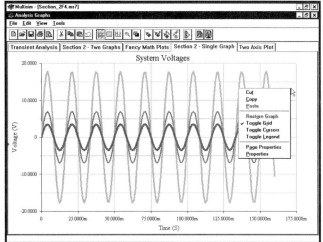

Select **Toggle Cursors**:

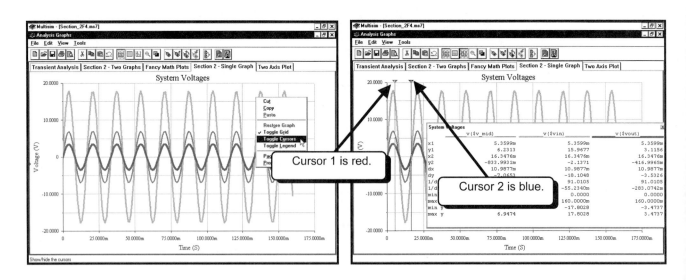

Two cursors are displayed on the page. Cursor 1 is shown in red and cursor 2 is shown in blue. Information for all traces is displayed on the cursor information dialog box. We will look closer at this dialog box:

Multisim 7 — The Postprocessor and the Grapher — 105

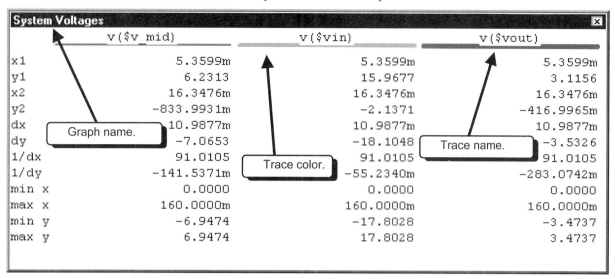

The cursor information is for the graph named **System Voltages**. Because this page contains a single graph, this information is obvious. If the page had more than one graph, this would be important information. We also see that the numerical information is displayed for all traces on this graph. If we want to display information only about a single trace with the cursors, we need to use the Graph Properties dialog box. Click the **RIGHT** mouse button on an empty area of the graph:

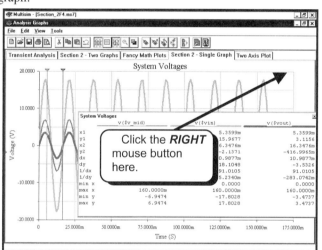
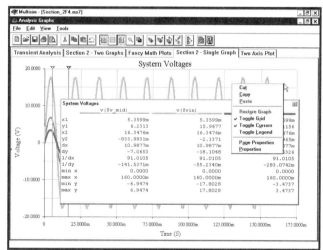

Select **Properties**:

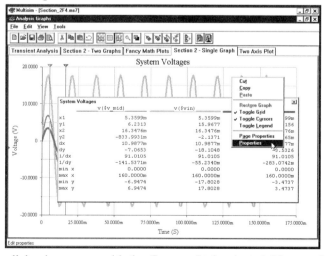
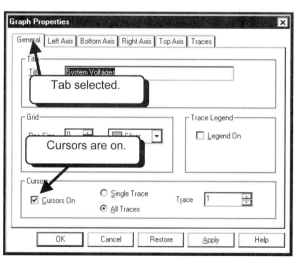

My dialog box opens with the **General** tab selected. If yours does not, click on the **General** tab to select it. Also note that the box indicates that the cursors are already on. This is because we turned on the cursors using the **RIGHT** mouse button. We can also use the **Cursors On** option to turn the cursors on or off.

This dialog box allows us to change the trace being displayed in the cursor information dialog box:

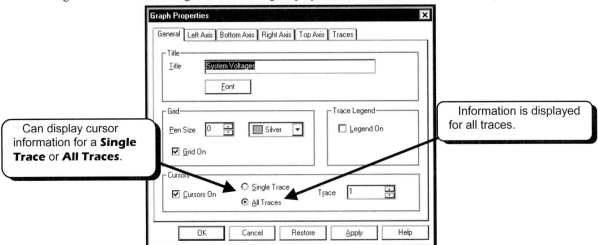

The cursor information dialog box can be set to display information for a single trace or all traces. Also, if we display information for only a single trace, we can select which trace to display. The options are currently set to display cursor information only for **All Traces**. As an example, we will display the cursor information for trace 1 only. Click the *LEFT* mouse button on the text **Single Trace** to select the option:

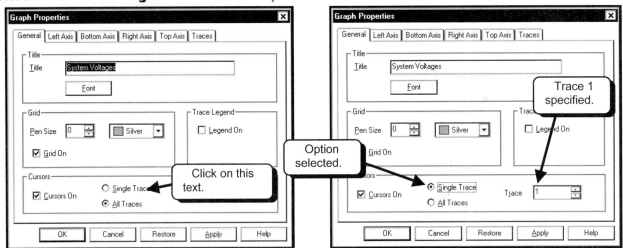

By default, **Trace 1** is specified. You can change this if you wish. Click on the **OK** button to accept the change. When you return to the Grapher, the cursor information dialog box will display information for **Trace 1** only:

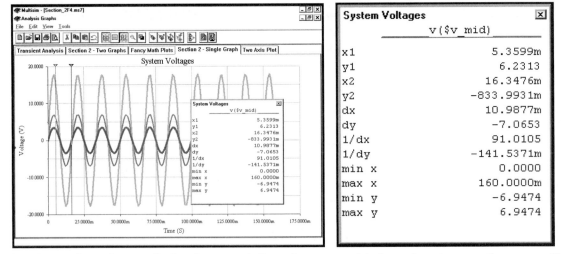

Now that we know how to display the cursor information, we need to know how to move the cursors. To move a cursor, drag the triangle at the top of the cursor. For example, to move cursor 1 (red), click and **HOLD** the *LEFT* mouse button on the red triangle as shown:

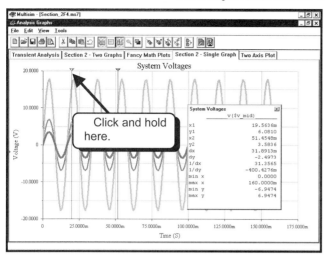
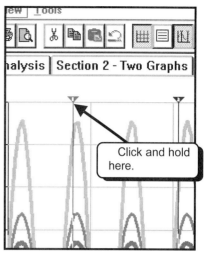

While continuing to hold down the **LEFT** mouse button, move the mouse. The cursor will move with the mouse and the numbers on the cursor dialog box will change. When you have found a suitable location, release the mouse button. To move cursor 2, click and **HOLD** the **LEFT** mouse button on the blue triangle and then move the mouse.

Instead of dragging the cursor to a specific location, you can use cursor search commands. If you click the **RIGHT** mouse button on a cursor, a menu will appear:

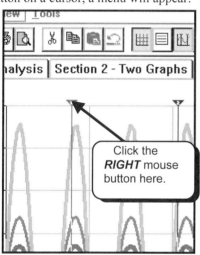
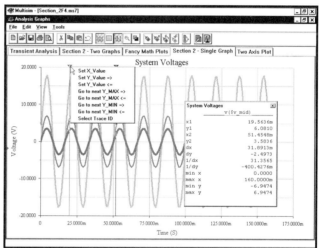

These menu commands will allow you to search along a trace and place the cursors at specific points along the trace. You can right-click on cursor 1 or cursor 2. Here we will show an example for cursor 1 only. The first thing you must do is select the trace you want the search to apply to. Select **Select Trace ID**:

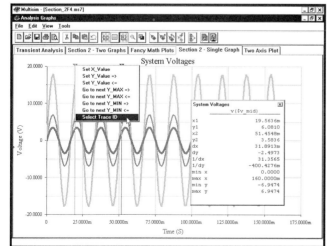
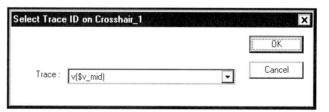

The down triangle allows us to select a trace to search:

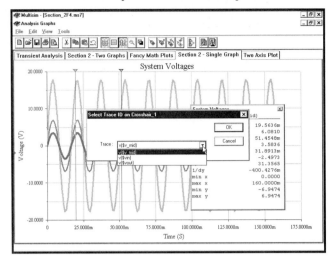

Select the input voltage trace and click the **OK** button. This tells the cursors to search the input voltage trace when they perform a search.

We will now search for a specific point on the input voltage trace. Click the **RIGHT** mouse button cursor 1 to obtain the pull-down menu:

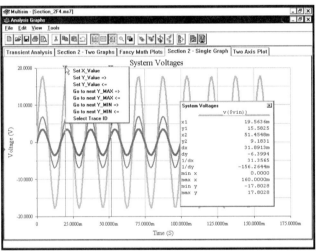

The specific search options are:
- **Set X_Value** — Set the x-coordinate of the cursor to a specific value.
- **Set Y_Value =>** — Find a point with a y-coordinate of the cursor with a specific value. Find the point closest to the present location of the cursor looking to the right.
- **Set Y_Value <=** — Find a point with a y-coordinate of the cursor with a specific value. Find the point closest to the present location of the cursor looking to the left.
- **Go to next Y_MAX =>** — Find the next local maximum. Find the point closest to the present location of the cursor looking to the right.
- **Go to next Y_MAX <=** — Find the next local maximum. Find the point closest to the present location of the cursor looking to the left.
- **Go to next Y_MIN =>** — Find the next local minimum. Find the point closest to the present location of the cursor looking to the right.
- **Go to next Y_MIN <=** — Find the next local minimum. Find the point closest to the present location of the cursor looking to the left.

For our example, we would like to find the next zero crossing looking to the right. Select **Set Y_Value =>**

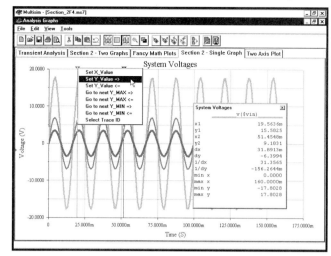
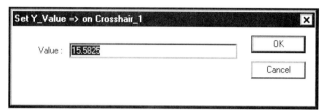

We are looking for a zero crossing, which would be a y-value of zero, so enter **0**:

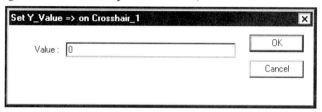

When you click on the **OK** button, the cursor will move to the next zero crossing to the right of the present cursor location:

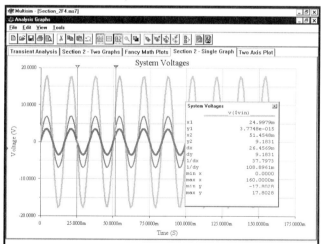
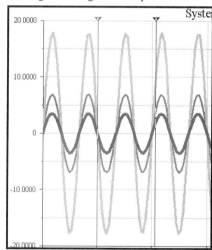

Next, we will find the next local maximum. Click the **RIGHT** mouse button on cursor 1 (red) and select **Go to next Y_MAX =>**

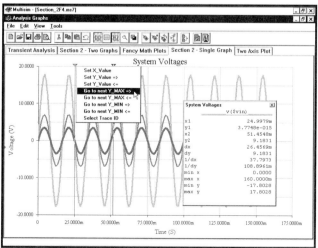

We see that the cursor jumps to the next local maximum going to the right:

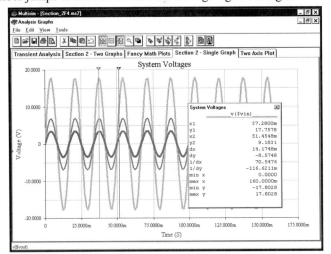 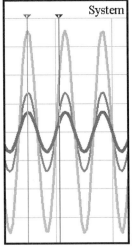

2.H. Zooming In and Out

There are two methods for zooming in on a trace in the Grapher. One method uses the mouse and the second method uses the Graph Properties dialog box. If you have a page with a single graph, the easiest way to zoom is to use the mouse. If you have a page with several graphs, the best way to zoom is to use the Graph Properties dialog box.

We will start with the single graph shown below:

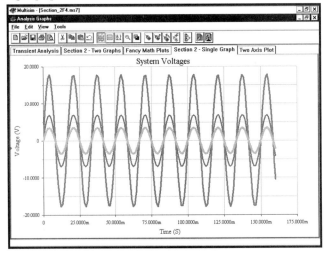

We will first zoom in using the mouse. In this method, we will draw a rectangle with the mouse and Multisim will zoom in to the area of the rectangle. Place the mouse pointer where you would like the upper left corner of the zoom rectangle to be:

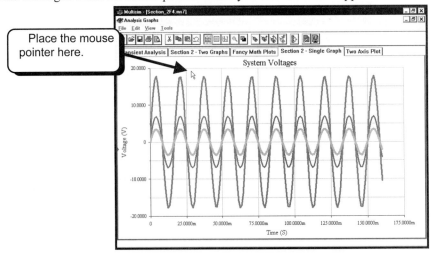

Click and **HOLD** down the *LEFT* mouse button. While continuing to hold down the *LEFT* mouse button, move the mouse. A dashed rectangle will be drawn:

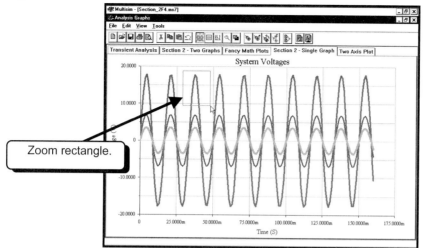

Draw the rectangle so that it encloses the area in which you want to zoom:

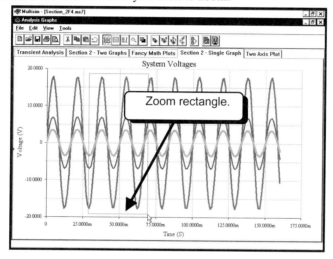

When you release the mouse button, Multisim will zoom in on the rectangle:

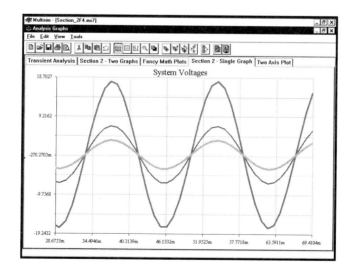

We can repeat the procedure to zoom in further:

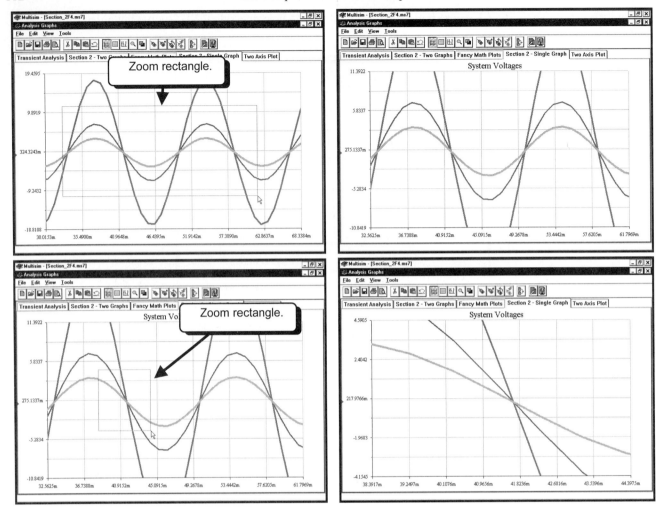

To go back to the original view, select **View** and then **Zoom Full** from the Grapher menus:

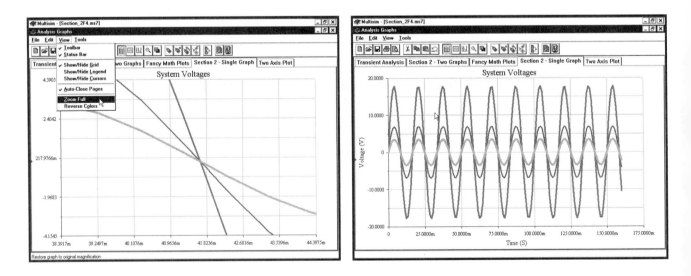

The second method we use for zooming is to change the x-axis and y-axis scales. This is tedious, but it allows us to zoom in to a precise location of the graph. Select **Edit** and then **Properties** from the Grapher menus to obtain the Graph Properties dialog box:

Multisim 7 — The Postprocessor and the Grapher

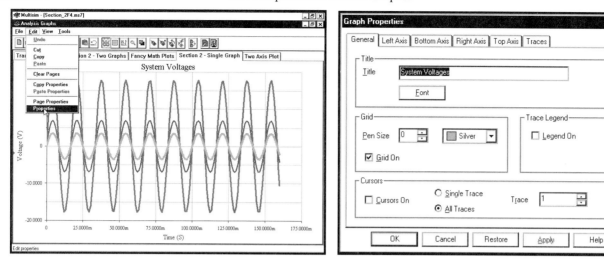

We will change the scale of the x-axis (**Bottom Axis**) first, so select the **Bottom Axis** tab:

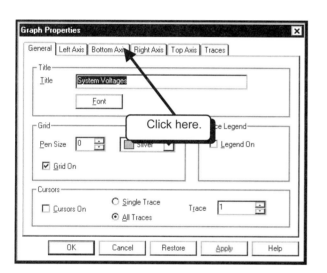
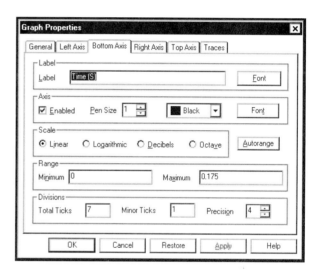

We would like to zoom in on the second cycle of the waveform, so change the **Range** of the x-axis to **0.016** to **0.032**:

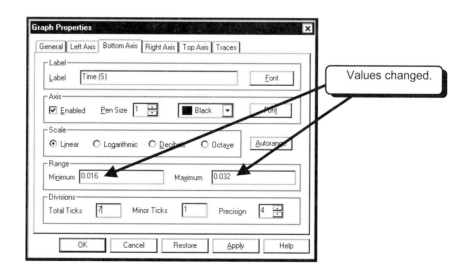

Next, we want to change the range of the y-axis (**Left Axis**), so select the **Left Axis** tab and change the **Range** to **-7.5** to **7.5**:

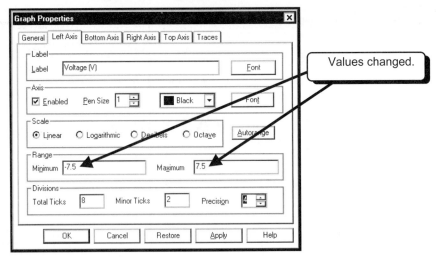

To zoom in, click on the **OK** button:

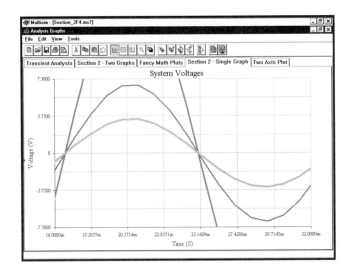

This second method may seem a bit cumbersome when compared to using the mouse. However, we must use this method when we have a page with several graphs on it. For our second demonstration, we will use the page with two graphs:

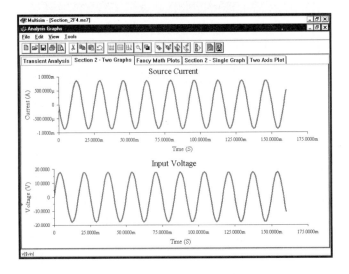

Next, use the mouse to zoom in on a portion of the top graph:

Multisim 7 — The Postprocessor and the Grapher — 115

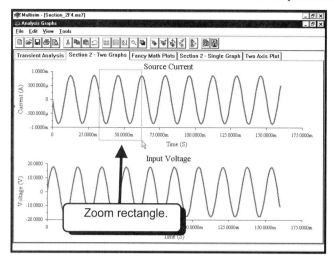
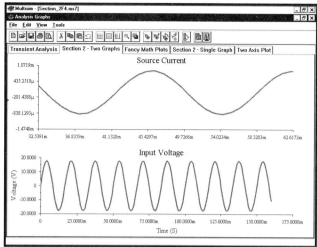

We see that Multisim only modifies the top graph. For a plot of this type we usually want the x-axes to be the same for both graphs. To accomplish this, the only thing we can do is use the Graph Properties dialog box for both graphs and set the bottom axis range to be the same for both plots. Before continuing, select **View** and then **Zoom Full** to start with the original graph.

Next, click the **RIGHT** mouse button on an empty space in the top graph and then select **Properties** from the menu:

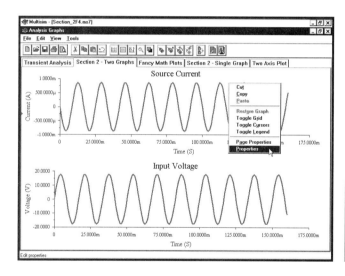
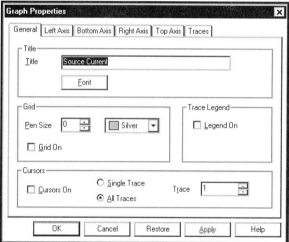

Select the **Bottom Axis** tab and change the range to `0.016` to `0.032`:

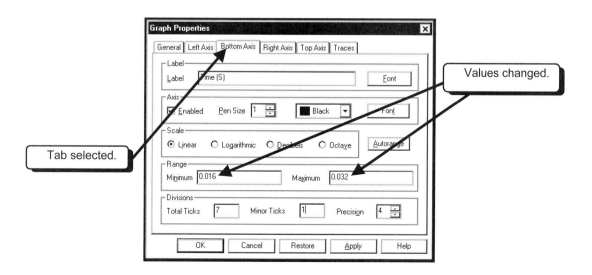

Click on the **OK** button to apply the changes to the top graph:

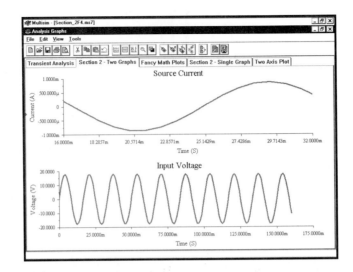

We must now repeat the procedure for the bottom graph. Click the ***RIGHT*** mouse button on an empty space in the bottom graph and then select **Properties** from the menu:

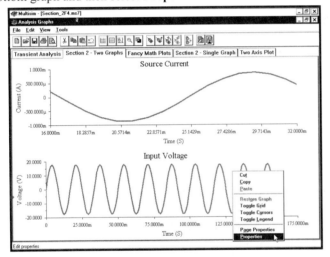
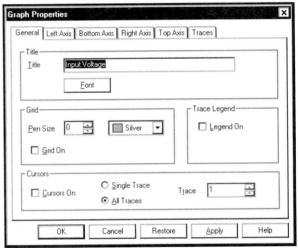

Select the **Bottom Axis** tab and change the range to **0.016** to **0.032**:

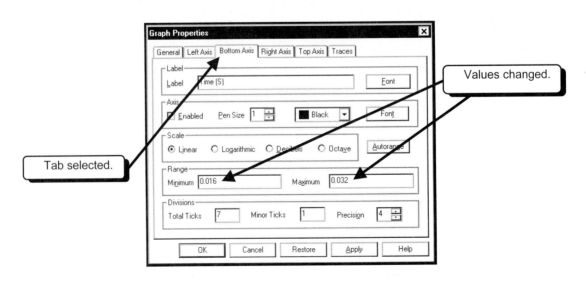

Click on the **OK** button to apply the changes to the bottom graph:

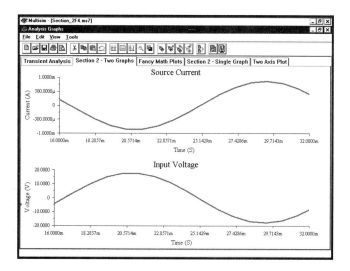

We see that the x-axis scale of both graphs is the same. However, the x-axes of the two graphs do not quite line up because the text on the left side of each graph takes different amounts of space.

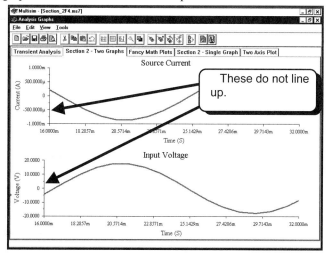

We can overcome this problem by changing the precision of the left axis on both graphs. Obtain the **Graph Properties** for the top graph and change the precision of the **Left Axis** of the top graph to 0:

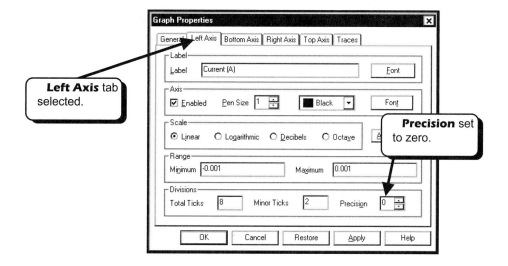

Click on the **OK** button and then obtain the **Graph Properties** dialog box for the bottom graph. Change the precision of the **Left Axis** of the bottom graph to 2:

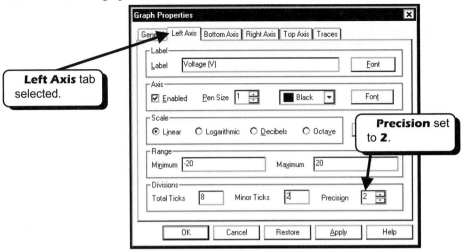

Click on the **OK** button to view the changes. The axes appear to line up. They are not aligned exactly, but they are close:

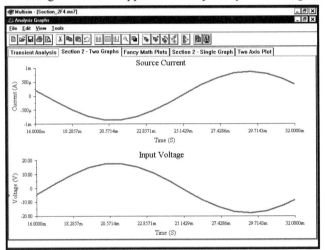

2.I. Saving and Opening Pages in the Grapher

The Grapher allows you to save and load graphs. This procedure allows you to select which graphs you would like to save. Only the graphs are saved; this method does not save any circuit information or Postprocessor information. Saving graphs does not save any information on how the graphs were created. This method just provides a quick way of saving all of the work we did in customizing the graphs.

To save the graphs, select **File** and then **Save** from the Grapher menus:

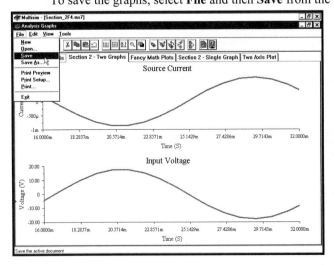 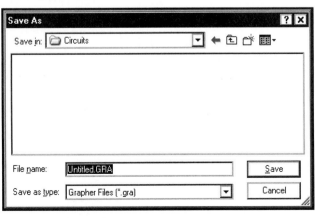

The Grapher file will be saved in the same directory as the circuit file and Postprocessor file unless you change the directory. Enter a name for the file:

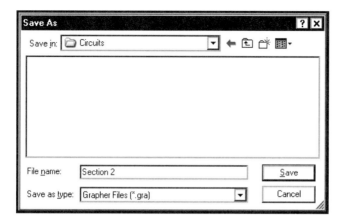

Multisim will automatically add the **.gra** extension. Click on the **Save** button to save the graphs:

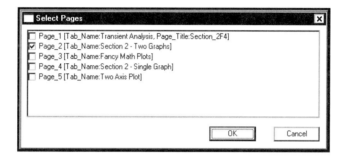

This dialog box allows us to select which of the graphs we would like to save. I will select all of them and then click the **OK** button. After clicking the **OK** button, the graphs are saved.

Close and then restart Multisim. You will be presented with an empty schematic window:

Type **CTRL-G** to open the Grapher. You will be presented with an empty Grapher window. Maximize the Grapher window so that it occupies the entire Multisim window:

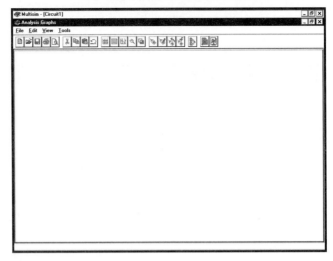

Select **File** and then **Open** from the Grapher menus and select the file we just created:

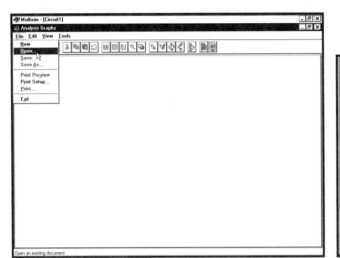
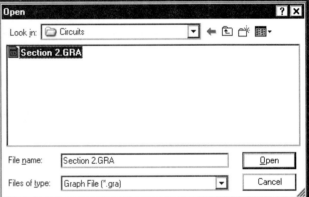

When you click on the **Open** button, the pages we saved earlier will be displayed:

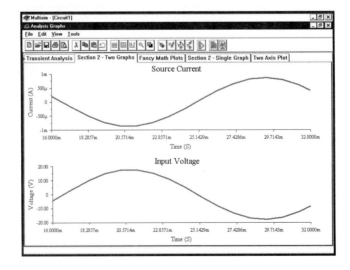

When we display the other pages, we see that they contain the changes we made using the Grapher facilities:

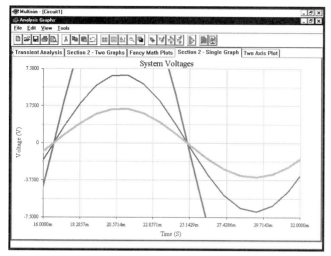
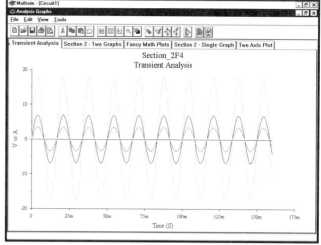

We cannot add any more traces to the plots, nor add any more pages to the Grapher with the file we just opened. If you want to add more traces for the circuit we simulated, you will need to open the circuit file, run the simulation, and then use the Postprocessor to create more graphs.

2.J. Problems

Problem 2-1: When you set up a SPICE simulation such as a Transient Analysis, what parameters must you specify so that you can view results using the Postprocessor and Grapher?

Problem 2-2: What does the **Add** button do?

Problem 2-3: What is the difference between an output variable such as $vin or $vout and an output variable such as vvx#branch?

Problem 2-4: What is the difference between the Grapher and the Postprocessor?

Problem 2-5: What is the difference between a page, a graph, and a trace?

Problem 2-6: Choose five functions available in the Postprocessor and describe their mathematical function. Example functions are abs, mean, and real.

Problem 2-7: What is the procedure for creating a new page with a single graph on it?

Problem 2-8: What is the procedure for adding a new trace to a page that has several graphs?

Problem 2-9: What are the two file types created by Multisim?

Problem 2-10: Under what circumstances would you add a second y-axis to a graph?

Problem 2-11: In the Grapher window below, how can you tell which graph is selected? This indicator was not covered in the manual.

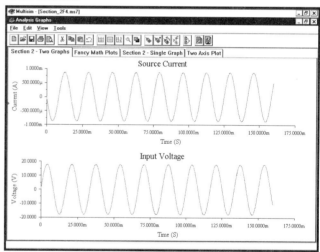

Problem 2-12: Specify the procedure for deleting a page in the Grapher window.

Problem 2-13: For this problem, use the circuit simulated in Part 2. With the Postprocessor, create a new page that has three graphs. The top graph plots the instantaneous power absorbed by Rx using the equation V^2/R, the middle graph plots the instantaneous power absorbed by Rx using the equation I^2R, and the bottom graph plots the instantaneous power absorbed by Rx using the equation $V \times I$.

Problem 2-14: For this problem, use the circuit simulated in Part 2. With the Postprocessor, create a new page that has one graph. Plot two traces. The first trace is the instantaneous power supplied by the source. The second trace is the instantaneous power absorbed by Rx plus the instantaneous power absorbed by Ry plus the instantaneous power absorbed by R1 plus the instantaneous power absorbed by R2. Show that the two traces are the same.

Problem 2-15: For this problem, use the circuit simulated in Part 2. With the Postprocessor, create a new page that has one graph. Plot two traces. The first trace is the average power supplied by the source. The second trace is the average power absorbed by Rx plus the average power absorbed by Ry plus the average power absorbed by R1 plus the average power absorbed by R2. Show that the two traces are the same.

Problem 2-16: What is the difference between saving information in Multisim and saving information in the Grapher?

Problem 2-17: Load the pages saved using techniques in section 2.D. On the page with three graphs, modify the page so that there is a grid on all three graphs, and a legend on the top two graphs but not the bottom graph. Turn in a printout of your solution.

Problem 2-18: In the Grapher page that contains three graphs, change the title of the middle plot to "My Middle Graph," change the y-axis label of the top graph to "Strawberries," and change the x-axis label to "Time – Warp." Turn in a printout of your solution.

Problem 2-19: What is the difference between selecting the **OK** button and selecting the **Apply** button in the Graph Properties dialog box of the Grapher?

For Problems 2-20 through 2-24, start with the graph below:

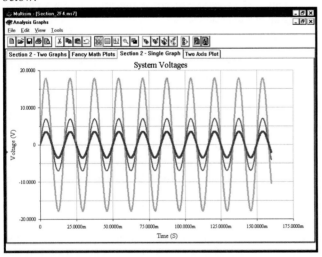

Problem 2-20: Modify the grid of a graph so that the pen size of the grid is 2 and the color of the grid is red. Turn in a printout of your solution.

Problem 2-21: Modify the left axis of a graph so that the pen size of the axis is 2 and the color is blue. Turn in a printout of your solution.

Problem 2-22: Modify the left axis of a graph so that the total number of ticks is 16 and the number of minor ticks is 4. Turn in a printout of your solution.

Problem 2-23: Modify the bottom axis of a graph so that the pen size of the axis is 3 and the color is green. Turn in a printout of your solution.

Problem 2-24: Modify the bottom axis of a graph so that the total number of ticks is 14 and the number of minor ticks is 2. Turn in a printout of your solution.

Problem 2-25: What is the maximum number of graphs you can put on a single Postprocessor page? What is the maximum number of graphs you can put on a single Postprocessor page and still have a printout that looks useful? Turn in printouts supporting your answer.

Problem 2-26: Specify three methods for showing or hiding cursors on a Grapher page.

Problem 2-27: Specify three methods for showing or hiding the grid on a Grapher page.

Problem 2-28: Specify three methods for showing or hiding the legend on a Grapher page.

Problem 2-29: Specify three ways to obtain the Graph Properties dialog box.

Problem 2-30: Describe the two methods for zooming in the Grapher.

Problem 2-31: List two ways of saving files and information in Multisim.

Problem 2-32: What are the advantages and disadvantages of using the two different zoom methods covered in section 2.H?

PART 3
DC Measurements

In this chapter we will find the DC values of currents and voltages in a circuit. For the analyses we show here, the DC voltages will not change. In Chapter 4 we will look at the DC sweep analysis where you can sweep a DC source through several values. The simulations in Chapters 3 and 4 are similar in that the values displayed are DC values. With a DC sweep, however, you can see how the DC values of a circuit change as a DC source in the circuit is changed. For example, we could plot how the bias current of a transistor changes for different values of the DC supply. In contrast, in this chapter we can only display the collector current for a single value of the supply. To find the bias current at a different value of the DC supply we would have to change the value of the supply and then rerun the simulation. We will show several ways to view the results, including using the current indicators, the Multimeter, and running a DC Operating Point Analysis.

3.A. Resistive Circuits

We will perform simulations of the circuit below:

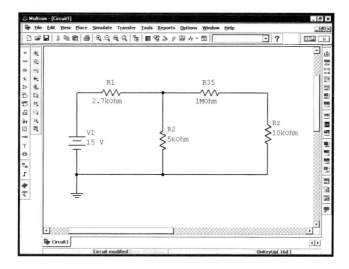

3.A.1. Measurements with Indicators

We would first like to find the current supplied by **V1** and the voltage across resistors **R2** and **Rz**. To measure current and voltage, we use the voltage and current indicators. Click the *LEFT* mouse button on the **Indicator** button

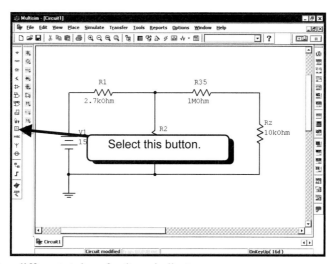

You will be presented with a few different styles of voltage indicators:

123

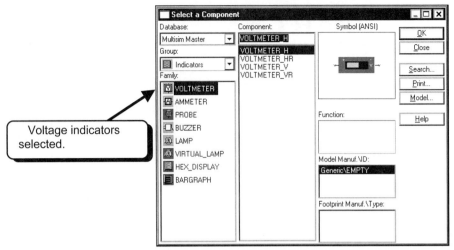

All of the voltage indicators perform the same function. Their terminals are arranged differently to make them more convenient to place in your circuit. Choose an indicator, place two of them in your circuit, and wire them as shown. Note that voltage is measured between two nodes in a circuit:

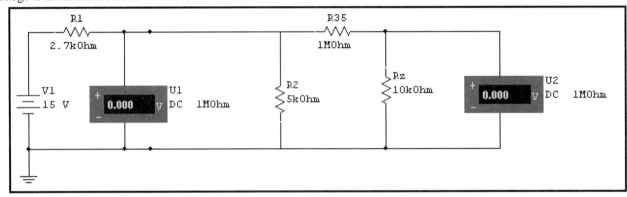

Note in the picture above that a resistance is shown next to the voltage indicators.

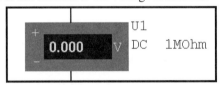

This is the equivalent parallel resistance of the indicators, and will be discussed later. The default value of this resistance is 1 MΩ.

To place a current indicator, click on the **Indicator** button . We are presented with the same dialog box as when we placed the voltage indicator, but now we must select a current meter. Click the *LEFT* mouse button on the text **AMMETER** to display the list of available current indicators:

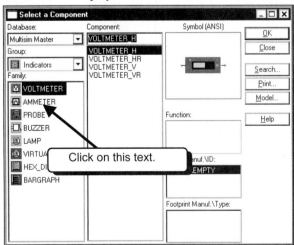
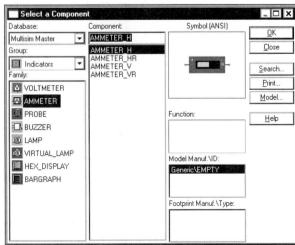

A current indicator must be placed in series with the element through which you want to measure the current. Choose a current meter and add it as shown:

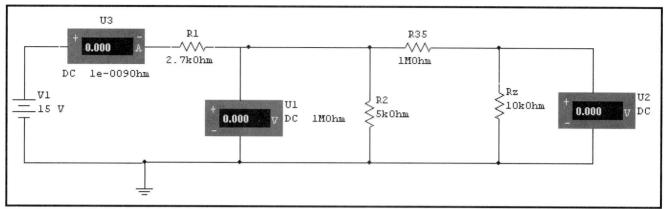

It is important to understand what the plus and minus signs mean in the current indicator. If the current enters the positive terminal and leaves the negative terminal, the monitor will read a positive number. That is, if the current flows in the direction shown by the arrow below, the current indicator will read a positive number:

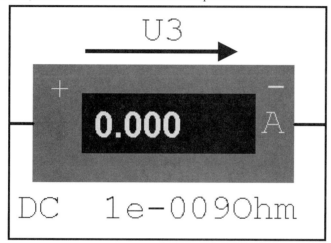

In the circuit shown above, the current indicator will measure the current supplied by V1 or the current flowing through R1. Also note that a resistance value is below the current indicator. This is the equivalent series resistance of the indicator and will be discussed later.

Before we simulate the circuit, we will look at the current and voltage indicators a little more closely. First, double-click the **LEFT** mouse button on the current indicator. This will open the properties dialog box for the indicator:

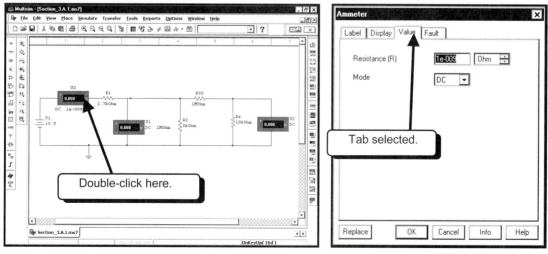

Select the **Value** tab if it is not already selected.

There are two important parameters for this part. First, the **Mode** is set to **DC**. This means that the indicator will display the **DC** or average value of the current it measures. Because all of our sources in this circuit are DC, this is the correct setting for this application. The other parameter is the series resistance of the current indicator. All real current meters are placed in series with the element through which they measure current, and all current meters have a small series resistance. Because the indicator is placed in series with another circuit element, the resistance of the indicator will affect the circuit by adding to the overall resistance of the circuit. To reduce the effects of the indicator a small series resistance is required. To provide an accurate model, Multisim allows you to specify the value of this series resistance.

There are two ways to choose a value for this resistance. If you are comparing this simulation to a real circuit where you measure the current with a real ammeter, you should set the resistance in the simulation to the resistance of your real ammeter. The manuals that accompany your meter should have a specification of its resistance. This will accurately model your ammeter in the circuit. If the ammeter has an effect on the circuit, your simulation will include this effect. The second method of choosing the current indicator resistance is to choose a small series resistance that does not affect the circuit. The default value of this series resistance is 1 nΩ (1e-009 ohm). This resistance is small compared to all other resistances in the circuit, so the current indicator resistance will have a negligible effect on the circuit. We are not comparing this circuit to a real-world measurement, so we will use the default value of 1 nΩ.

We note that both the **Mode (DC)** and the ammeter series **Resistance (1e-009Ohm)** are displayed on the schematic. You may or may not find this helpful. We will now show how to hide these attributes. Select the **Display** tab. We note that the option **Use Schematic Global Setting** is selected:

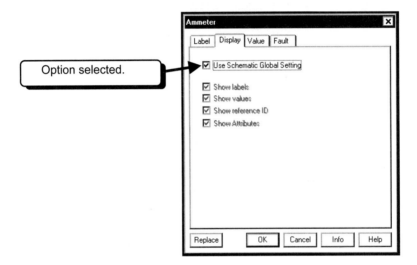

With this option selected, the part will display the four categories of information listed in this dialog box on the schematic. To hide some of the information displayed on the schematic, you must deselect the option **Use Schematic Global Setting**:

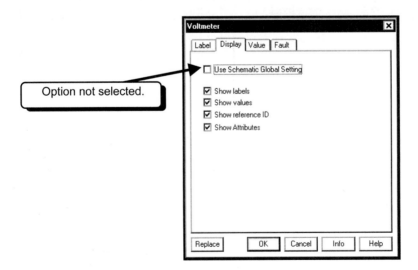

The ammeter has no labels or Attributes that are displayed. The **reference ID** for my ammeter is U3. The **Mode** and the **Resistance** are both values. If we uncheck the **Show values** and the **Show reference ID** options, the information will not be displayed on the schematic.

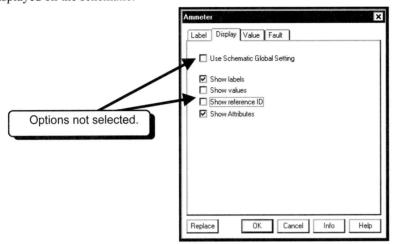

Click on the **OK** button to close the dialog box. We see that the reference ID (U3), and the mode (DC) and ammeter resistance (1e-009Ohm) are no longer displayed:

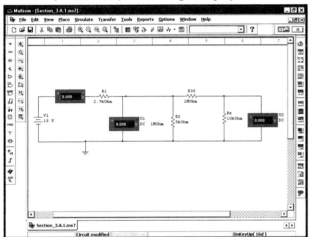

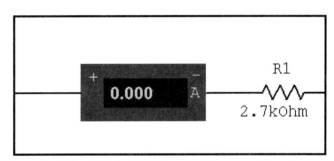

Next, we will look at the properties of the voltage indicator. Double-click the *LEFT* mouse button on the voltage indicator. This will open the properties dialog box for the indicator:

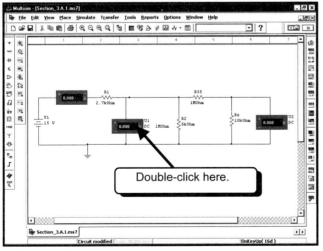

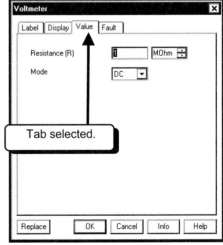

Select the **Value** tab if it is not already selected.

As with the current indicator, you can set the voltage indicator to measure AC or DC voltages. We will leave the **Mode** set at **DC**. The other parameter is the parallel resistance of the voltage indicator. All real voltage meters are placed in parallel with the element across which they measure voltage. In order to measure the voltage, the meter must draw a small current from the nodes to which it is connected. The current the meter draws is modeled as a resistance between the meter terminals. Because the voltage indicator is connected between two nodes, the resistance of the indicator will draw current and affect the circuit. To reduce the effects of the voltage indicator, a large resistance is required. To provide an accurate model, Multisim allows you to specify the value of this resistance. The default is 1 MΩ. If the current that the meter draws is small compared to other elements in the circuit, the meter will not affect the circuit very much. If the meter resistance is comparable to that of other elements in the circuit, the act of making a measurement may cause a large change in the circuit, and the measurement will be inaccurate. For now, we will assume that 1 MΩ is large enough and we will not change the value. We will check later to see if this is indeed the case.

We note that both the **Mode** (**DC**) and the voltmeter parallel **Resistance** (**1MOhm**) are displayed on the schematic. You may or may not find this helpful and it takes up a large amount of space on the schematic. We will now show how to hide these attributes. Select the **Display** tab. We note that the option **Use Schematic Global Setting** is selected:

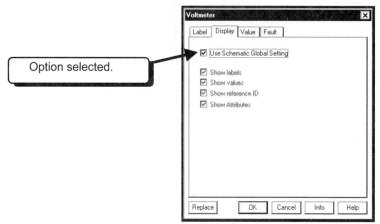

With this option selected, the part will display the four categories of information listed in this dialog box on the schematic. To hide some of the values displayed on the schematic, you must deselect the option **Use Schematic Global Setting**:

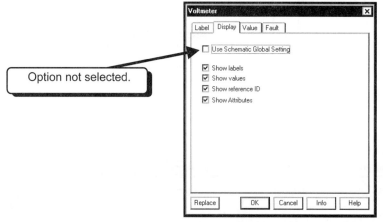

This part has no labels or attributes that are displayed. The **reference ID** for my voltmeter is U1. The **Mode** and the **Resistance** are both values. If we uncheck the **Show values** and the **Show reference ID** options, the information will not be displayed on the schematic.

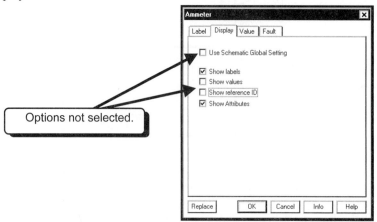

Click on the **OK** button to close the dialog box. We see that the reference ID (U1), and the mode (DC) and voltmeter resistance (1MOhm) are no longer displayed:

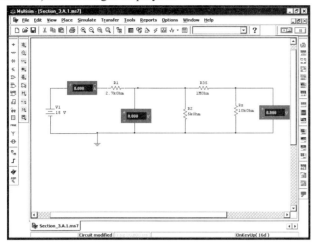
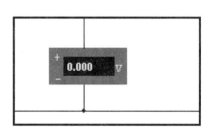

Removing this information will allow me to clean up my schematic a bit and to enlarge it for better screen captures. A slightly enlarged view is shown below:

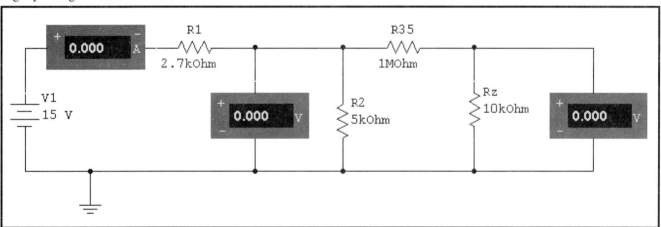

We are now ready to simulate the circuit. Select **Simulate** and then **Run** from the Multisim menus, press the **F5** key, click the **Run/stop simulation** button, or click on the simulate button:

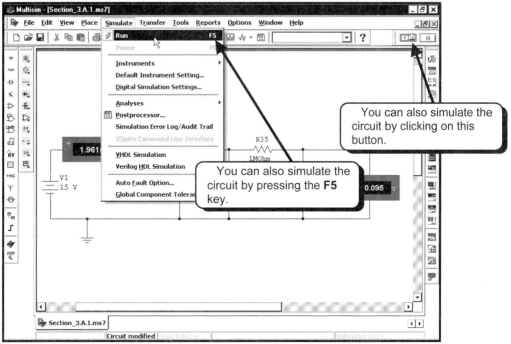

When you simulate the circuit, the indicators will display their measured values:

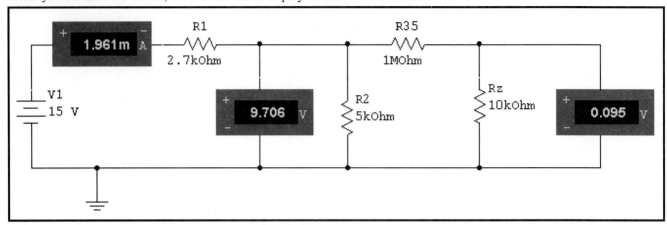

Next, we will check to see if the resistance of the voltage indicators had an effect on the readings. The first thing we need to do is to stop the simulation. Select **Simulate** and then **Run** from the Multisim menus, press the **F5** key, click the **Run/stop simulation** button, or click on the simulate button to toggle the simulation off. You can tell that the simulation has stopped running because the time indicator at the bottom of the screen should stop counting:

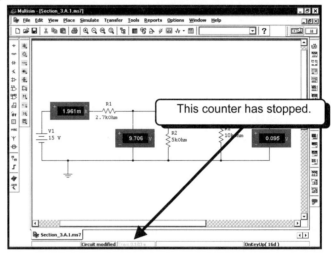

Double-click the *LEFT* mouse button on a voltage indicator and select the **Value** tab. Change the resistance to 1 GΩ as shown below:

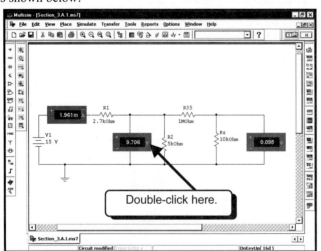
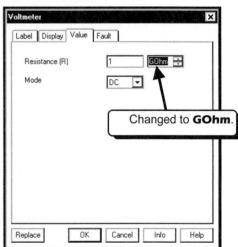

Note that 1 GΩ equals $1 \times 10^9 \Omega$. This should be large enough compared to other elements in the circuit so that it draws a negligible amount of current. Click on the **OK** button to close the dialog box. Modify the other voltage indicator in the circuit and change its resistance to 1 GΩ as well. When you are done, the circuit will look unchanged:

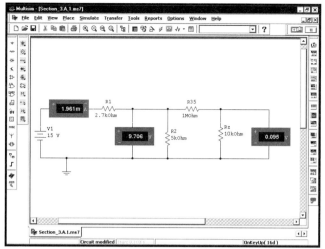

Note that the numbers displayed by the indicators will not change until we run the simulation. Select **Simulate** and then **Run** from the Multisim menus, press the **F5** key, click the **Run/stop simulation** button, or click on the simulate button to start the simulation. In a few moments, the indicators may change their display:

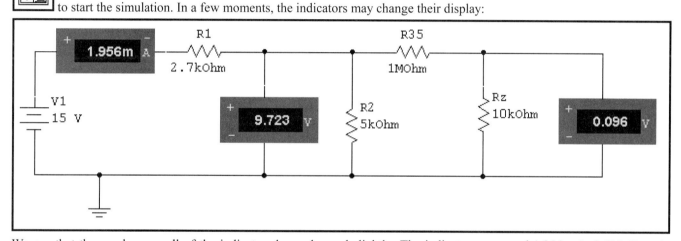

We see that the numbers on all of the indicators have changed slightly. The indicators now read 1.956 mA, 9.723 V, and 0.096 V. These values are all slightly different from the original readings of 1.961 mA, 9.706 V, and 0.095 V. These changes may or may not be significant depending on your application. The important thing to note is that whenever you measure a quantity, the act of making that measurement may affect the quantity you are measuring. Multisim accurately models these effects by giving a current indicator a series resistance and a voltage indicator a parallel resistance. The theoretical value of the current is 1.954 mA, so we see that even the high-resistance meters we created do affect the circuit behavior slightly.

Note that an ideal current indicator would have no series resistance, and an ideal voltage indicator would have an infinite parallel resistance. Be advised that if you make the resistance of a current indicator too small, it will not read the correct current. The screen capture below shows the result of setting the current indicator resistance to 1 pΩ, or 10^{-12} Ω:

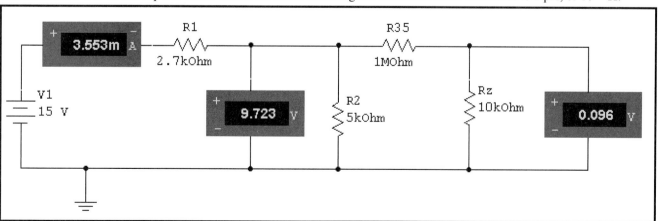

The current is improperly displayed as 3.553 mA. These examples show us that you should be aware of the effects of using an indicator, and that you cannot adjust its resistance to make it ideal either. Note that a series resistance of 1 nΩ is close to ideal for the current monitor while a parallel resistance of 1 MΩ for a voltage indicator is on the small side. When you use a voltage monitor, you should consider increasing its resistance to 1 GΩ so that it has less effect on your circuit.

EXERCISE 3-1: Find the unknown node voltages and the current through R1.

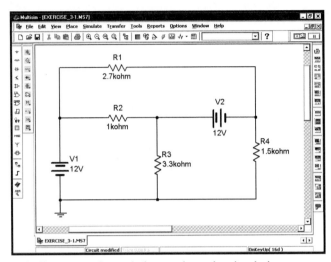

SOLUTION: Add voltage and current indicators as shown below and run the simulation.

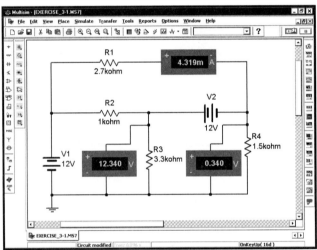

EXERCISE 3-2: Find the DC node voltages for the circuit below and the DC current through R1:

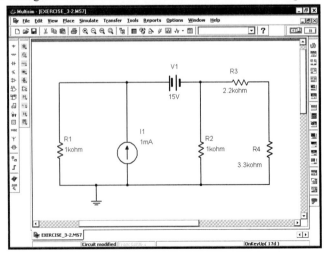

HINT: A DC current source is located in the Signal Source toolbar:

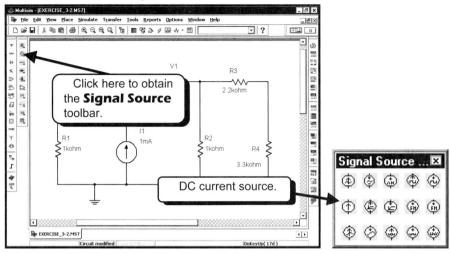

Click on the indicated button to place a DC current source.

SOLUTION: Add indicators as shown and run the simulation:

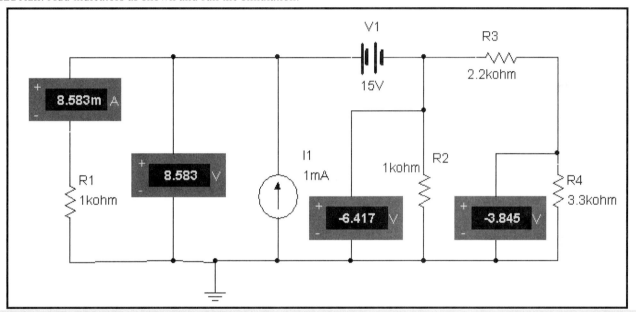

3.A.2. Measurements with the Multimeter

You can also measure voltage and current using a model of a Multimeter. We will start with the original circuit, which is repeated below:

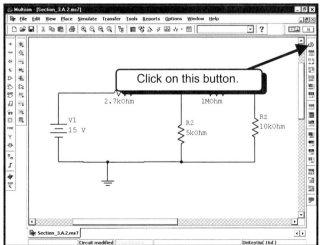

To place a Multimeter, click the *LEFT* mouse button on the Multimeter button as shown above. A meter will become attached to your mouse pointer.

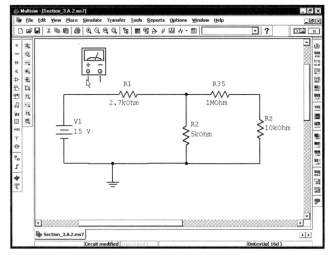

Place the meter as you would any part. We will measure a voltage and a current, so we will need two meters in our circuit:

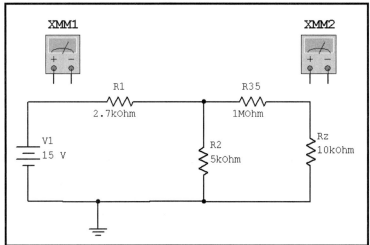

We will measure the current through **R1** using meter **XMM1** and the voltage across **R2** using **XMM2**. When you wire up a current meter, you must place the meter in series with the element through which you want to measure the current. When you measure a voltage, you place the meter in parallel with the element across which you want to measure the voltage. Wire the meters as shown:

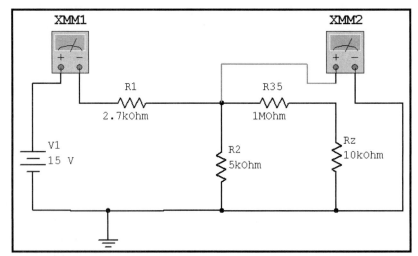

It is important to understand what the plus and minus signs mean when you measure current with a Multimeter. If the current enters the positive terminal and leaves the negative terminal, the monitor will display a positive number. That is, if the current flows in the direction shown below, the current indicator will display a positive number:

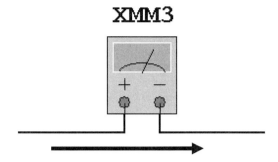

To view the display and function of a meter, double-click the **LEFT** mouse button on the meter. A window will open displaying the meter:

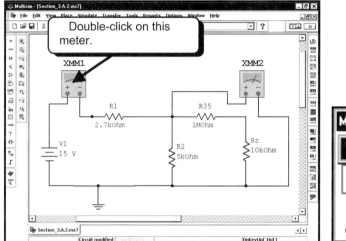

We see that we can use the meter to measure current, voltage, resistance, and decibels. We can also select between AC and DC measurements. On my monochrome screen captures it is not easy to see which mode is selected. If you click on the **Set** button on the meter, you will see the model parameters for this meter:

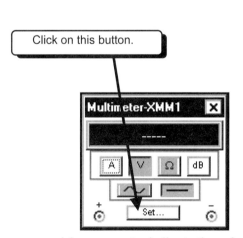
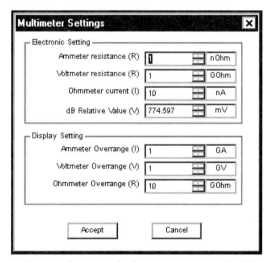

Here we see model components similar to those for the voltage and current indicators on page 125. When the meter is set to measure current, it has a resistance of 1 nΩ. When the meter is set to read voltage, the meter looks like a large resistance of 1 GΩ. When the meter is used to measure resistance, it passes a 10 nA current through the device it is connected to, and it reads the voltage across that element. These resistances make the meters close to ideal, and we will not change them. It is important to note that if a series resistance of 1 nΩ or a parallel resistance of 1 GΩ is significant in your circuit, you will need to account for these resistances in your circuit analysis. The lower pane of the window allows you to specify error conditions for the meter. When the meter reads quantities that are above the specified value, the meter will indicate an error. Click on the **Accept** button to close the **Multimeter Settings** dialog box.

We can view both meters at the same time, so double-click the *LEFT* mouse button on the other Multimeter to open its window. You should have both Multimeters opened as shown below:

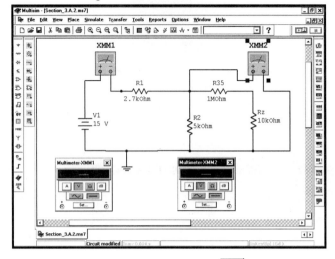

Select the left meter to measure current by clicking on the **A** button . Select the right meter to measure voltage by clicking on the **V** button .

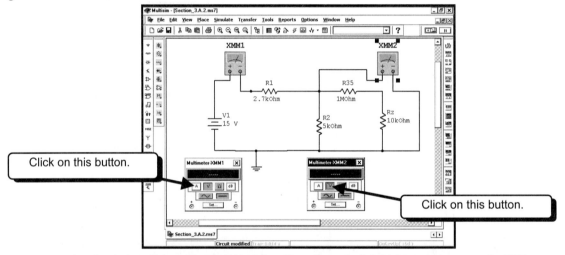

We can now start the simulation. Select **Simulate** and then **Run** from the Multisim menus, press the **F5** key, or click on the simulate button to start the simulation. The meters will display their measured values:

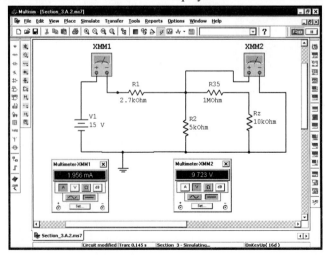

Note that you can press the meter buttons while the simulation is running. I will change **XMM1** to read voltage:

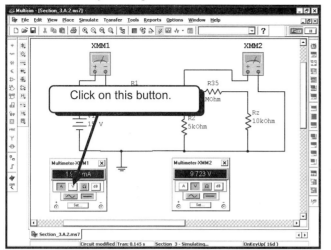 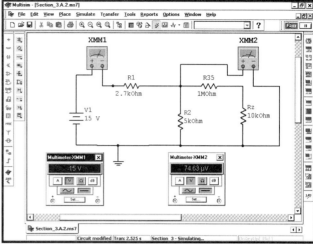

After a few moments the meters change their values, but we notice that the value displayed by **XMM2** also changes. This is because when we changed **XMM1** to read voltage, the model of the meter changes to a 1 GΩ resistor. This large resistance dominates the circuit and causes all of the readings to change. This is obviously not the correct way to use this meter, so I will change **XMM1** back to reading current.

We can also use these meters to read AC or DC waveforms. Switch the meters to AC by clicking on the AC button as shown below:

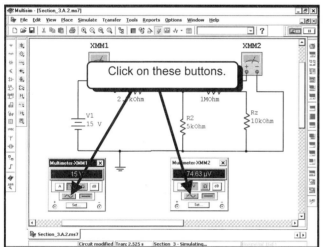 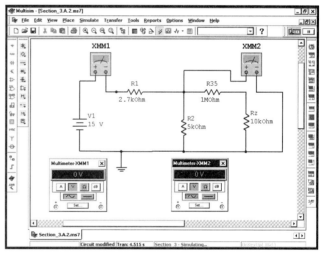

We see that both meters change their display after a few moments and read 0. This is because there are no AC sources in the circuit, so the measured items have no AC component and therefore read 0. You can close the meter displays by clicking on the ☒ in the upper right corner of the Multimeter windows:

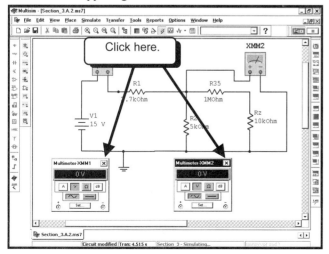 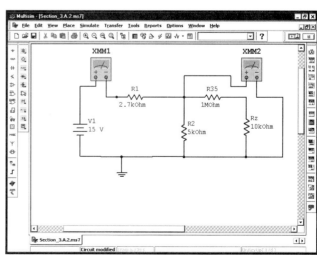

Remember to press the **F5** key to stop the simulation.

EXERCISE 3-3: Find the unknown node voltages and the current through R1 using Multimeters.

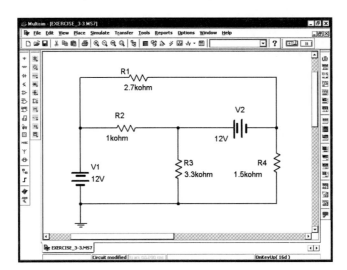

SOLUTION:

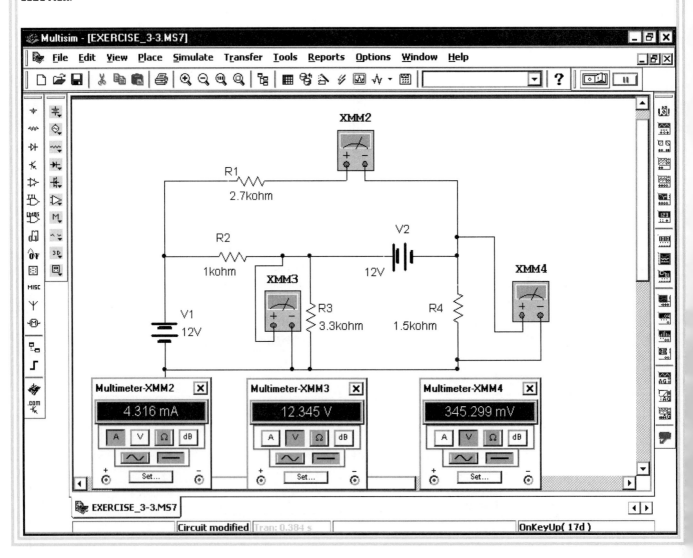

EXERCISE 3-4: Find the DC node voltages for the circuit below and the DC current through R1 using meters.

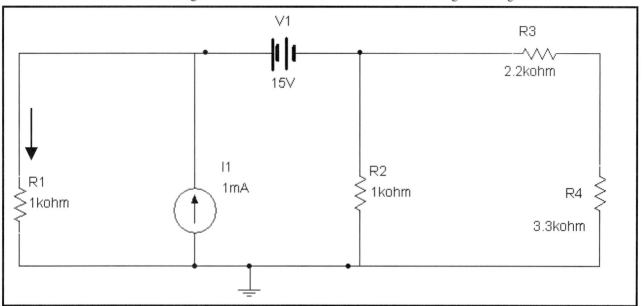

SOLUTION:

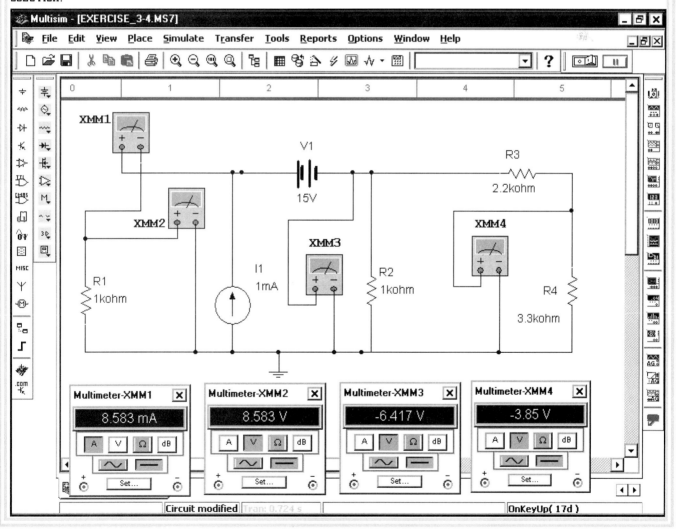

3.A.3. Using the Power Meter

Another meter we can use in the circuit of the previous section is the power meter. We will start with the circuit below:

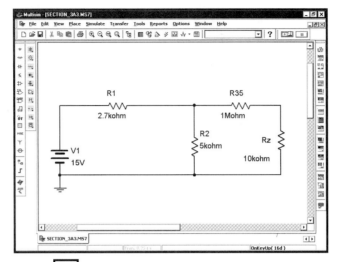

Double-click on the Wattmeter button [icon] in the instruments toolbar. A wattmeter will become attached to your mouse pointer:

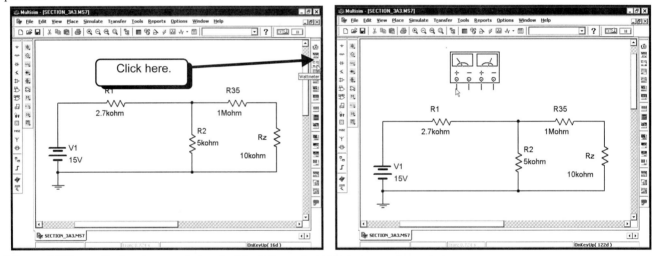

We would like to display the power absorbed by R1, so place the meter as shown below:

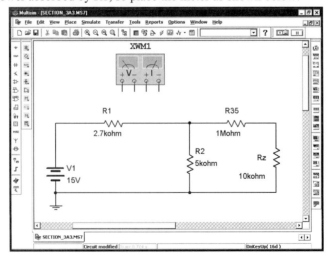

To measure the power absorbed by R1, we must measure the current through the resistor and the voltage across the resistor. Measuring current with the wattmeter is the same as measuring current with a Multimeter or with a current indicator; if current enters the positive terminal and leaves the negative terminal, the meter will interpret it as a positive current. Wire the meter as shown:

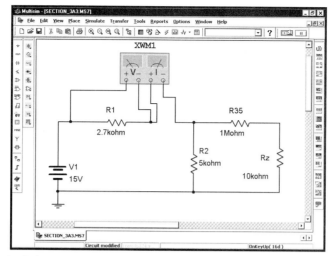

As shown, the meter measures the left terminal of R1 as the positive voltage reference, and current is positive flowing left to right. Thus, the positive direction of current is entering the positive voltage terminal, and this is the definition of power, P = VI. Note that P = VI applies only if you follow the convention that the positive direction of current enters the positive voltage reference terminal.

Double-click the **LEFT** mouse button on the power meter to open its window:

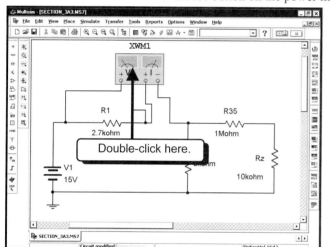 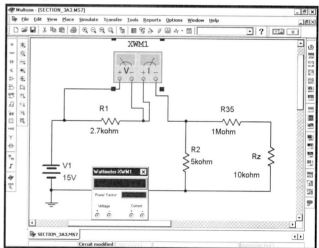

This meter works for AC or DC measurements. Select **Simulate** and then **Run** from the Multisim menus, press the **F5** key, click the **Run/stop simulation** button, or click on the simulate button to start the simulation.

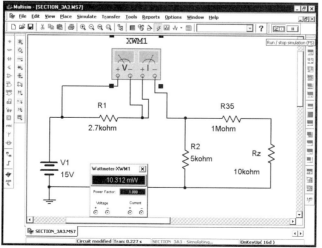

We see that R1 absorbs **10.312 mW** of power.

EXERCISE 3-5: Find the power absorbed by R3 in the circuit below:

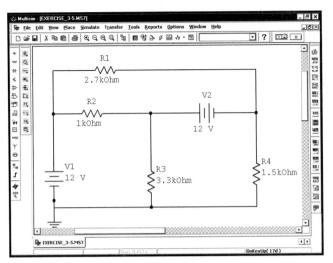

SOLUTION: Add the power meter as shown:

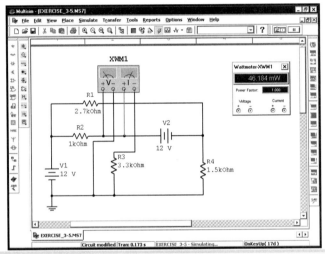

EXERCISE 3-6: Find the power supplied by I1:

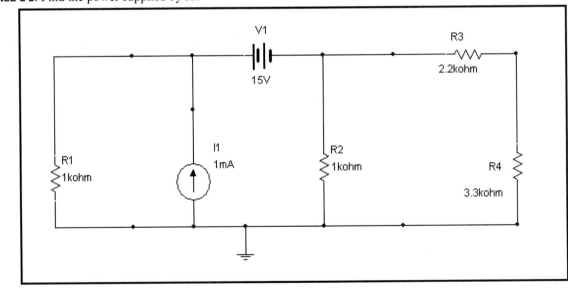

SOLUTION: Redraw the circuit to make the placement of the power meter more convenient.

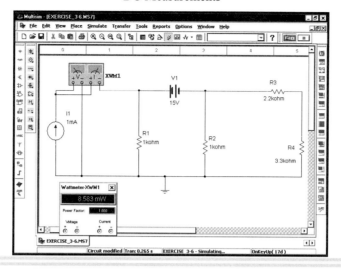

3.A.4. DC Operating Point Analysis

The final way of simulating a circuit and finding the DC node voltages we will look at is the DC Operating Point Analysis. We will start with the example circuit used in previous sections:

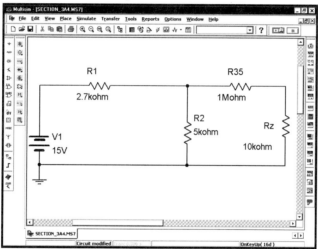

This analysis gives us the voltages at all nodes of the circuit. However, we do not know how the nodes are numbered. We can display the node names in two ways. One way is to double-click on a wire. For example, double-click the *LEFT* mouse button on a wire connected to the middle node:

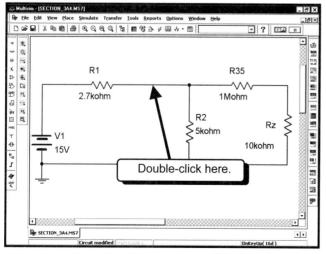
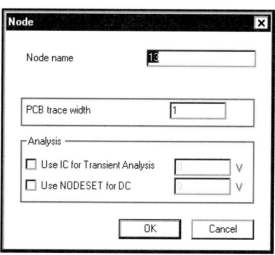

This dialog box tells us that this node is numbered 13 (your name may be different) and it also allows us to change the node name. I will change the name to **Vmid**:

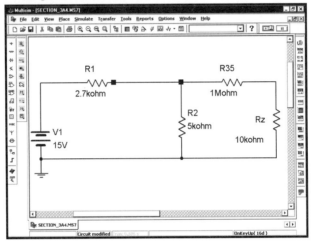

Click on the **OK** button to save the change:

We now know the name of this node, but it is still not displayed on the circuit.

To display the node names on the schematic, select **Options** and then **Preferences** from the Multisim menus:

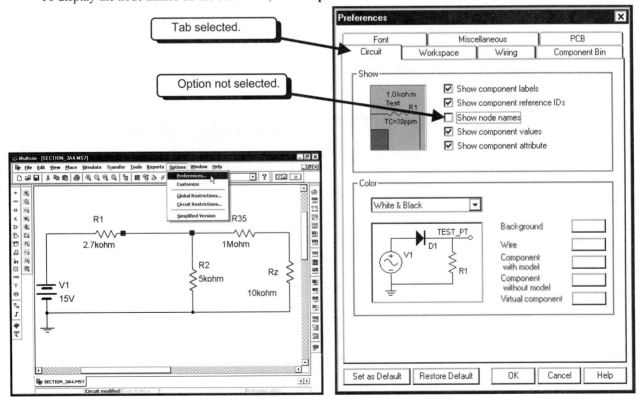

Select the **Circuit** tab if it is not already selected. Note in the dialog box above that the option to display node names is not selected. To display node names on the circuit, we need to select this option. Click the *LEFT* mouse button on the empty box ☐ next to the **Show node names** option to enable it. When the option is enabled, the box will have a checkmark ☑ :

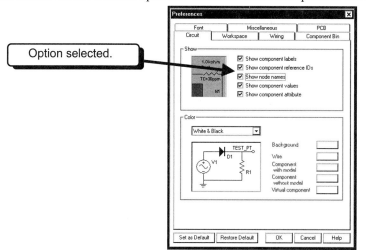

When you click on the **OK** button, you will return to the schematic and the node names will be displayed.

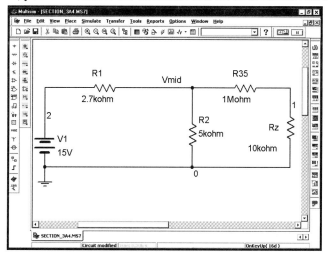

The node numbering of your circuit may be different. You can rename the other nodes (except node 0) if you want to. We now know the names and can set up the DC Operating Point simulation. Select **Simulate**, **Analyses**, and then **DC Operating Point** from the Multisim menus:

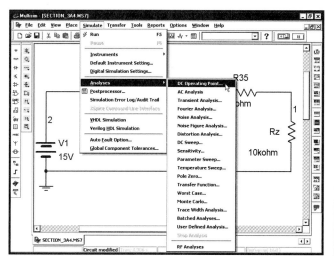
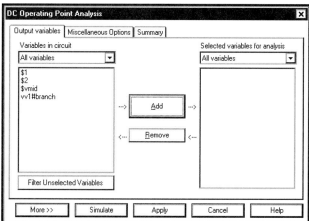

The only thing we need to do is to specify the nodes we are interested in. To select a node, click the **LEFT** mouse button on the node to select it. For example, click the **LEFT** mouse button on the text $2 as shown. It will become highlighted, indicating that it has been selected:

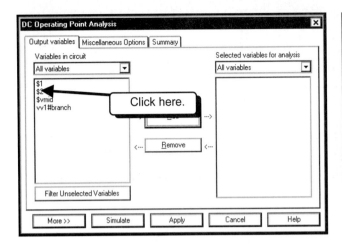 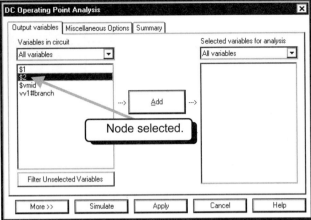

Click on the **Add** button. The node will be added to the list of nodes whose result is displayed:

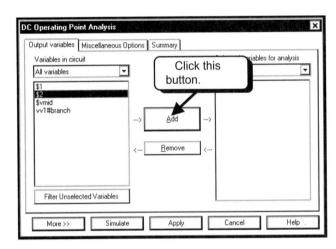 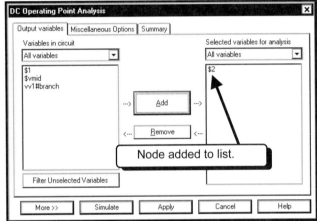

Repeat the procedure to add nodes **$1** and **$vmid**:

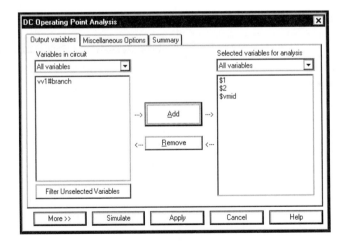

We are now ready to run the simulation, so click on the **Simulate** button. The results are displayed in a new window:

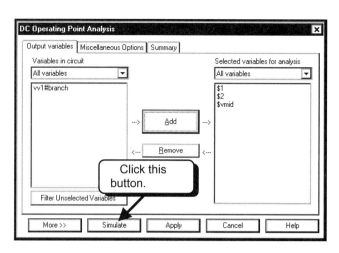
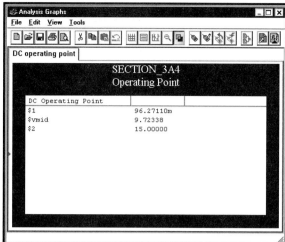

We see that the voltage at node **$1** is **96.27** mV, the voltage at node **Vmid** is **9.72** volts, and the voltage at node **2** is **15** volts. These results agree with the results of the previous analyses.

3.B. Nodal Analysis with Dependent Sources

To illustrate an example with dependent sources, we will find the node voltages in the circuit below. Create the circuit below.

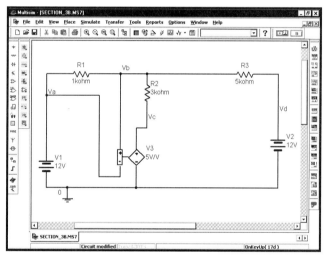

The new part in this circuit is the voltage-controlled voltage source. The way this element is wired, the voltage at node **Vc** is 5 times the voltage across **R1**: $V_c = 5(V_b - V_a)$. If you zoom in on the voltage-controlled voltage source, you will see the graphic shown in Figure 3-1. The plus (+) and minus (−) terminals on the voltage-sensing element (the left element) are open circuited. These connections draw no current and only sense the voltage of the nodes to which they are connected. The right half of the graphic contains the dependent source. The voltage of this source is the gain 5V/V times the voltage at the sensing nodes.

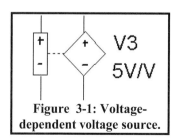

Figure 3-1: Voltage-dependent voltage source.

We can use any of the methods presented in section 3.A to measure the node voltages. We will use a number of voltage indicators to measure various voltages. You can use the DC Operating Point Analysis or the meters to measure the voltages if you wish. My modified circuit is shown below:

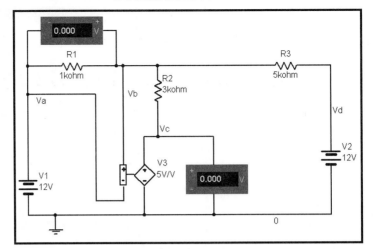

When you have made the changes to your circuit, click the **LEFT** mouse button on the **Run/stop simulation** button to simulate the circuit. The indicators will be updated and display the DC node voltages:

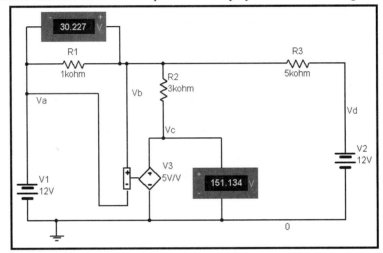

It turns out that this circuit is fairly sensitive to the parallel impedance of the voltage indicators. We will change the input impedance of the indicators from the default value of 1 MΩ to 1 GΩ. First, we must stop the simulation. Click the **LEFT** mouse button on the **Run/stop simulation** button to turn the simulation off. Next, double-click the **LEFT** mouse button on one of the voltage indicators as shown below:

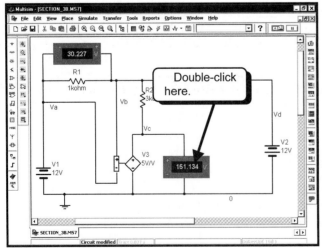
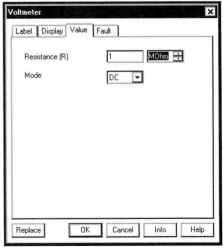

We see that the default value of the parallel resistance is 1 MΩ. To compare the simulated results to calculated results, we need to increase the parallel resistance. Change this value to 1 GΩ or higher and then click on the **OK** button. Increase the resistance of all voltage indicators in your circuit to 1 GΩ or higher. Note that my meters do not display the value of the parallel resistance nor the text **DC** on the schematic. You can also hide this text by following the procedure shown on pages

126-129. Click on the **Run/stop simulation** button to simulate the circuit and update the displays of the indicators:

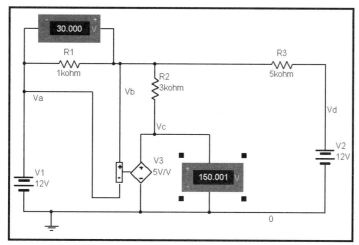

We see that the numbers displayed now are different than before. For most of your simulations, you should increase the voltage indicator parallel resistance to 1 GΩ or higher, or the voltage indicators will affect the operation of the circuit.

EXERCISE 3-7: Find the DC node voltages for the circuit below:

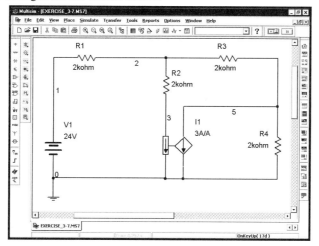

HINT: I1 is a current-controlled current source. Note that node 3 is connected to ground with a wire. Thus, the voltage of node 3 should be zero volts. Node 3 is necessary because it joins the lower terminal of R2 to the current-sensing terminal of I1.

SOLUTION: Use meters or voltage indicators to measure the voltages of the unknown nodes. The voltage at node 1 is known and is equal to V1.

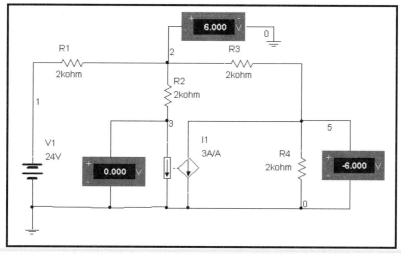

EXERCISE 3-8: Find the DC node voltages for the circuit below:

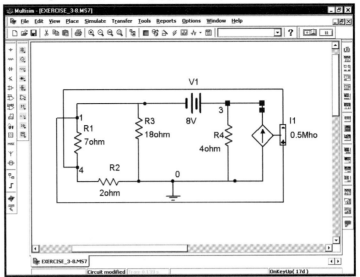

HINT: I1 is a voltage-controlled current source. The current through **I1** is 0.5 times the voltage at node 1 minus the voltage at node 4.

SOLUTION: Add voltage indicators to the circuit and change their parallel resistance to 1 GΩ.

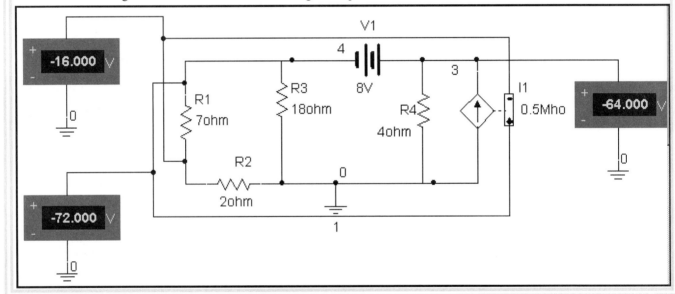

3.C. Diode DC Current and Voltage

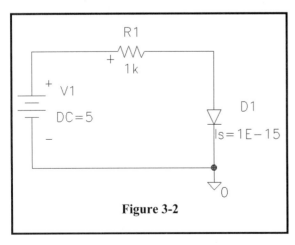

Figure 3-2

We will now use Multisim to find the diode current and voltage in the circuit of Figure 3-2. The diode current is given as $I_D = I_S[\exp(V_D/\eta V_T) - 1]$. I_S is the diode saturation current and is 10^{-15} amp for this example. V_T is the thermal voltage and is equal to 25.8 mV at room temperature. η is the emission coefficient for the diode and its default value is 1. Multisim automatically runs all simulations at room temperature by default.

When you use a diode in a circuit, you will have to specify a model for the diode. In our case, the model will tell Multisim the value of I_S for our diode. The libraries have a number of predefined models that are usually used in a classroom environment. However, the model for this diode is not in our libraries, so we will have to define a new model for it. The part for the diode you should use is **DIODE_VIRTUAL**. Draw the circuit below:

DC Measurements

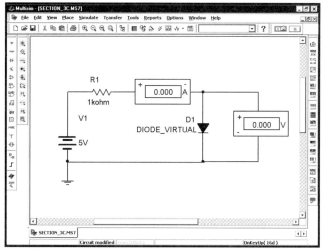

The **DIODE_VIRTUAL** model is located in the Diodes group. To display the available diodes, click on the **Diode** button as shown:

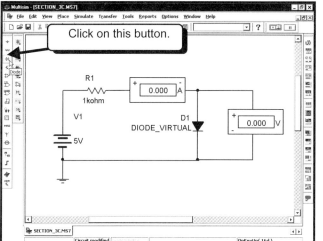
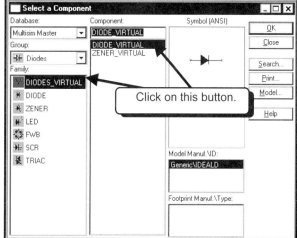

On my screen, a virtual diode is selected by default. Make sure that you have a virtual diode selected and click the **OK** button to place the part in your circuit.

Parts that are referred to as "virtual" by Multisim are ideal parts, or parts that are easy to specify by a value or model. The virtual diode implements the basic diode equation $I_D = I_S[\exp(V_D/\eta V_T) - 1]$ with η equal to 1. All other model parameters are set to their default.

After placing the diode in your circuit, we need to change the model parameter I_S. Double-click on the diode graphic:

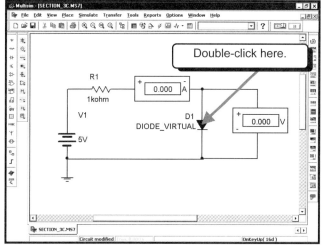
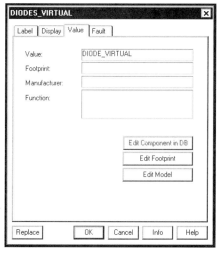

Click on the **Edit Model** button:

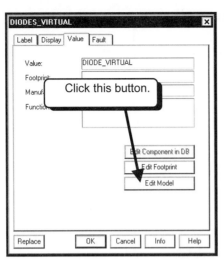
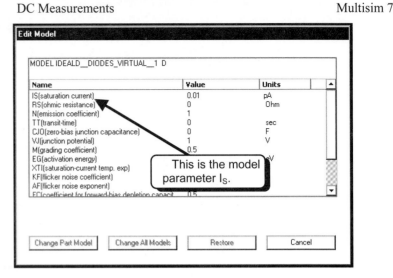

We see that the default value for the saturation current is **0.01 pA** or $I_S = 10^{-14}$ amp. Change the value to **0.001 pA**:

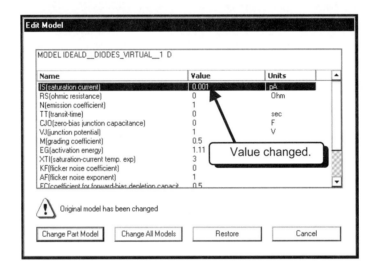

Click on the **Change Part Model** button and then click on the **OK** button. You will return to the schematic.

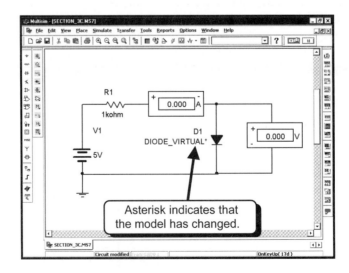

The schematic displays an asterisk next to the part name to indicate that we have changed the model. Run the simulation by clicking on the **Run/stop simulation** button. The diode current and voltage will be displayed:

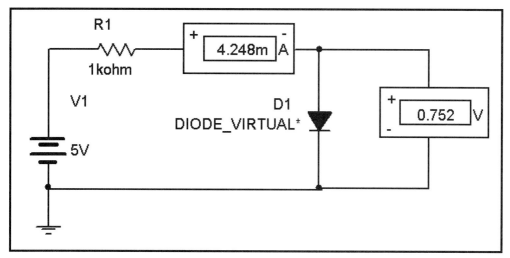

We see that the diode current is 4.248 mA, and the diode voltage is 0.752 volts.

EXERCISE 3-9: Find the diode current and voltage in the circuit below:

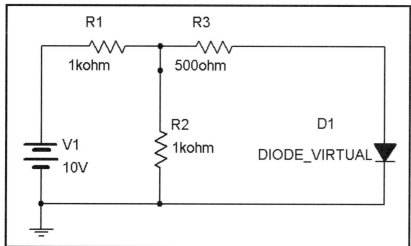

SOLUTION: Use the same diode parameters as in the previous example. Add indicators and then simulate the circuit.

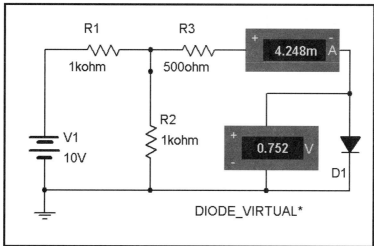

The diode voltage is 0.752 V, and the diode current is 4.248 mA. The diode voltage and current of this circuit are the same as the diode voltage and current of the previous example. This result should be expected since the Thevenin equivalent of V1, R1, R2, and R3 in this exercise is exactly the same as the circuit of the previous example.

3.C.1. Changing the Temperature of the Simulation

In the last simulation we found the diode voltage and current at the default temperature of 27°C. Suppose we want to simulate the circuit at a different temperature? This can easily be done by changing the default instrument settings. From the Multisim menus select **Simulate** and then **Default Instrument Setting**:

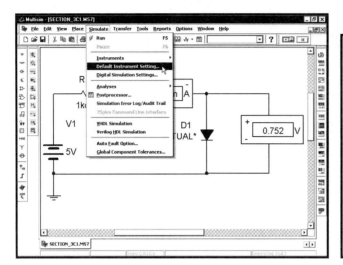
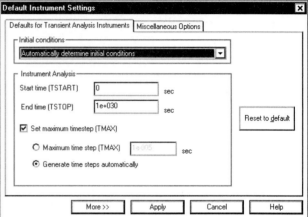

Select the **Miscellaneous Options** tab:

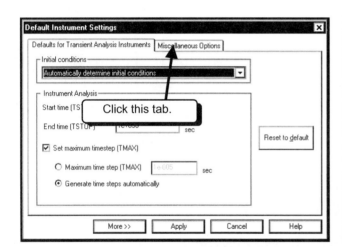
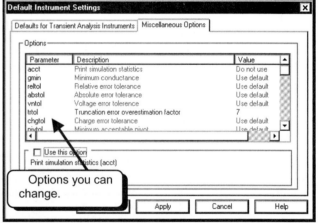

This dialog box lists most of the options you can change. Scroll down the list until you see the **temp** option:

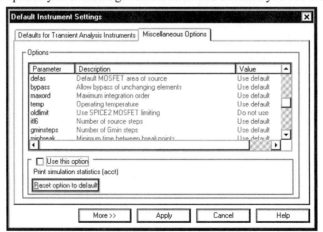

Click the *LEFT* mouse button on the text **temp** to select the option:

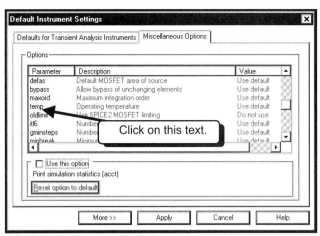 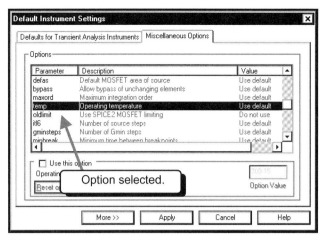

To modify the value of this option, click the ***LEFT*** mouse button on the square ☐ next to the text **Use this option**. The box should become filled with a checkmark ☑, indicating that the option is selected:

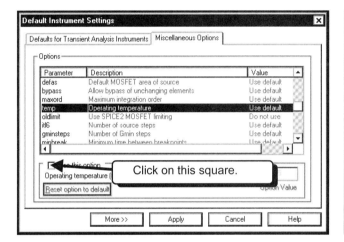 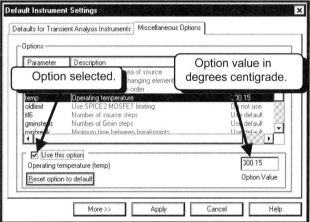

When you select the option, a value appears in the **Option Value** text field. For temperature, the units of this option are degrees centigrade. By default, all simulations are run at 27°C unless you specify a different temperature. For this run we will find the diode voltage and current at 125°C. Change the **Option Value** as shown:

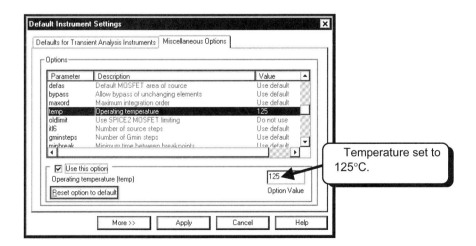

Click on the **Apply** button to accept the value and return to the schematic. Run the simulation by clicking on the **Run/stop simulation** button. The diode current and voltage will be displayed for a temperature of 125°C:

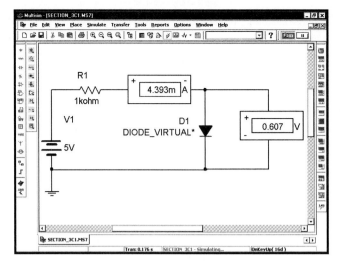

3.D. Finding the Thevenin and Norton Equivalents of a Circuit

Multisim can be used to easily calculate the Norton and Thevenin equivalents of a circuit. The method we will use is the same as if we were going to find the equivalent circuits in the lab. We will make two measurements, the open-circuit voltage and the short-circuit current. The Thevenin resistance is then the open-circuit voltage divided by the short-circuit current. This will require us to create two circuits, one to find the open-circuit voltage, and the second to find the short-circuit current. In this example, we will find the Norton and Thevenin equivalent circuits for a DC circuit. This same procedure can be used to find the equivalent circuits of an AC circuit (a circuit with capacitors or inductors). However, instead of finding the open-circuit voltage and short-circuit current using DC sources, we would use AC sources.

For this example, we will find the Thevenin and Norton equivalent circuits for the circuit attached to the diode in **EXERCISE 3-9**. The circuit is repeated below:

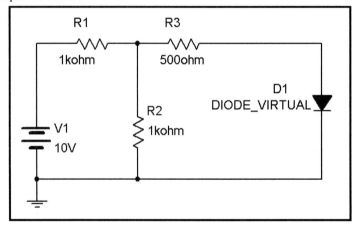

This circuit is difficult because it contains a nonlinear element (the diode) and a complex linear circuit. If we could replace V1, R1, R2, and R3 by a simpler circuit, the analysis of the nonlinear element would be much easier. To simplify the analysis of the diode, we will find the Thevenin and Norton equivalent circuits of the circuit connected to the diode; that is, we will find the Thevenin and Norton equivalents of the circuit below:

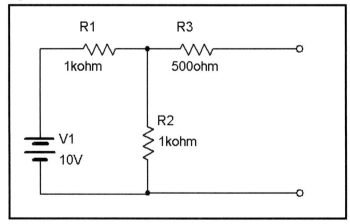

We will convert this circuit into the Thevenin equivalent:

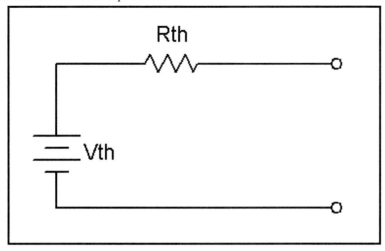

Once we find numerical values for Vth and Rth, the entire circuit of **EXERCISE 3-9** reduces to:

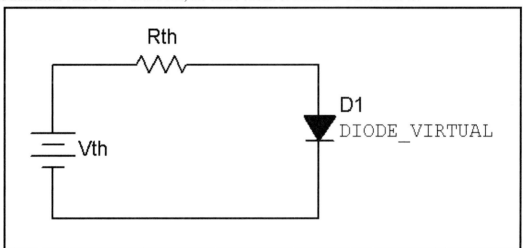

For determining the diode voltage and current, this circuit is much easier to work with than the original. This example is concerned with finding the numerical values of the equivalent circuit. The analysis of the circuit above for finding the diode voltage and current was covered in Section 3.C. We will now find the Thevenin and Norton equivalent circuits of the circuit shown below:

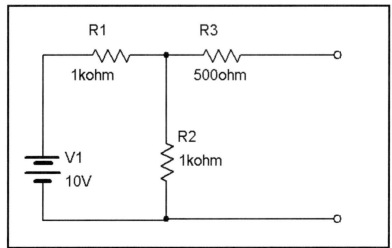

We will first find the open-circuit voltage. This is just the voltage across the two terminals in the circuit shown above. We will add a ground to the circuit and a voltage indicator:

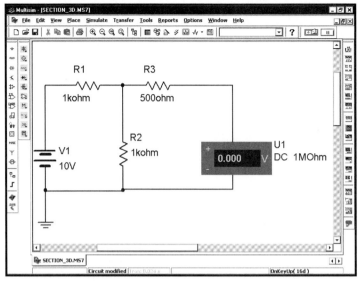

The lower terminal is now at ground potential, or zero volts. Before we run the simulation, we need to change the attributes of the voltage indicator. Double-click the **LEFT** mouse button on the voltage indicator graphic to edit its properties:

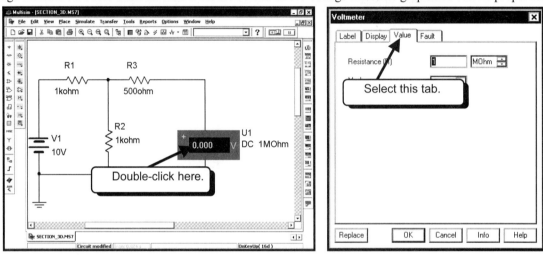

Select the **Value** tab as shown above and then change the resistance to 1 TOhm (10^{12} Ω).

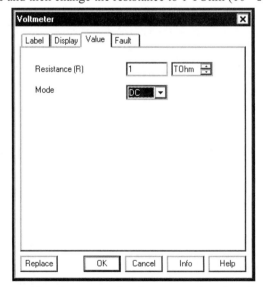

This resistance is large enough to be considered an open circuit compared to the other resistors in the circuit. Click on the **OK** button when you have changed the value. Run the simulation by pressing the **F5** key:

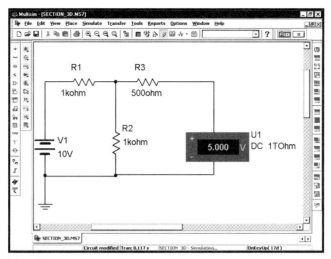

The open-circuit voltage is **5.000** volts.

Next we must find the short-circuit current. All we need to do is replace the voltage indicator with a current indicator:

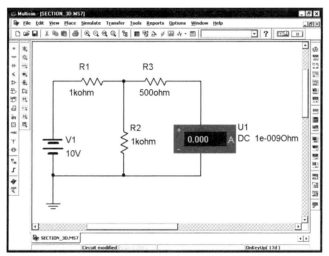

Current measuring devices have a very low resistance, so they have a negligible effect on the current they are measuring. The current indicator in Multisim has a series resistance of 1 nΩ, or 10^{-9} Ω. If we place a current indicator across two terminals, as shown above, we are essentially placing a 1 nΩ resistor between the two terminals. This resistance is small enough to be considered a short circuit. Thus, the current indicator shown above measures the short-circuit current. All we need to do now is run the simulation:

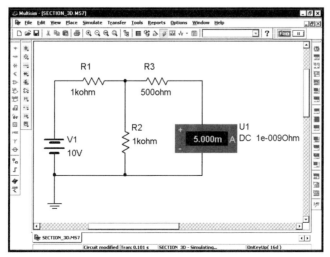

We see that the short-circuit current is **5.000** mA.

We can now find the Thevenin resistance by dividing the open-circuit voltage by the short-circuit current:

$$R_{th} = \frac{Voc}{Isc} = \frac{5.000 \text{ V}}{5.00 \text{ mA}} = 1000 \, \Omega$$

Our Thevenin and Norton equivalent circuits are shown below:

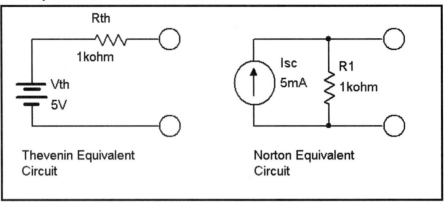

EXERCISE 3-10: Find the Thevenin and Norton equivalent circuits for the circuit below:

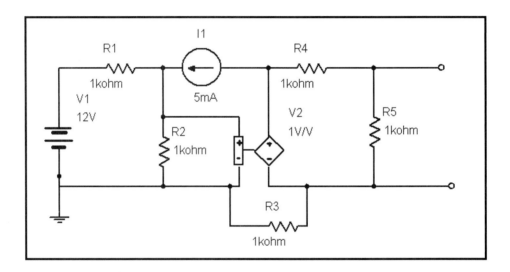

SOLUTION: Voc = 4.25 V, Isc = 8.5 mA, Rth = 500 Ω. Use the circuits below:

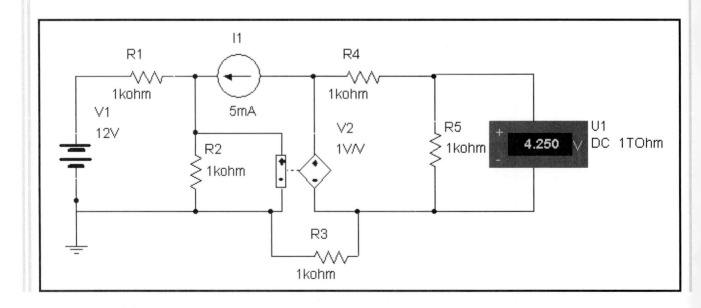

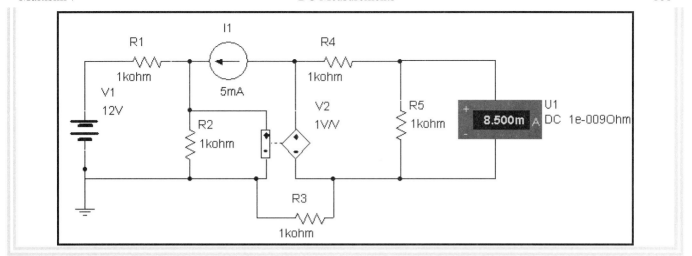

3.E. Transistor DC Operating Point

One of the first things you should do when you are simulating a transistor amplifier circuit is to check the transistor operating point, also referred to as the transistor DC Operating Point. If the transistor operating point is incorrect, none of the other analyses will be valid. If another analysis does not make sense, check the operating point to see if all active devices are biased correctly. When Multisim finds the operating point, it assumes that all capacitors are open circuits and that all inductors are short circuits.

For this example we will demonstrate the DC Operating Point of an NPN BJT. The same procedures could be used for a PNP BJT, a jFET, or a MOSFET. We will show two ways to view the data: using the audit trail, and displaying values in a simulation output window. We will illustrate the analysis with the circuit below:

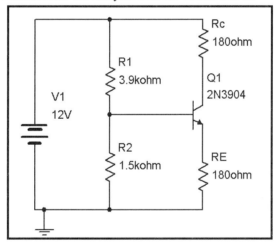

The 2N3904 transistor is located in the **Transistors** group. Click on the **Transistor** button as shown below:

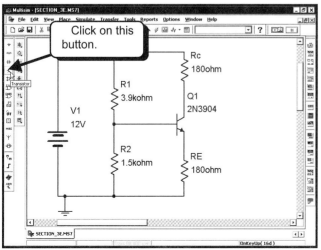
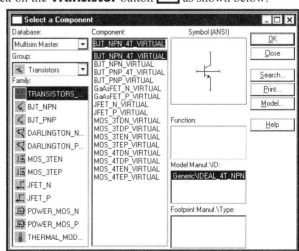

The **Transistors Group** is split into the different types of transistors. A 2N3904 is an NPN bipolar junction transistor (BJT). Select the **BJT_NPN** family of transistors:

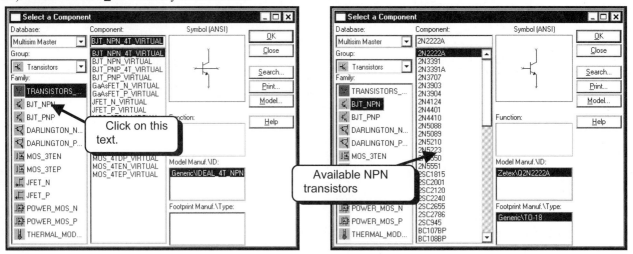

The dialog box now lists the available NPN bipolar junction transistors. Select the **2N3904** and click on the **OK** button to place the part. When you finish drawing your circuit, you can continue with the procedure below.

We must first set up the DC Operating Point simulation. Select **Simulate**, **Analyses**, and then **DC Operating Point** from the menus:

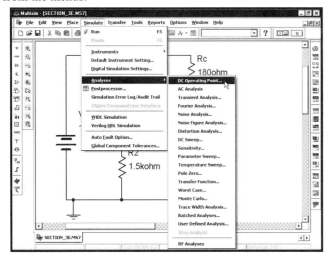
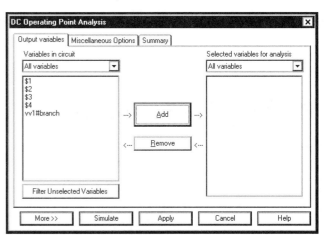

The first method we will use to display the operating point is the audit trail. Click on the **More>>** button:

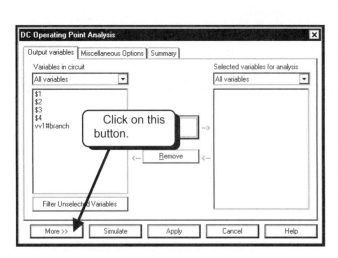
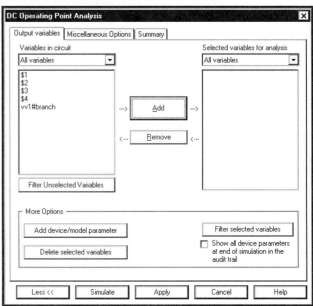

We need to enable the option **Show all device parameters at end of simulation in the audit trail**. Click on the ☐ next to this option to enable it. The square should fill with a checkmark ☑:

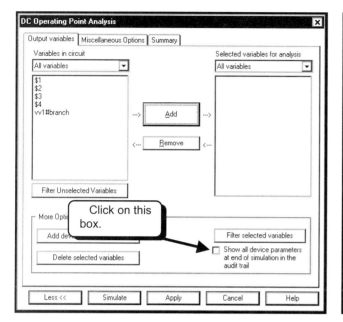
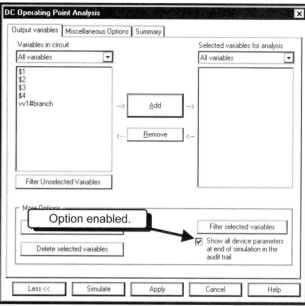

Although we are not interested in any of the node voltages, we need to specify an output variable in order to run the simulation. Click on the text **$1** as shown to select the output variable:

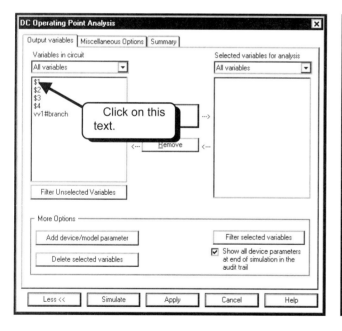
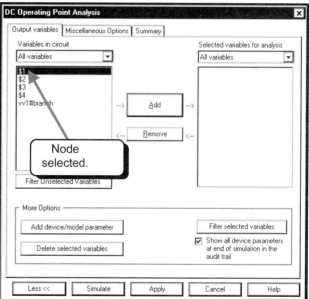

Click on the **Add** button to add the node to the list of output variables that will be displayed during the simulation:

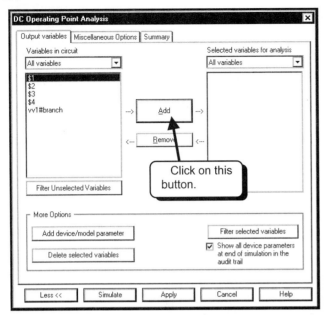 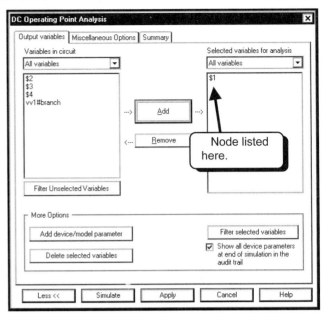

Now that we have an output variable listed, we can run a simulation. Click on the **Simulate** button to run the simulation:

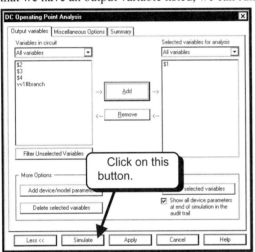 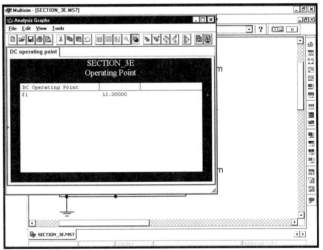

A new window pops up and displays the results of the simulation. The only value shown is the DC voltage at node $1. The information we seek is displayed in the audit trail, so we can close this window. Click on the ✖ in the upper right corner of the window to close it:

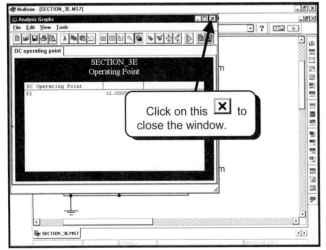 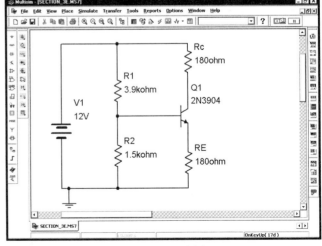

To view the audit trail, select **Simulate** and then **Simulation Error Log/Audit Trail** from the Multisim menus:

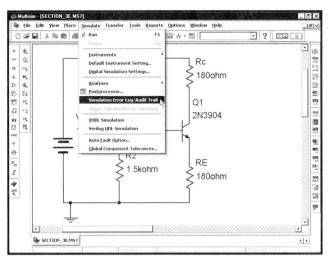
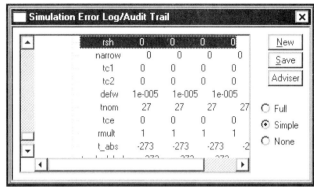

Scroll up or down in the log until you see the information about the BJT:

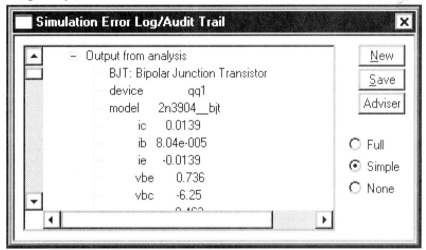

We see that the collector current (**ic**) is 13.9 mA, the base current (**ib**) is 80.4 µA, **vbe** is **0.736** volts, and **Vbc** is **-6.25** volts. Vce is not directly displayed, but Vce is equal to –Vbc + Vbe, which is equal to 6.986 volts.

If we scroll down the list, we can see the parameters for the hybrid-π small-signal model:

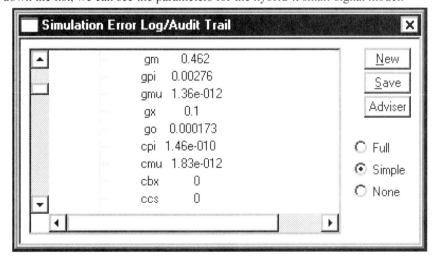

In the hybrid-π model we are used to resistances such as r_π, r_o, r_x, and r_μ. The model parameters shown here are conductance parameters with $r_\pi = 1/$**gpi**, $r_o = 1/$**go**, $r_x = 1/$**gx**, and $r_\mu = 1/$**gmu**. There is more information displayed in the audit trail, but we will not look at it now. Click on the ⊠ in the upper right corner of the window to close it:

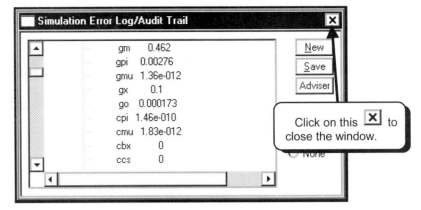

Next, we will run the DC Operating Point Analysis, but we will display the results differently. From the Multisim menus, select **Simulate**, **Analyses**, and then **DC Operating Point** from the menus:

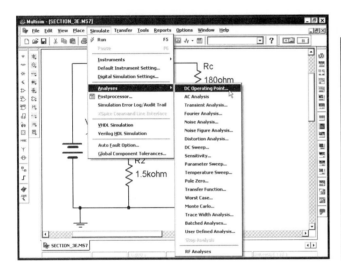
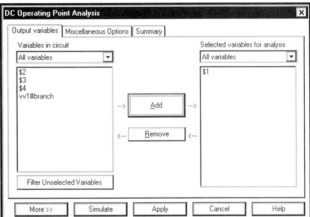

Node **$1** was selected in our last exercise. You can remove it if you wish. We would like to display the transistor currents, voltages, and some small-signal quantities found in the DC operating point simulation. A problem is that these quantities are not yet displayed in this dialog box and we need to add them. Click on the **More>>** button:

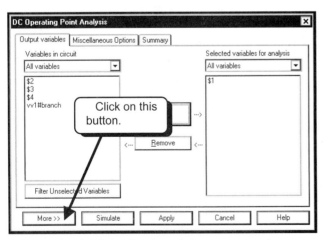
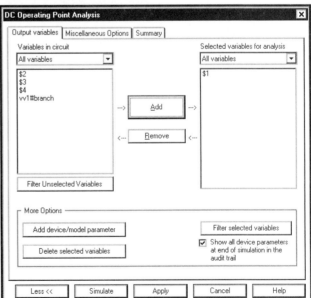

Click on the **Add device/model parameter** button:

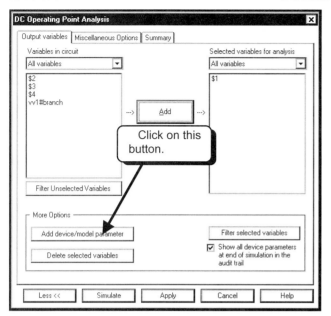
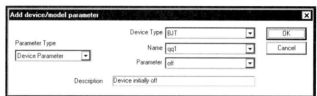

There is only one BJT device in this circuit, so this dialog box selects that device and refers to it as **qq1**. We can display the device model parameters or the device parameters. A model parameter would be something like a diode's saturation current or a transistor's current gain, β. A device parameter would be something like a transistor's collector current or base-emitter voltage. We are interested in the transistor's currents, voltages, and small-signal model parameters, and these quantities are listed as device parameters. By default this dialog box is set to display the device parameters of **qq1**, so all we need to do is select those parameters. Click the *LEFT* mouse button on the down triangle ▼ as shown:

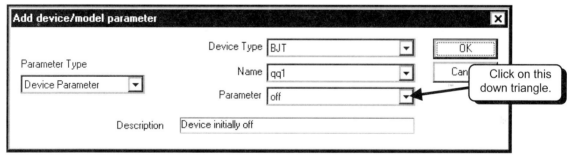

A list of device parameters will be displayed:

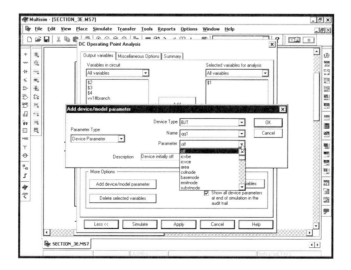

Scroll down the list until you see the parameter **ic**. This is the transistor's collector current:

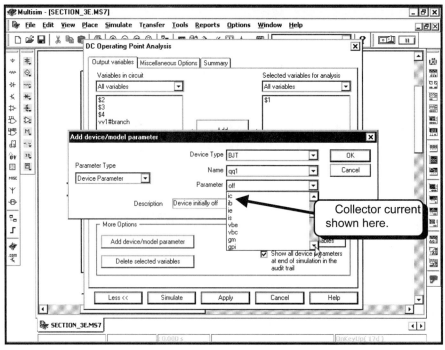

We see several parameters that we are interested in. Click the **LEFT** mouse button on the text **ic** to select it:

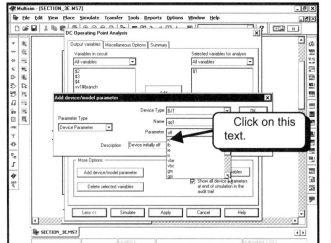

The dialog box tells us that this parameter is the collector current. Click on the **OK** button to add the parameter to the list:

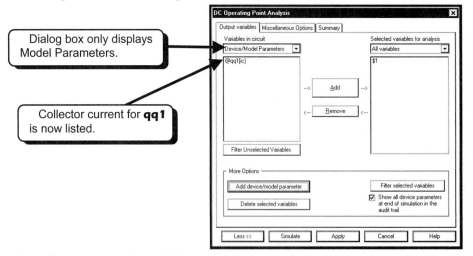

The collector current for **qq1** is now listed, but we have not yet selected it to be displayed when the simulation runs. We need to specify that we want this parameter displayed in the output. Click the **LEFT** mouse button on the text **@qq1[ic]** to select it:

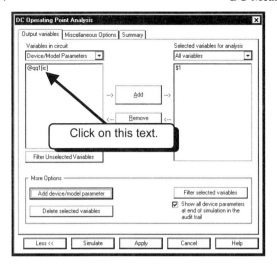 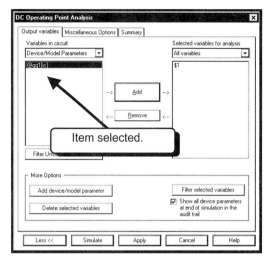

Click the *LEFT* mouse button on the **Add** button:

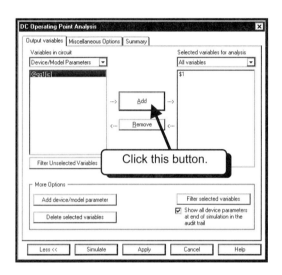 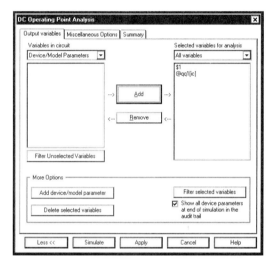

The bias collector current will now be displayed when the DC Operating Point Analysis runs. Repeat the procedure to display the base current, Vbe, Vbc, gpi, and gm for this device:

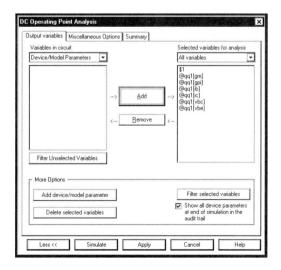

We are now ready to run the simulation. Click on the **Simulate** button. The results are displayed in the **Analysis Graphs** window:

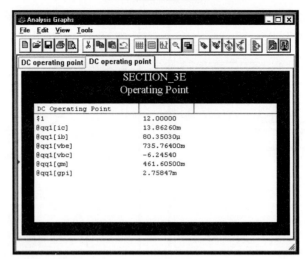

These numbers agree with the information displayed in the audit trail.

EXERCISE 3-11: Run the DC Operating Point Analysis for the transistor circuit and display the collector current, base current, Vbc, Vbe, gpi, and gm for the BJT.

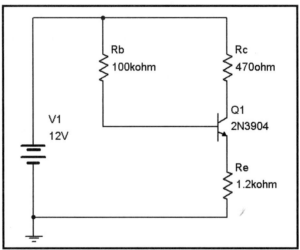

SOLUTION:

EXERCISE 3-12: Run the DC Operating Point Analysis for the circuit below and display the drain current, gate current, Vgs, Vgd, and gm for the jFET.

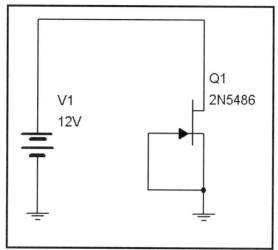

SOLUTION: You will need to click on the **Filter Unselected Variables** button and enable all of the options to display the required parameters. You must enable the options because the model for the 2N5486 is a subcircuit and these items are not usually displayed.

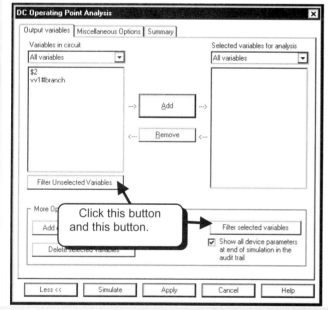
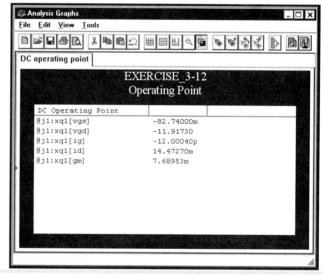

3.F. Problems

Problem 3-1: For the circuit shown below, find the unknown node voltages and the current through R3 using indicators.

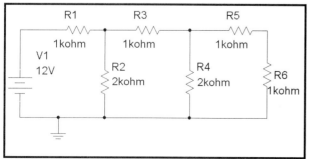

Problem 3-2: For the circuit of Problem 3-1, find the unknown node voltages and the current through R3 using Multimeters.

Problem 3-3: For the circuit of Problem 3-1, find all node voltages using the SPICE DC Operating Point Analysis.

Problem 3-4: For the circuit of Problem 3-1, find the power absorbed by R6 using the power meter.

Problem 3-5: For the circuit of Problem 3-1, find the current through all resistors using the indicators.

Problem 3-6: For the circuit of Problem 3-1, find the power absorbed by R4, R5, and R6 using the power meter.

Problem 3-7: For the circuit of Problem 3-1, find the power supplied by V1 using the power meter.

Problem 3-8: For the circuit of Problem 3-1, find the unknown node voltages and the current through R2 using indicators.

Problem 3-9: For the circuit of Problem 3-1, find the unknown node voltages and the current through R4 using Multimeters.

Problem 3-10: For the circuit of Problem 3-1, use the indicators to show that Kirchhoff's voltage law holds for all three loops of this circuit.

Problem 3-11: For the circuit of Problem 3-1, use the indicators to show that the current through R1 equals the current through R2 and R3 (verifying Kirchhoff's current law).

Problem 3-12: For the circuit shown below, find the unknown node voltages and the current through R3 using indicators.

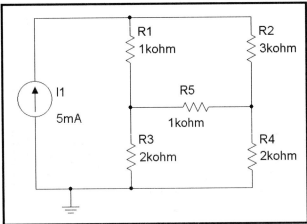

Problem 3-13: For the circuit of Problem 3-12, find the unknown node voltages and the current through R3 using Multimeters.

Problem 3-14: For the circuit of Problem 3-12, find all node voltages using the SPICE DC Operating Point Analysis.

Problem 3-15: For the circuit of Problem 3-12, find the power absorbed by R5 using the power meter.

Problem 3-16: For the circuit of Problem 3-12, find the current through all resistors using the indicators.

Problem 3-17: For the circuit of Problem 3-12, find the power absorbed by R4 using the power meter.

Problem 3-18: For the circuit of Problem 3-12, find the power supplied by I1 using the power meter.

Problem 3-19: For the circuit of Problem 3-12, find the unknown node voltages and the current through R2 using indicators.

Problem 3-20: For the circuit of Problem 3-12, find the unknown node voltages and the current through R4 using Multimeters.

Problem 3-21: For the circuit of Problem 3-12, use the indicators to show that Kirchhoff's voltage law holds for all three loops of this circuit.

Problem 3-22: For the circuit of Problem 3-12, use the indicators to show that the current through I1 equals the current through R1 and R2 (verifying Kirchhoff's current law).

Problem 3-23: For the circuit shown below, find the unknown node voltages and the current through R3 using indicators.

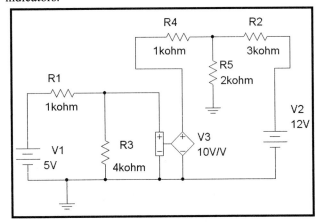

Problem 3-24: For the circuit of Problem 3-23, find the unknown node voltages and the current through R3 using Multimeters.

Problem 3-25: For the circuit of Problem 3-23, find all node voltages using the SPICE DC Operating Point Analysis.

Problem 3-26: For the circuit of Problem 3-23, find the power absorbed by R5 using the power meter.

Problem 3-27: For the circuit of Problem 3-23, find the current through all resistors using the indicators.

Problem 3-28: For the circuit of Problem 3-23, find the power absorbed by R4 using the power meter.

Problem 3-29: For the circuit of Problem 3-23, find the power supplied by V1 using the power meter.

Problem 3-30: For the circuit of Problem 3-23, find the unknown node voltages and the current through R2 using indicators.

Problem 3-31: For the circuit of Problem 3-23, find the unknown node voltages and the current through R4 using Multimeters.

Problem 3-32: For the circuit used in section 3.C, find the diode voltage and current when the diode saturation current is 10^{-13}amp, 10^{-14}amp, 10^{-15}amp, and 10^{-16}amp. By hand, plot the diode current versus the saturation current, and plot the diode voltage versus the saturation current.

Problem 3-33: Find the diode current and voltage in the circuit below.

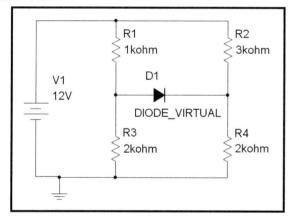

Problem 3-34: Find the diode current and voltage in the circuit below.

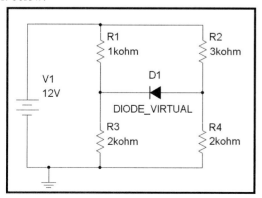

Problem 3-35: Find the diode current and voltage in the circuit below.

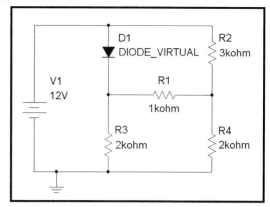

Problem 3-36: Find the diode current and voltage in the circuit below.

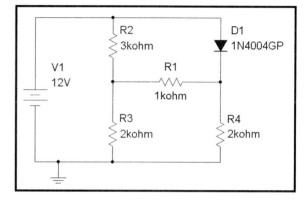

Problem 3-37: Find the diode current and voltage in the circuit below.

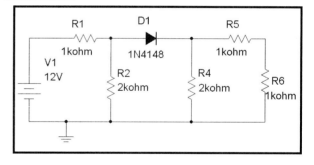

Problem 3-38: Find the diode current and voltage in the circuit below.

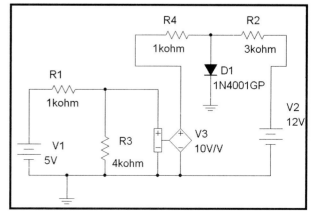

Problem 3-39: Find the Thevenin equivalent circuit for the circuit below at the indicated terminals:

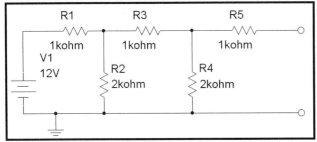

Problem 3-40: Find the Thevenin equivalent circuit for the circuit below at the indicated terminals:

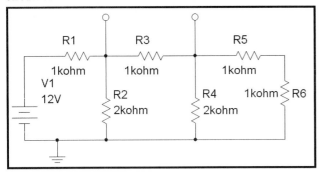

Problem 3-41: Find the Thevenin equivalent circuit for the circuit below at the indicated terminals:

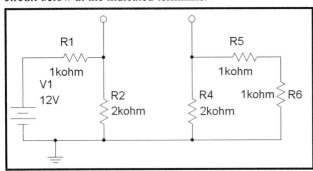

Problem 3-42: Find the Thevenin equivalent circuit for the circuit below at the indicated terminals:

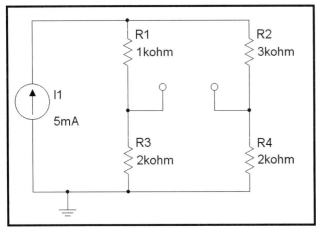

Problem 3-43: Find the Thevenin equivalent circuit for the circuit below at the indicated terminals:

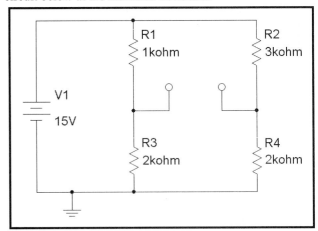

Problem 3-44: Find the Thevenin equivalent circuit for the circuit below at the indicated terminals:

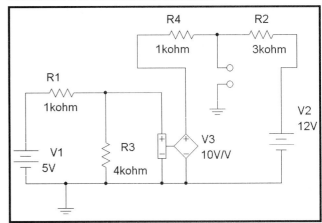

Problem 3-45: Find the Thevenin equivalent circuit for the circuit below at the indicated terminals:

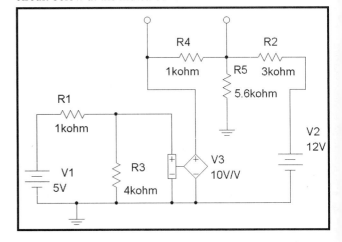

Problem 3-46: Find the transistor operating point I_C and V_{CE} in the circuit below:

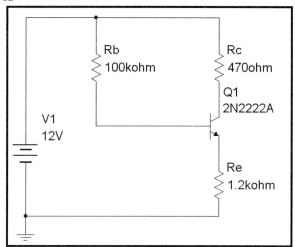

Problem 3-47: Find the transistor operating point I_D and V_{DS} in the circuit below:

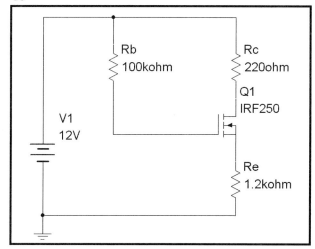

Problem 3-48: Find the transistor operating point I_C and V_{CE} in the circuit below:

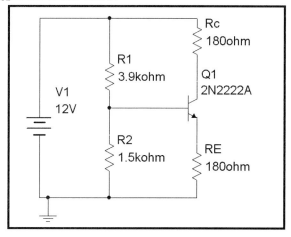

Problem 3-49: Find the transistor operating point I_D and V_{DS} in the circuit below:

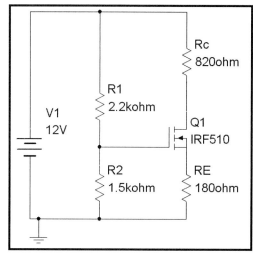

Hint: You may want to add a zero-volt DC source in series with the drain as shown below so that you can easily measure the drain current:

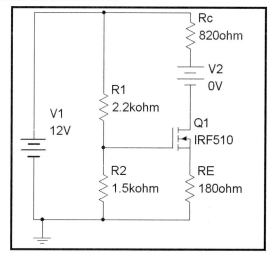

Problem 3-50: Find the transistor operating point I_C and V_{CE} in the circuit below:

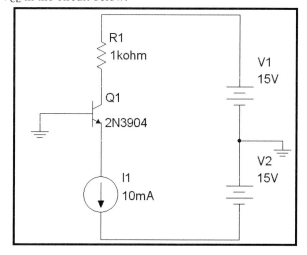

PART 4
DC Sweep

The DC Sweep can be used to find all DC voltages and currents in a circuit. The DC Sweep is similar to the analyses of Part 3 but adds more flexibility. The added flexibility is the ability to allow DC sources to change voltages or currents. For example, the circuit on page 123 will give us results only for the single value of Vx = 15 V if the DC Operating Point Analysis is used. For each different value of Vx we are interested in, we must run the simulation again. If we use the DC Sweep, we can simulate the circuit for several different values of Vx in the same simulation. How node voltages vary for changing source voltages or how a BJT's bias collector current changes for different DC supply voltages would be example applications of the DC Sweep. As in the node voltage analysis, all capacitors are assumed to be open circuits, and all inductors are assumed to be short circuits.

An important note about this part is that the simulations we perform can only be done using the DC Sweep analysis. In Part 3, we found that we could use the DC Operating Point simulation to get the same results as we found using the current and voltage indicators, or the Multimeter. In the AC Analysis part, we will see that we can use the AC Sweep analysis or the Bode plotter to view similar results. In the Time Domain part, we can use the Oscilloscope or the Transient Analysis to view similar results. In most of the other parts, we can either run an analysis and plot results with the Postprocessor, or view results using one of the instruments. The DC Sweep simulation differs in that there is no standard instrument that will allow us to view the same results as we can with the DC Sweep analysis. Thus, in this part we will not demonstrate similar simulations using both the analyses and the instruments to view results. Instead, we will only run the DC Sweep analysis.

4.A. Basic DC Analysis

We will start with a modification of the circuit discussed in section 3.A:

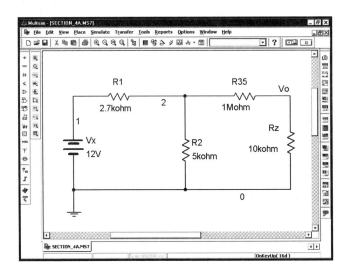

The question we will ask is: How does the voltage at **Vo** vary as **Vx** is raised from 0 to 25 volts? We will also view some of the currents through the components. Since this is a DC Sweep, all capacitors are assumed to be open circuits, and all inductors are assumed to be short circuits. We will now set up the DC Sweep. From the menu bar select **Simulate**, **Analyses**, and then **DC Sweep**:

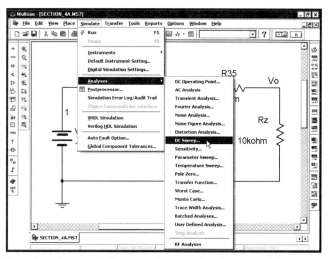 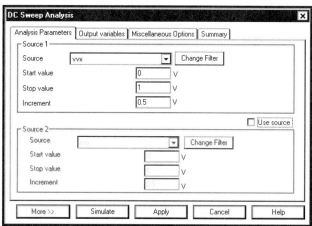

There is only one DC source in our circuit, so Multisim selects this source as the one to be swept. Note that we called the source **Vx** while Multisim refers to the source as **vvx**. This is done because in the SPICE convention for naming voltage sources, the name must start with the letter v. In Multisim, a voltage source can start with any letter because Multisim attaches the letter v as a prefix for voltage sources. The adding of prefixes applies to other components such as resistors, capacitors, inductors, and current sources. We would like to sweep **Vx** from 0 volts to 25 volts in 0.01-volt increments. Fill in the dialog box as shown:

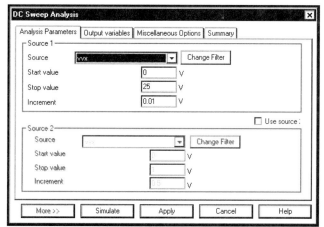

Note that we are setting the parameters for sweeping **Source 1**. Since we only have a single source this is all we can sweep. However, we are able to set up sweeps for two sources so that we can run nested sweeps. See section 4.D for an example of a nested sweep.

The parameters for the sweep are now set up. Before we can simulate the circuit we must specify output variables to plot during the simulation. Click on the **Output variables** tab:

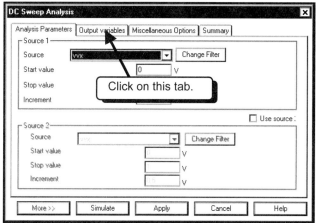 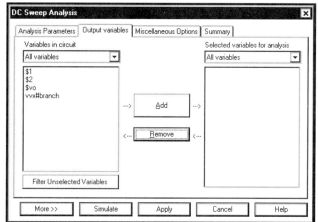

We must select at least one variable to plot during the simulation. See Part 2 for information on displaying traces with the Postprocessor. When the simulation is complete, we can display traces that incorporate any node voltage in the circuit.

Displaying currents, however, is a bit more difficult. Here, we will illustrate how to plot the current through a few of the resistors using the Postprocessor.

First, we will select the voltage at node **Vo** to be plotted during the simulation. We need to select at least one variable to be plotted during the simulation. If we are interested in other variables after the simulation is complete, we can display them using the Postprocessor. Click the **LEFT** mouse button on the text **$Vo** as shown:

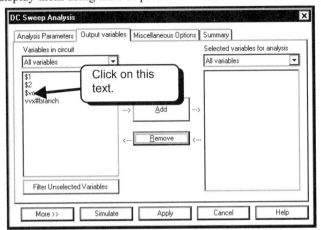 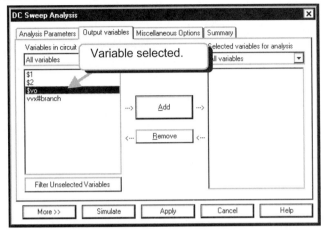

Click on the **Add** button.

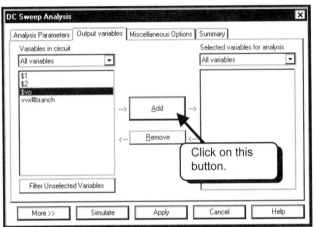 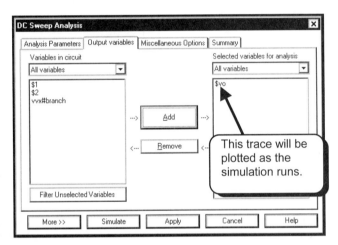

Vo will now be plotted as the simulation runs.

We are now ready to run the simulation. Click on the **Simulate** button. The **Analysis Graphs** window will open and display the selected output voltage:

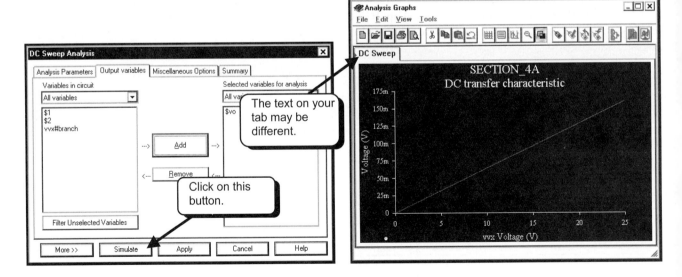

The graph plots the selected variable versus the sweep variable, **Vx** in this example.

We can now use the Postprocessor to view other voltages in the circuit. To access the Postprocessor, select **Simulate**, and then **Postprocess** from the Multisim menus:

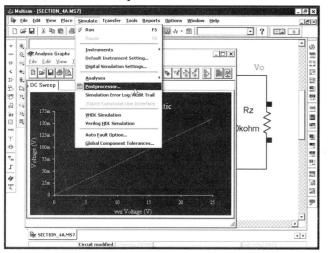
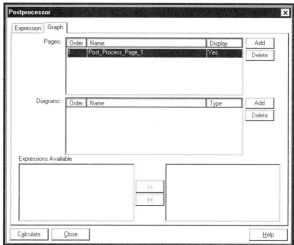

Select the **Expression** tab:

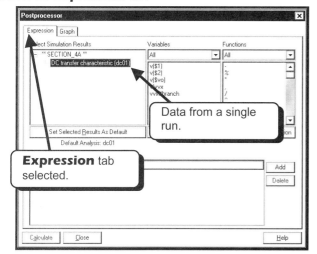
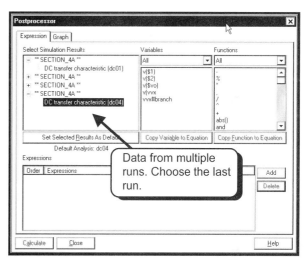

Note that in the **Select Simulation Results** section, all analyses you have run in this Multisim session are listed. In my case as shown in the above left screen capture, I have only run a single analysis, this DC sweep. If you have run the simulation more than once, or have run other simulations before you ran the simulation for this section, your dialog box will list more than the single analysis shown in my dialog box. An example is shown in the above right screen capture where the data from four DC sweep simulations are listed. Select the data set you wish to use, usually the last data set because it is the most recent simulation.

We would like to create a plot that displays the voltage at nodes 1 and 2. The first thing we must do is make the traces available to the Postprocessor. Click the **Add** button:

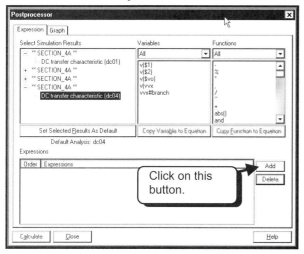
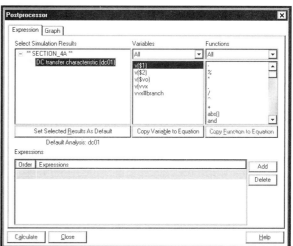

We must now create the trace **Expressions** we wish to plot. We notice that we can only plot node voltages such as **v($1)**, **v($2)**, **vo**, and the voltage of the DC source **Vx** (indicated by **vvx** in this dialog box). The only current listed is current through voltage sources. The line **vvx#branch** specifies the branch current through voltage source **Vx**. We note that the currents through the resistors are not listed. To view these, we need either to add a current monitor to the circuit (a current-dependent voltage source) or to specify the current during the analysis setup. We will show both later in this section. For now we are limited to the node voltages and the current through the DC source.

Double-click the *LEFT* mouse button on the text **v($1)** to add the trace to the list:

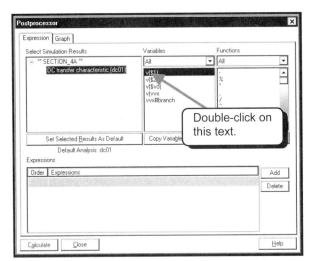 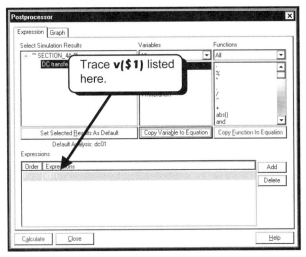

We must now create the **Expression** for the voltage at node 2. Click the **Add** button:

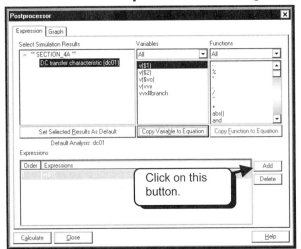 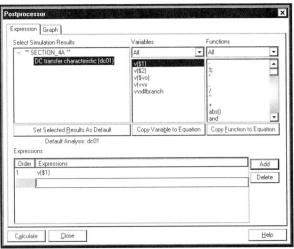

Double-click the *LEFT* mouse button on the text **v($2)** to add the trace to the list:

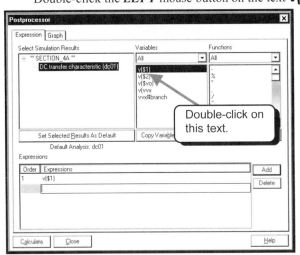 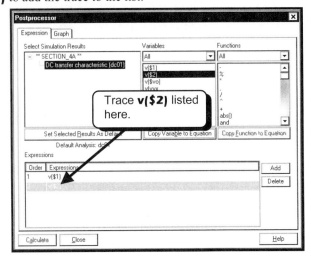

We now need to create the graph on which the traces reside. Select the **Graph** tab:

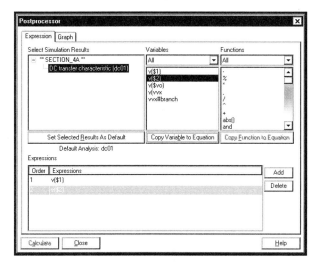 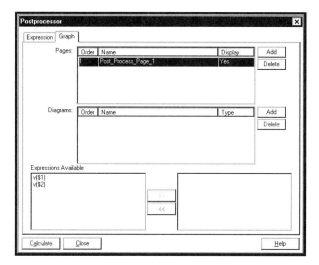

The first thing we will do is name the graph page we are about to create. Click the *LEFT* mouse button on the text **Post_Process_Page_1** as shown below to select the text field:

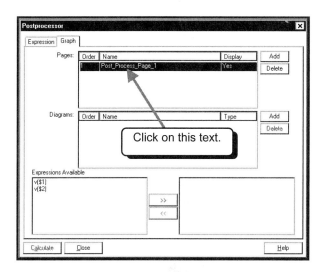 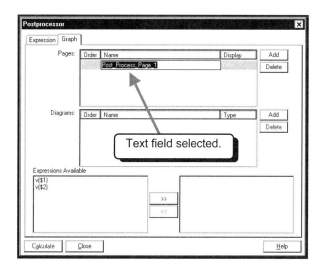

Type the new name of the Grapher page. I will rename the page **Node Voltages**:

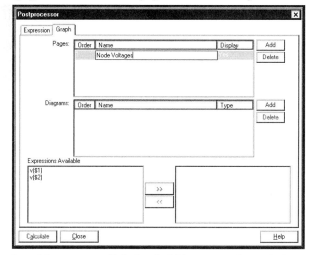

Next, we want to add a single graph to this page, so click the **Add** button as shown below:

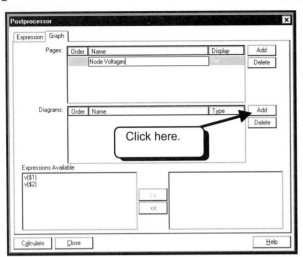 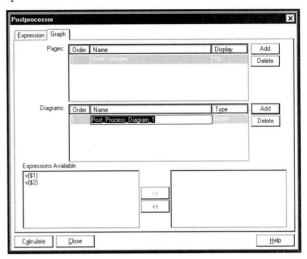

Type the name of this new graph. I will call it **Node Voltages** also:

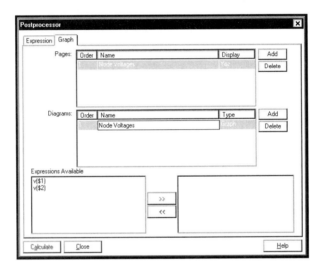

We now have a single graph on this page, and we can add traces to this graph. Click the **LEFT** mouse button on the text **v($1)** to select the text:

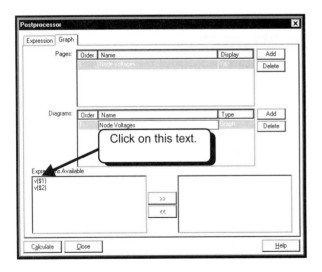 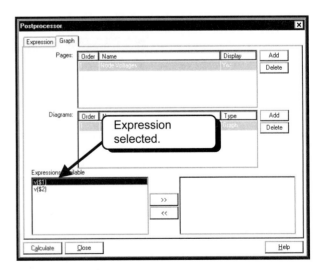

Click the **LEFT** mouse button on the [>>] button to add the trace:

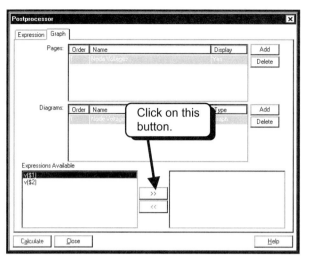 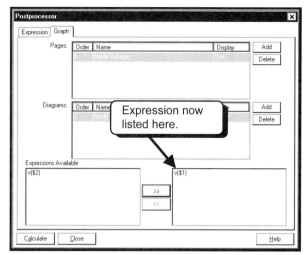

Repeat the procedure to add **v($2)** to the right-side window pane:

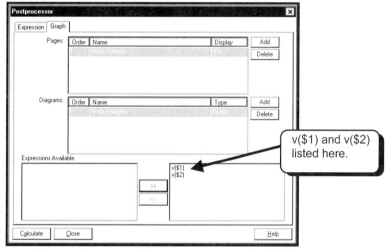

We are now ready to generate the plots. Click the **Calculate** button to view the graphs and display the node voltages at nodes 1 and 2:

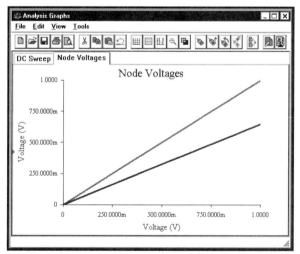

To show currents, we must either add a current monitor, or specify the current traces as outputs when we set up the DC Sweep analysis. We will show both here. Close the **Analysis Graphs** window. We can monitor a current by placing a current-controlled voltage source or a zero-volt voltage source in series with the element we wish to measure the current through. We will show both methods here. We will measure the current through R1 with a current-controlled voltage source and the current through R35 with a zero-volt voltage source. We already know how to place a DC voltage source, so we will show how to place a dependent voltage source. We need to place a part called a **Current Controlled Voltage Source** in our circuit. Click on the **Source** button:

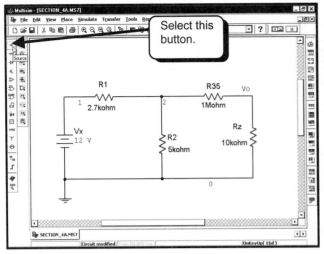
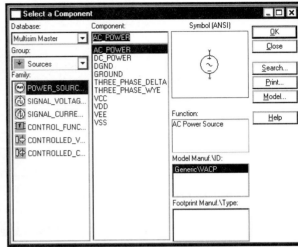

Select the **Controlled_V**oltage source group:

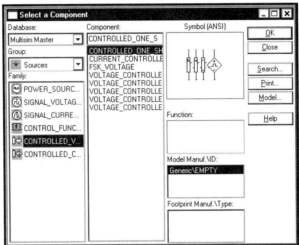

This dialog box now lists all of the available controlled voltage sources. We want to use a current-controlled voltage source, so select the **CURRENT_CONTROLLED** voltage source. An enlargement of the current-controlled voltage source part is shown on the right below.

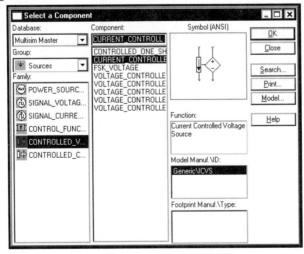
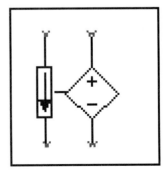

Click the **OK** button to place the part in your circuit.

We would like to monitor the current through **R1**. To add the monitor, we will have to break the circuit and insert the current monitor between **R1** and node **2** as shown:

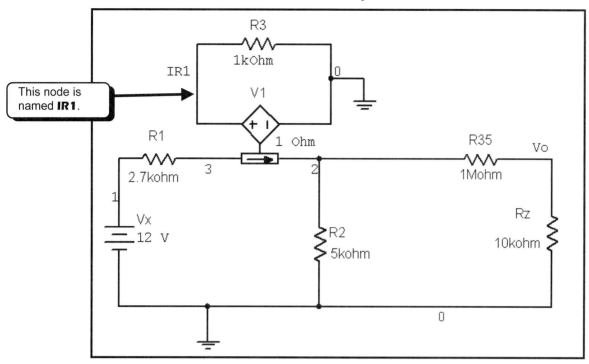

The current-dependent voltage source is named **V1**. Its gain is **1 Ohm**. This means that the output voltage of the source is 1 ohm times the current it measures, or the transfer ratio of this current monitor is 1 volt per amp. Also note that we added a resistor in parallel with the source. The value and name of this resistor are not important. It was added so that we can easily name the node between the two elements. In my circuit the node is named **IR1**. We need to name the node so that we can easily view its voltage. To name the node, double-click the **LEFT** mouse button on the wire between the resistor and the voltage source.

Next, we will add a DC source between R35 and Rz so that we can measure the current through these two resistors. If we set the voltage of this source to zero, it will not affect the circuit. However, Multisim allows us to display the current through voltage sources, so this allows us to plot this current. Add a DC voltage source as shown in the circuit below:

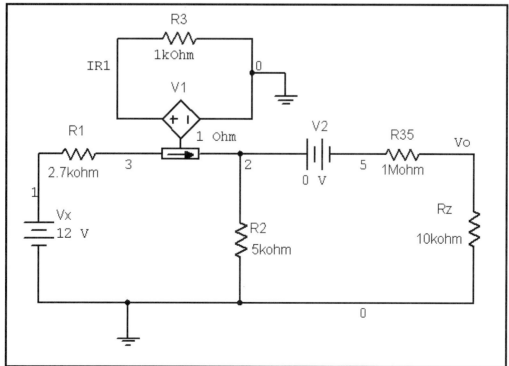

In our circuit diagram, the name of the DC source is V2 and the value of the source is zero volts. Note that for a voltage source, current is defined as positive when it enters the positive terminal. Thus, positive current is indicated by the arrow shown in the diagram below:

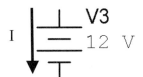

When you have finished changing the circuit, select **Simulate**, **Analyses**, and then **DC Sweep**:

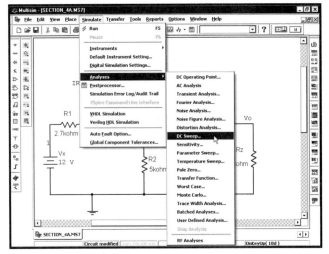
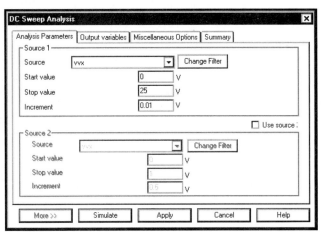

If your dialog box does not have the settings shown above, modify the settings.

We will now show a second way of viewing the current through R1. Click on the **Output variables** tab:

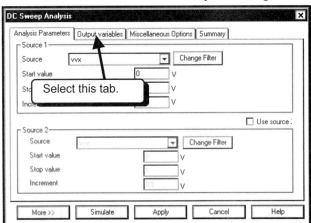
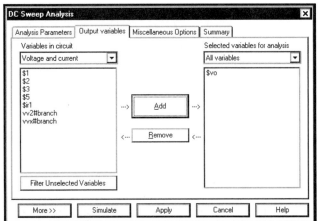

Click on the **More** button:

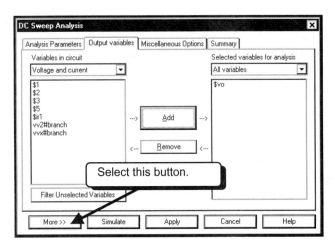
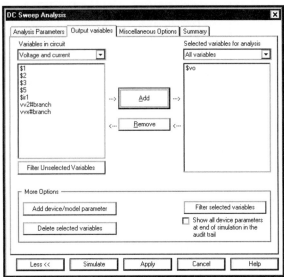

Click on the **Add device/model parameter** button.

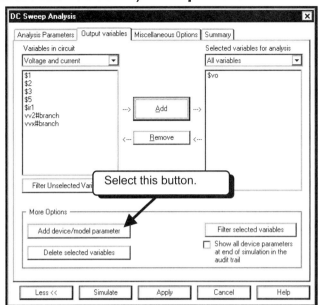

We would like to view the current through R1. We need to specify the **Device Type** as a resistor and then select R1. Click the *LEFT* mouse button on the down triangle ▼ for the **Device Type** field:

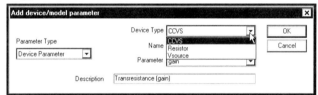

Select **Resistor**:

In the **Name** field of my dialog box, resistor R1 is already selected (**rr1**). If your dialog box does not specify resistor R1, click the *LEFT* mouse button on the down triangle ▼ for the **Name** field and select **rr1**. The last thing we must do is specify what parameter of the resistor we would like to plot. Click the *LEFT* mouse button on the down triangle ▼ for the **Parameter** field to display the available parameters:

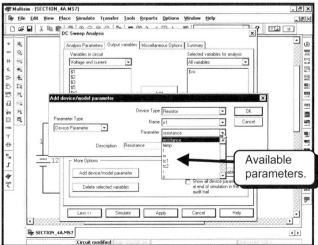

Select current by clicking on the lower case **i**. Be careful not to select **l**, the length of the resistor. If you select the correct parameter, the text **Current** will be displayed in the **Description** field:

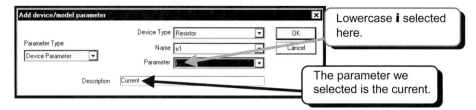

Click on the **OK** button. The resistor current will now be displayed as a variable we can plot:

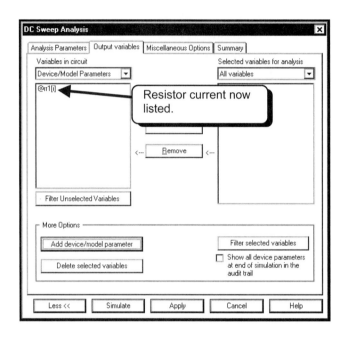

We must now add this trace to the traces that are plotted during the simulation. Click the **LEFT** mouse button on the text **@rr1[i]** to select it:

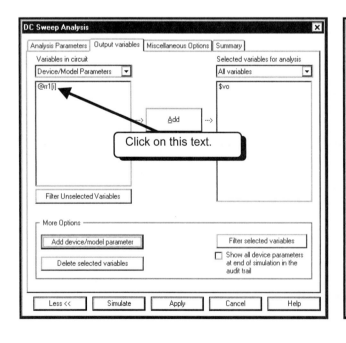
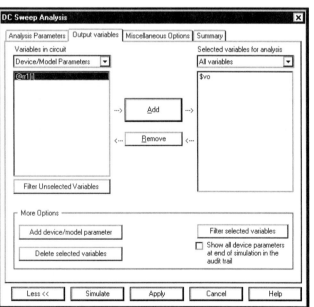

Click on the **Add** button:

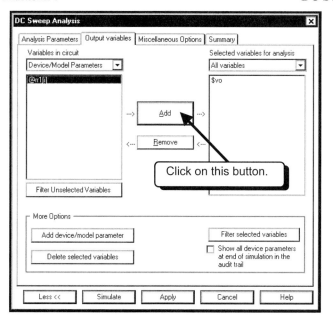 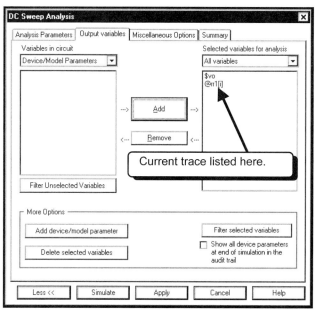

The trace will now be displayed as the simulation is running, and we can plot it with the Postprocessor as well. Remember that trace **$vo** was added at the beginning of this example.

Click on the **Simulate** button to simulate the circuit. The voltage at node **Vo** and the current through **R1** will be displayed in the **Analysis Graphs** window:

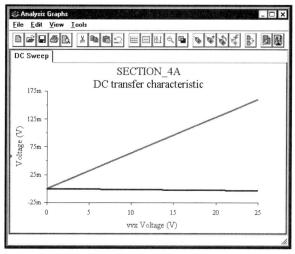

We will now use the Postprocessor to plot the current traces. To access the Postprocessor, select **Simulate** and then **Postprocess** from the Multisim menus:

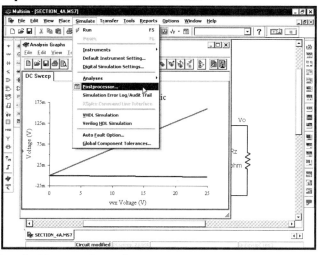 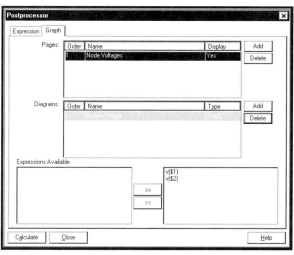

Select the **Expression** tab:

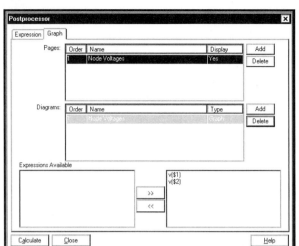 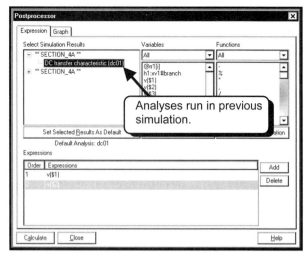

The data set currently selected is from the previous DC Sweep. We need to select the data set from the analysis we just ran. Click the **LEFT** mouse button on the ⊞ as shown. The tree will expand to show the available data sets:

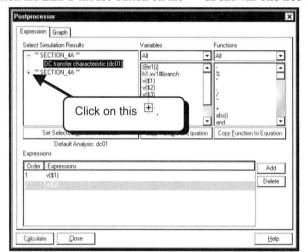 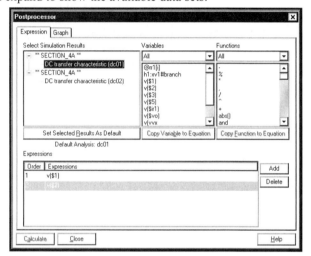

Click on the text **DC transfer characteristic (dc02)** to select the data set:

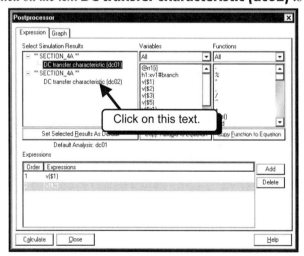 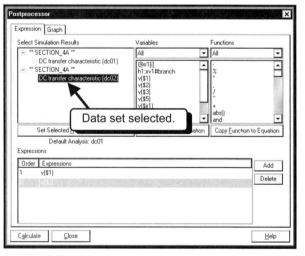

We must now create the expressions we wish to plot. We want to plot the following traces:

- @rr1[i] – The current through resistor R1.
- v($ir1) – This is the voltage at node IR1. This is the voltage at the output node of the current sensor we created.
- vv2#branch – this is the current through source v2, which is equal to the current through resistors R32 and Rz.

To add an **Expression**, click the **Add** button:

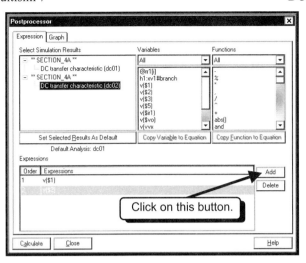 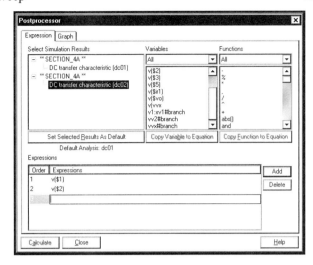

We will demonstrate by adding the expression @rr1[i]. Click the **LEFT** mouse button on the text @rr1[i] to select the text:

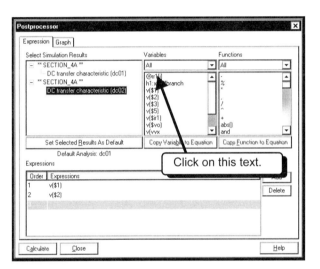 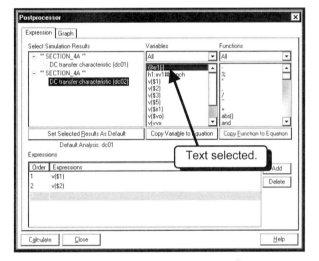

Click on the **Copy Variable to Equation** button to add the variable to the expression:

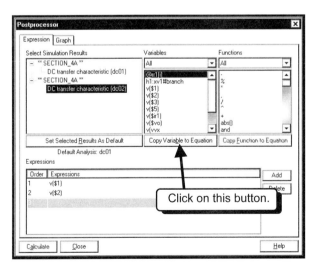 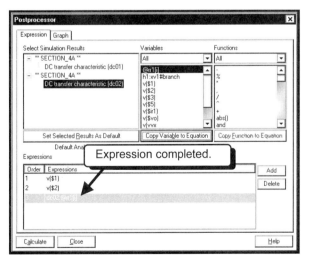

The trace's expression is now completed. To begin the next expression, click the **Add** button. Repeat the procedure and add two more expressions for **v($ir1)** and **vv2#branch**:

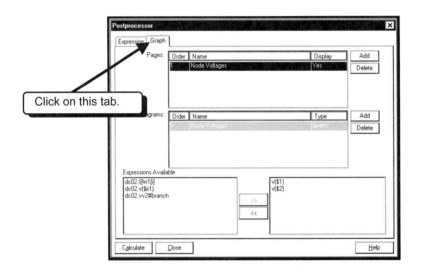

Note that the traces we just added have the prefix **dc02** before the traces, for example, **dc02.v($ir1)**. This is because we have run multiple simulations and the Postprocessor allows us to plot traces from any of those simulations. The text **dc02** identifies which of the simulations the data is associated with.

We are now ready to switch to the Graph tab and create our plots. Select the tab:

The Postprocessor presently has one plot created from our last example, which we named **Node Voltages**. We need to create a new page and add a new graph to that page. Click the *LEFT* mouse button on the top **Add** button to create a new Grapher page:

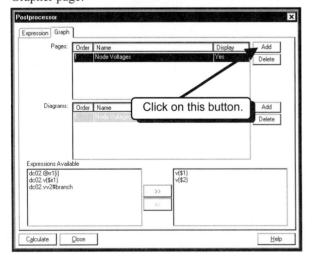
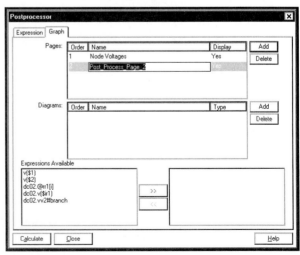

Type the name of the new page, **Resistor Currents**, for example:

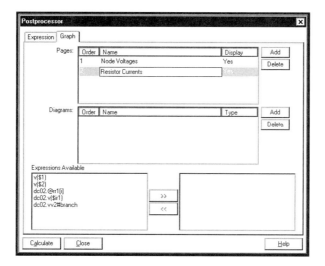

Next, we need to add two graphs to this page. Click the *LEFT* mouse button on the lower **Add** button to add a graph to this page:

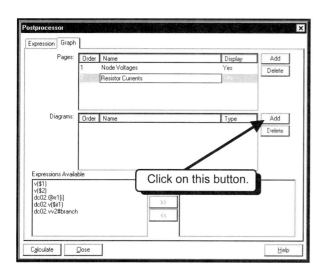
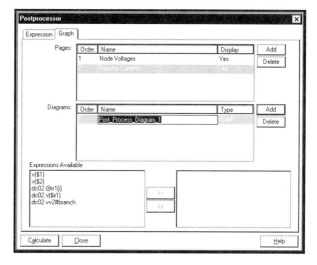

Type the name of the graph, **Current Through R1**, for example:

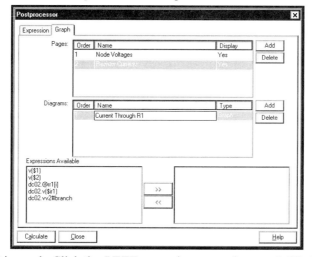

We will now add the traces for this graph. Click the *LEFT* mouse button on the text **dc02.@rr1[i]** to select the trace:

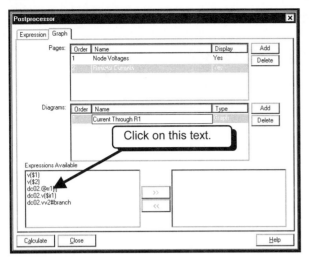 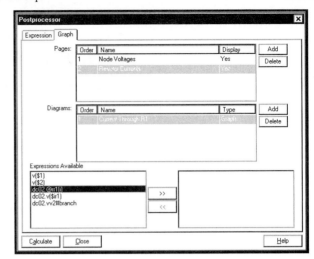

Click the >> button to add the trace to the graph:

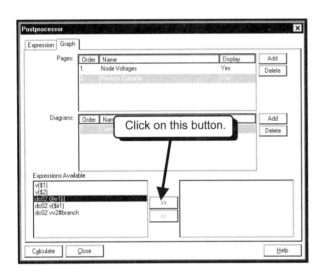 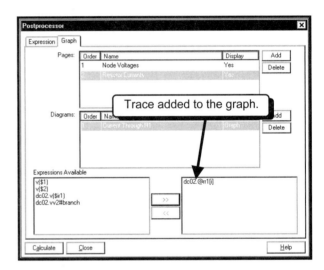

Repeat the procedure to add the trace **dc02.v($ir1)** to the graph:

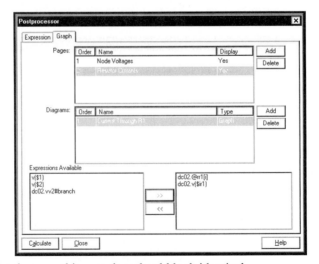

Note that both of these traces plot the same thing, so they should look identical.

Next, we will add a second graph to this page for the current through resistors R35 and Rz. Click the lower **Add** button to create a second graph:

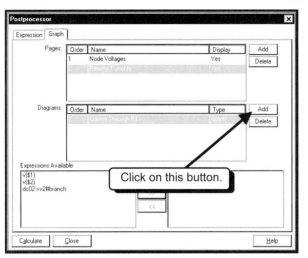 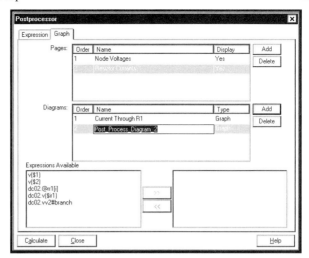

Name the graph **`Current Through R35 and Rz`** and then add the trace **`dc02.vv2#branch`** to it:

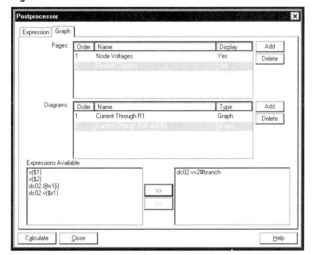

Click the **Calculate** button to create the graphs:

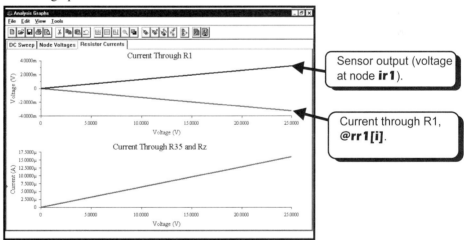

We see that the two traces for the current through R1 are the same except that one is the negative of the other. The problem with using a parameter such as **@rr1[i]** is that we do not know which direction Multisim considers to be the positive direction of current. It turns out that if I rotated R1 in the schematic by 180 degrees, the current would then match up with the sensor output. We also know that the y-axis for **@rr1[i]** should be current rather than voltage. We can easily change the label of the axis if we wish. This procedure was covered in section 2.F.3. With a sensor we know which direction is positive (left to right in my schematic) because the arrow in the part graphic indicates the positive direction.

Because I know that the current through R1 is left to right, I can always plot the trace -@rr1[i] to correct for the sign error:

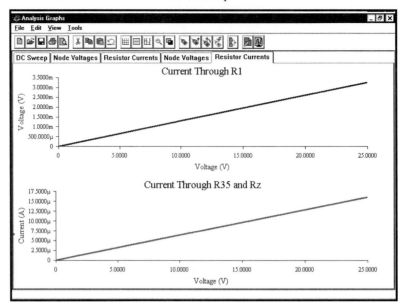

We see that the two methods for displaying the current through R1 now produce the same result.

EXERCISE 4-1: Find the voltages at nodes **1**, **2**, and **3** if the source voltage **V1** is swept from a DC voltage of 6 volts to a DC voltage of 36 volts.

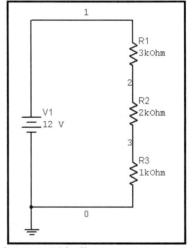

SOLUTION: The results of the DC Sweep are shown graphically:

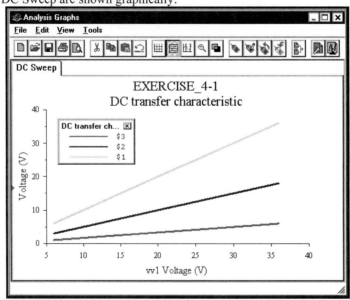

4.B. Diode I-V Characteristic

We would now like to use SPICE to obtain the I-V characteristic of a semiconductor diode. Wire the circuit shown below:

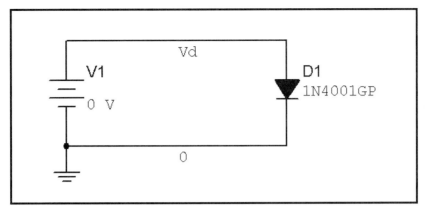

Note that we labeled the top node as **Vd** rather than a number. To label a wire, double-click on the wire.[1] To place the 1N4001GP diode, click on the **Diode** button ⤻ :

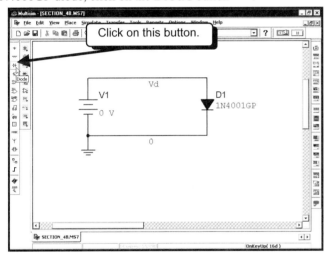
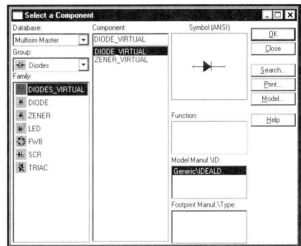

Select the **DIODE** family to view the list of available diode models in Multisim 7:

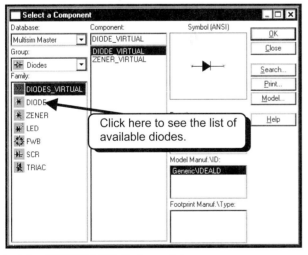
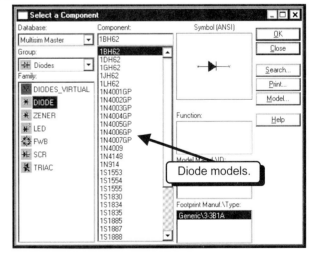

Select the **1N4001GP** diode model:

---

[1] See section 1.G to label a wire or give a node a name.

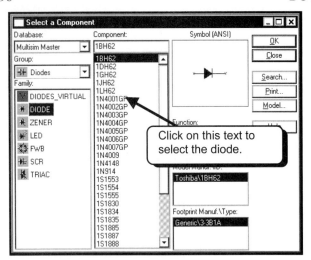 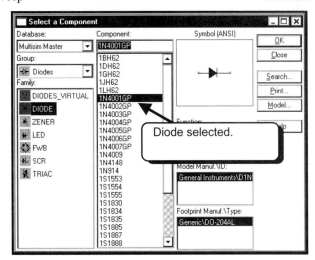

Click the **OK** button to place the part.

After creating the circuit, we need to set up a DC sweep. Select **Simulate**, **Analyses**, and then **DC Sweep** and fill in the dialog box as shown:

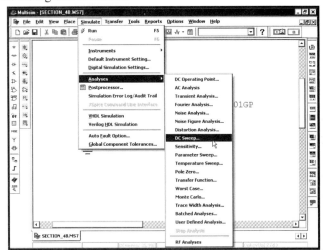 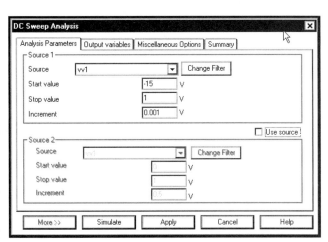

The settings in the **DC Sweep Analysis** dialog box specify that voltage source **V1** is to be swept from **-15** volts to **1** volt in **0.001** volt steps.

To view the output current, we need to add the diode current to the list of output variables. Click the **LEFT** mouse button on the **Output variables** tab:

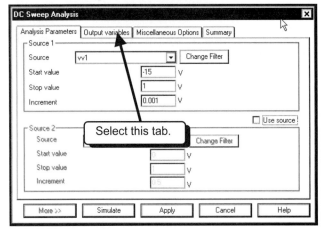 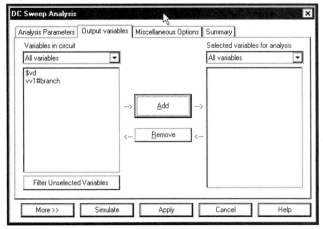

Click on the **More** button and then select the diode current as an output variable. See pages 186-189 for a detailed explanation of how to specify a device current as an output variable.

Multisim 7 — DC Sweep

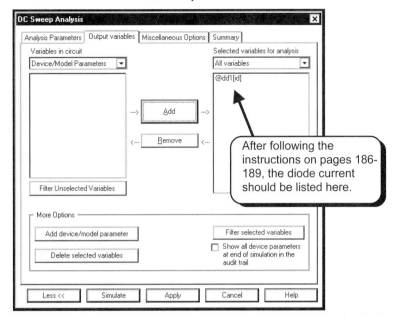

Since the diode current is listed as an output variable, it will be plotted automatically when the simulation runs.

Click on the **Simulate** button to run the simulation. The **Analysis Graphs** window should pop up and display the I-V characteristic:

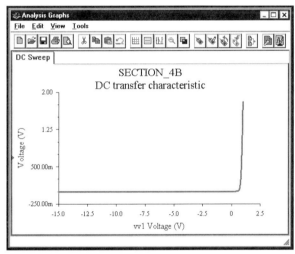

To modify the plot settings, click the ***RIGHT*** mouse button on the graph:

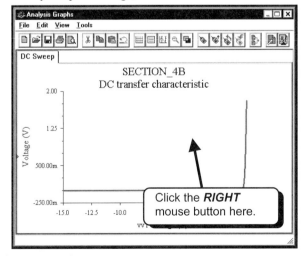
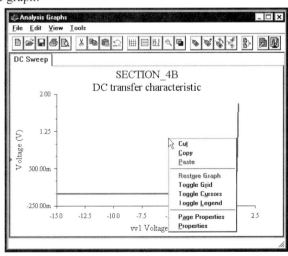

Select **Properties**:

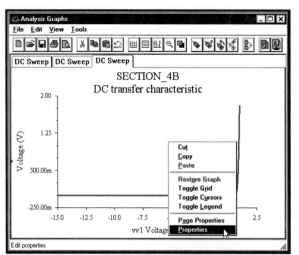
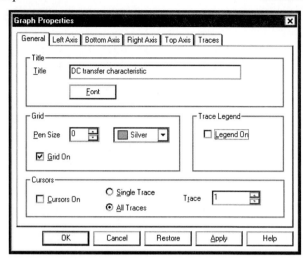

Change the properties of the plot so that it looks like the plot below:

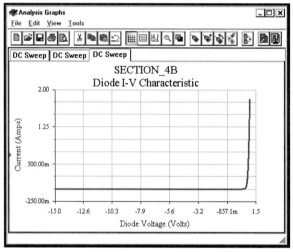

Note that the diode current shown here is too large for this device to handle. A 1N4001 can handle only 1 amp of current or less. The current goes to nearly 2 amps because we applied a large voltage across the diode, 1 V or more, and this is the current the model produces with a 1 volt diode drop. For a typical 1N4001, the diode voltage should be closer to 0.7 volt.

EXERCISE 4-2: Display the I-V characteristic of the Multisim diode_virtual part:

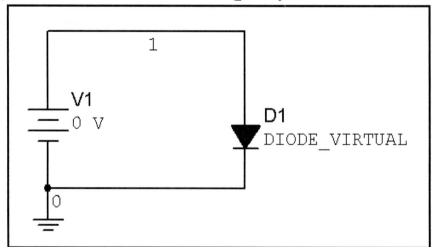

SOLUTION: Use the DC Sweep analysis and sweep the voltage from 0 to 1 volts. After plotting the trace, modify the y-axis to only display currents from 0 to 1 amp.

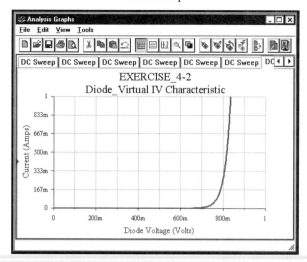

4.B.1. Diode I-V Characteristic Using the IV Plotter

We will now generate the same I-V plot of the 1N4001GP diode using the IV plotter instrument provided by Multisim 7. This is a new instrument added in version 7 of Multisim. To place the instrument, click the *LEFT* mouse button on the **IV-Analysis** button. The instrument will become attached to the mouse pointer and move with the mouse:

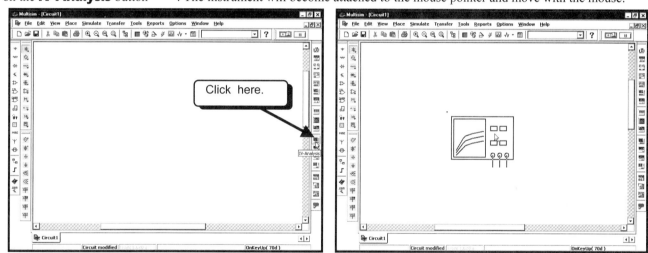

Place the instrument in your circuit and then place a 1N4001GP diode in your circuit:

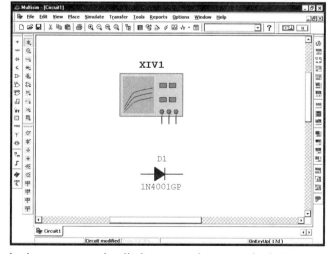

To understand how to connect the instrument to the diode, we need to open the instrument. This instrument can be used to generate IV plots for two- and three-terminal devices. Double-click the *LEFT* mouse button on the instrument to open it:

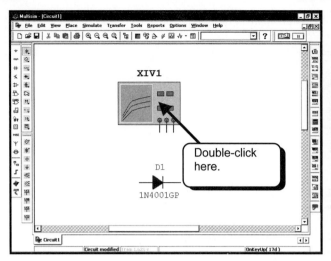 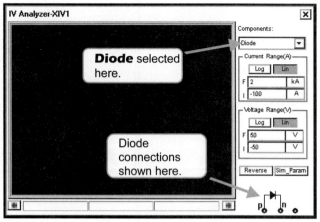

The bottom right portion of the screen shows us how to connect the diode to the instrument. Also, make sure that the **Components** field specifies a **Diode** as the type. If not, change it to a **Diode**. Now that we know how to connect the diode, close the dialog box and connect the diode to the instrument:

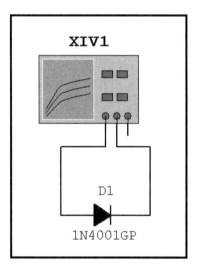

We are now ready to set up the parameters for the instrument. Double-click on the instrument to open it up again and then click on the **Sim_Param** button:

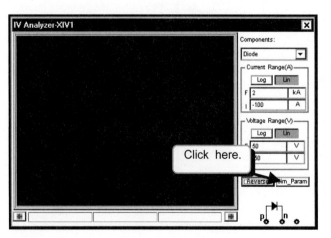 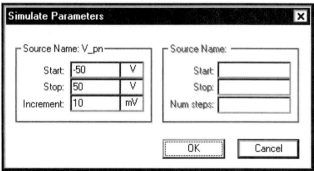

This dialog box specifies the parameters for the IV plot. In the plot, we sweep the voltage across the diode just like we did for the DC sweep earlier. We will specify the same parameters here as we did for the DC sweep on page 198:

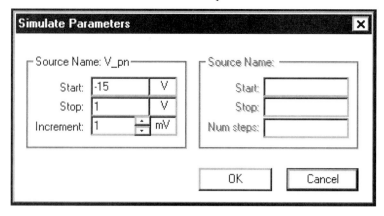

We will sweep the diode voltage from –15 volts to 1 volt in 1 mV steps. Click the **OK** button to accept the settings. Press the **F5** button to start the simulation and generate the IV plot:

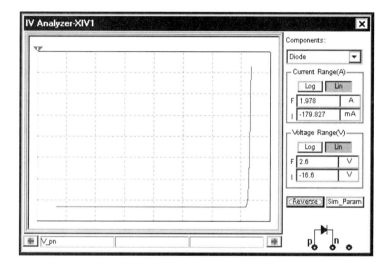

When the curve is complete, you should press the **F5** key again to stop the real-time simulation. You can now use the cursor to view points along the curve:

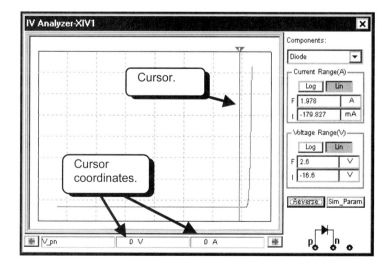

You can also type **CTRL-G** to open the Grapher and view the IV plot with the Grapher:

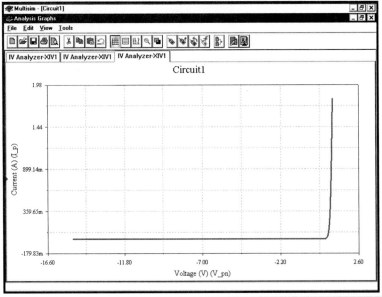

EXERCISE 4-2A: Display the I-V characteristic of the Multisim diode_virtual part using the IV plotter.

SOLUTION:

4.C. DC Transfer Curves

One of the more useful functions of the DC Sweep is to plot transfer curves. A transfer curve plots an output versus an input. A DC transfer curve plots an output versus an input, assuming all capacitors are open circuits and all inductors are short circuits. In a DC Sweep, all capacitors are replaced by open circuits and all inductors are replaced by short circuits. Thus the DC Sweep is used for DC transfer curves.

4.C.1. Zener Clipping Circuit

The circuit below should clip positive voltages at the Zener breakdown voltage and negative voltages at the diode cut-in voltage, approximately 0.6 to 0.8 volt:

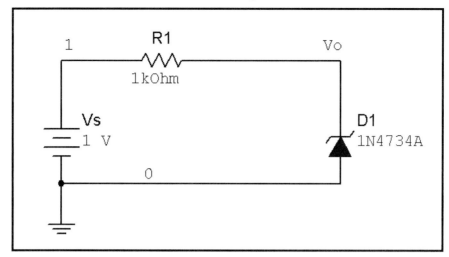

We would like to sweep **Vs** from −15 volts to +15 volts and plot the output. Select **Simulate**, **Analyses**, and then **DC Sweep** and fill in the dialog box as shown:

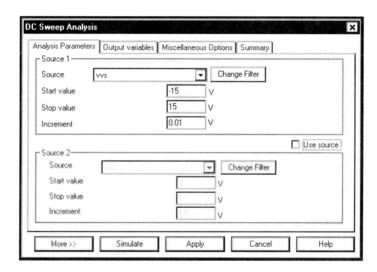

Next, select the **Output variables** tab and plot **Vo** during the simulation:

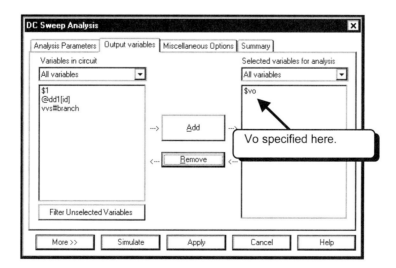

Click on the **Simulate** button to run the simulation. The result will be displayed in the **Analysis Graphs** window:

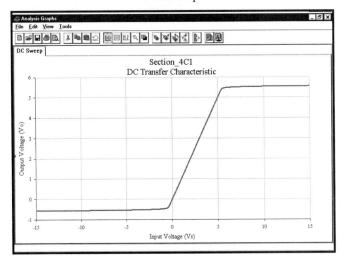

I manually modified the x- and y-axis titles. We see that this circuit limits the output at the Zener breakdown voltage of 5.6 volts and –0.7 volt, minus the diode turn-on voltage.

EXERCISE 4-3: Find the transfer curve of the circuit below:

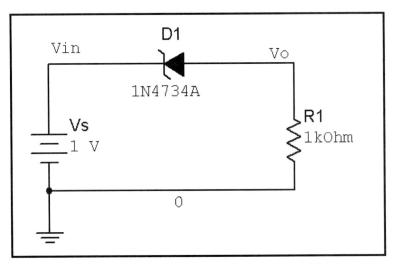

SOLUTION: Sweep **V1** from –15 V to +15 V as in the previous example and specify Vo as the variable to plot during the simulation.

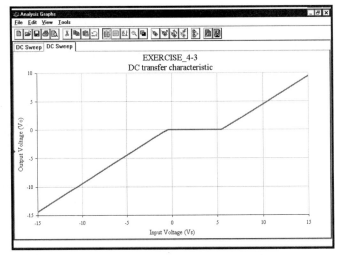

Notice that the output is zero for input voltages from approximately –0.7 V to about 5.6 V.

4.C.2. NMOS Inverter Transfer Curve

We would like to plot the transfer curve **Vo** versus **Vin** for the NMOS inverter below:[2]

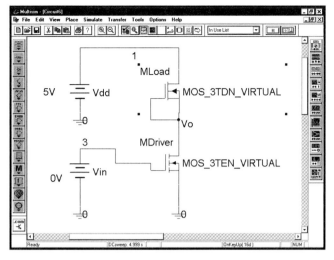

This circuit is an NMOS inverter with a depletion MOSFET load. The driver transistor is an N-type enhancement MOSFET and the load transistor is an N-type depletion MOSFET. In SPICE, both use the same model parameters. The way SPICE distinguishes between enhancement and depletion devices is that the threshold voltage is positive for enhancement devices and negative for depletion devices. Both types of transistors use the SPICE NMOS model. In the screen capture above, Multisim uses different symbols to emphasize that one is a depletion type and the other is an enhancement type. To create the circuit above, you need to place a part called **MOS_3TDN_VIRTUAL** and one called **MOS_3TEN_VIRTUAL**. Both are located in the Transistors toolbar.

Before we create the models for the two transistors, we need to discuss the basic operation of the devices and how the standard SPICE equations may differ from convention. Many textbooks describe MOSFET operation by the following equations:

$$I_D = K(V_{GS} - V_T)^2(1 + \lambda V_{DS}) \qquad \text{saturation region}$$

$$I_D = K(2(V_{GS} - V_T)V_{DS} - V_{DS}^2)(1 + \lambda V_{DS}) \qquad \text{linear region}$$

For our example, the load MOSFET is a depletion-mode NMOS transistor. Its parameters are K = 20 µA/V^2, V$_T$ = −1.5 V, and λ = 0.05 V^{-1}. The driver is an enhancement-mode NMOS transistor. Its parameters are K = 20 µA/V^2, V$_T$ = +1.5 V, and λ = 0.05 V^{-1}. We note that the only difference between the two is the threshold voltage, V$_T$.

First, we need to create models to describe these transistors. The standard SPICE MOSFET model uses the following equations to describe MOSFET operation:

$$I_D = \frac{K_P}{2}(V_{GS} - V_{TO})^2(1 + \lambda V_{DS}) \qquad \text{saturation region}$$

$$I_D = \frac{K_P}{2}(2(V_{GS} - V_{TO})V_{DS} - V_{DS}^2)(1 + \lambda V_{DS}) \qquad \text{linear region}$$

The conversion between the two sets of equations is shown in Table 4-1. The SPICE model parameter names are close to the standard names used to represent MOSFET operation in many textbooks. One difference is that the SPICE model parameter Kp is twice the value of K. Thus, in our model we should set K$_P$ = 40 µA/V^2. All other model parameters will be the same.

After drawing the circuit, the next thing we need to do is modify the models used by the virtual components. First, double-click on the driver MOSFET graphic to change its properties:

Table 4-1	
SPICE Model Parameter	Equation Variables
Kp	2K
VTO	V$_T$
lambda	λ

---

[2] This section is left as a Multisim 2001 example because there is an error in the MOSFET library. When the error is corrected, this section will be updated to Multisim 7 and will be available on the author's website.

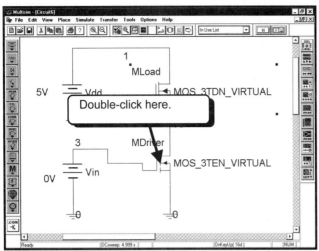 

Select the **Value** tab if it is not already selected. To modify the SPICE model, click the *LEFT* mouse button on the **Edit Model** button:

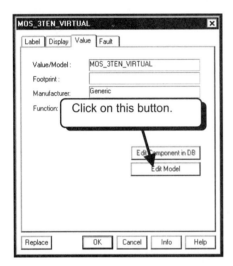 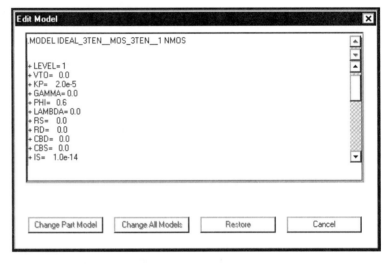

We are not allowed to change the name of the model, but we can change the values of the model parameters. Change the value of **VTO** to 1.5, **KP** to 40e-6, and **LAMBDA** to 0.05:

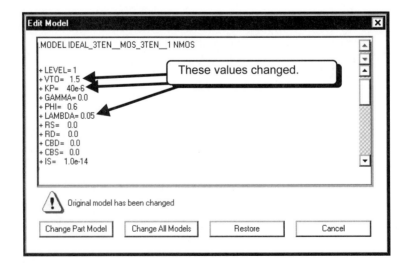

Click on the **Change Part Model** button to apply the changes to the part we double-clicked on.

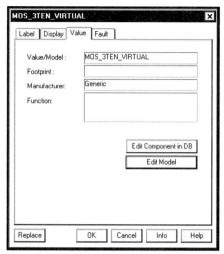

Click on the **OK** button to return to the schematic:

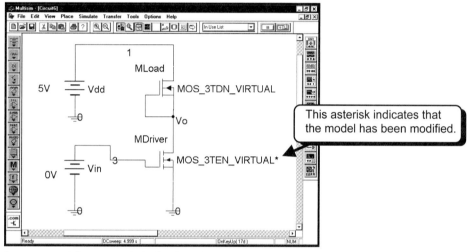

Note that the model name now has an asterisk next to it indicating that the model it uses has been changed. Use the same procedure to change the model for the load transistor. For that transistor, change the value of **VTO** to –1.5, **KP** to 40e-6, and **LAMBDA** to 0.05. Below I show the model editor window and the resulting schematic:

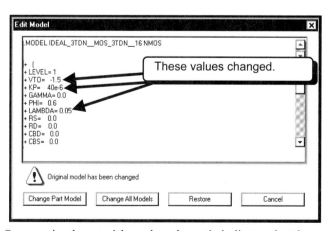
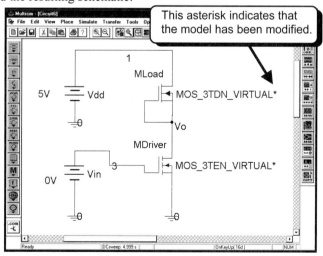

Once again, the asterisk on the schematic indicates that the model has been modified.

We are now ready to set up the DC Sweep. Select **Simulate**, **Analyses**, and then **DC Sweep** and fill in the dialog box as shown. The dialog box specifies that **Vin** should be swept from **0** volts to **5** volts in 1 mV steps. Under the **Output variables** tab, we have selected to display the voltage at node **Vo**:

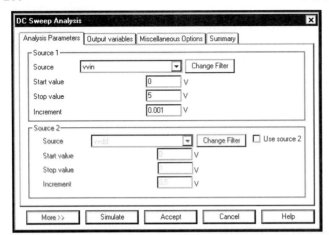 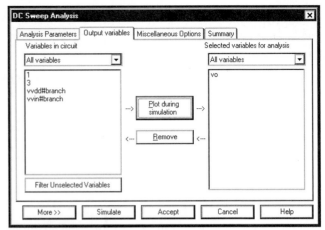

Click on the **Simulate** button to run the simulation. In the screen capture below, I have modified the axis labels and added a grid to the plot:

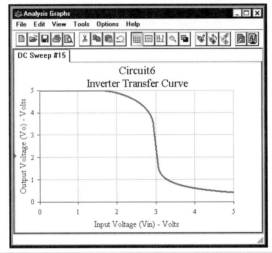

EXERCISE 4-4: Find the transfer curve for a CMOS inverter. The top transistor is a PMOS enhancement device, and the bottom transistor is an NMOS enhancement device. For the NMOS transistor, let Kp = 24 µA/V^2, Vto = 1.5 V, and lambda = 0.01 V^{-1}. For the PMOS transistor, let Kp = 24 µA/V^2, Vto = −1.5 V, and lambda = 0.01 V^{-1}. In most processes, Kp for PMOS devices is less than Kp for NMOS devices, and the difference between the two values is equalized by changing the width-to-length ratio of the PMOS and NMOS devices to make the effective value of Kp the same. The effective value of Kp is $\left(\dfrac{W}{L} K_P\right)$. This Multisim model does not let us specify a width-to-length ratio, so we will just specify that the Kp values for both transistors are the same.

SOLUTION: Create the circuit below:

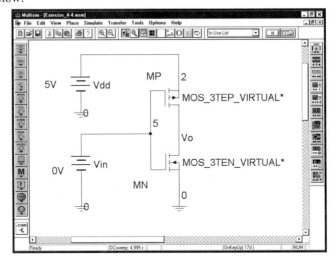

Create the following models for the two transistors:

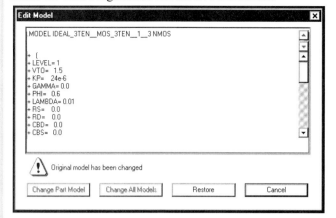 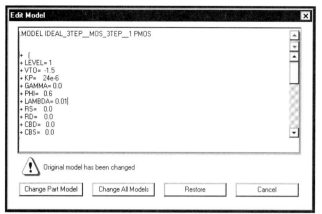

Run a DC Sweep to sweep **Vin** from 0 to 5 V in 0.001 V increments. A small increment is required for a CMOS inverter because the slope of the transfer curve is very steep. Specify **Vo** as the output variable:

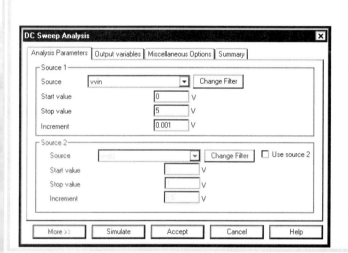 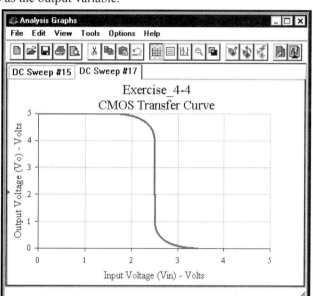

4.D. Nested DC Sweep — BJT Characteristic Curves

The nested DC Sweep can be used to generate characteristic curves for transistors. We will illustrate generating these curves using a BJT. Wire the circuit below:

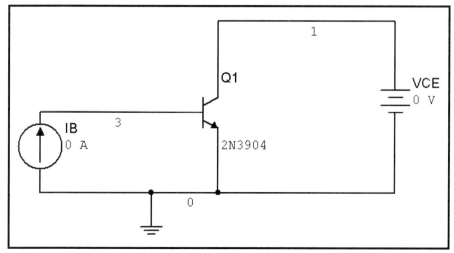

We need to set up a nested DC sweep. The inner loop will sweep VCE from 0 to 15 volts, and the outer loop will step IB from 0 to 100 µA in 10 µA steps. Select **Simulate**, **Analysis**, and then **DC Sweep** from the Multisim menus and fill in the dialog box as shown:

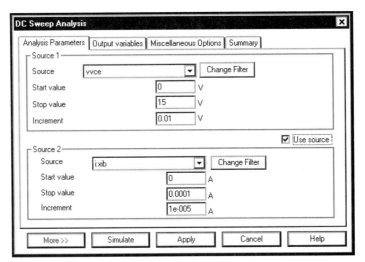

To select the current source **i:xib** in **Source 2** as shown above, you must select the **Change Filter** option and then select all of the options. After selecting all of the options, the choice **i:xib** will become available.

The way the above dialog box is set up, for each value of source **IB**, **VCE** will be swept from 0 volts to 15 volts in 10 mV steps. In other words, **IB** will be set to 0 A and then **VCE** will be swept from 0 volts to 15 volts; then **IB** will be set to 10 µA and **VCE** will be swept from 0 volts to 15 volts; then **IB** will be set to 20 µA and **VCE** will be swept from 0 volts to 15 volts, and so on. This nested DC sweep will contain eleven I_C versus V_{CE} curves.

The next thing we must do is specify the output variable. We are interested in the collector current I_C of **Q1**. We can view this current in two ways. (1) We can use the procedure on pages 186-189 to add the collector current of **Q1**, which is referred to as a device parameter. (2) The second way is to note that the collector current is also equal to the current supplied by the source **VCE**. The definition of positive current for a voltage source is that current is positive if it flows into the positive voltage terminal. This direction is the opposite of that of positive collector current. Thus, to plot the collector current, we can specify the output variable to be the negative of the current through source **VCE**. We cannot specify functions until we use the Postprocessor, so we cannot plot the negative of the source current during the simulation. We would have to run the simulation and then generate a new plot with the Postprocessor. I will plot the collector current of Q1 since this is the correct direction for the current. Select the **Output variables** tab and follow the procedure on pages 186-189 to add the collector current of **Q1**:

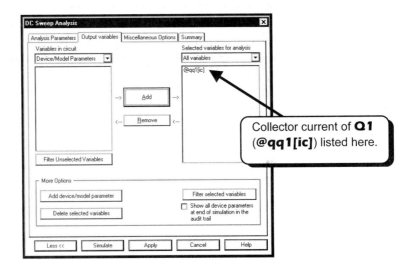

Click on the **Simulate** button to run the simulation and plot the collector current. In the screen capture below, I modified the titles after the simulation was complete:

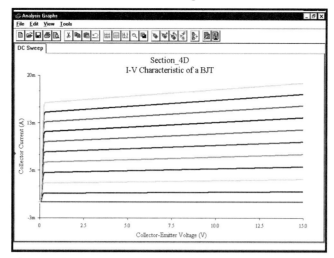

This is the characteristic family of curves for a 2N3904 NPN bipolar junction transistor.

EXERCISE 4-5: Use the nested DC Sweep to find the characteristic curves of a 2N3821 jFET.

SOLUTION: Draw the circuit below:

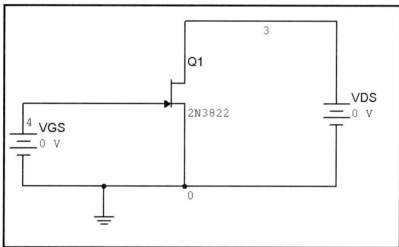

For each value of **VGS**, we want to sweep **VDS** from 0 to 15 volts. Thus, we want **VDS** to be the inner loop or source 1 and **VGS** to be the outer loop or source 2. Note also that for jFETs we must sweep **VGS** from zero to a negative value. Fill in the DC Sweep Analysis dialog box as shown:

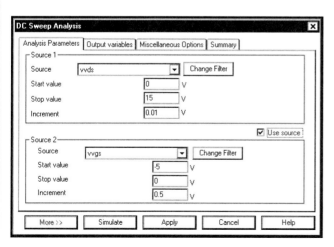
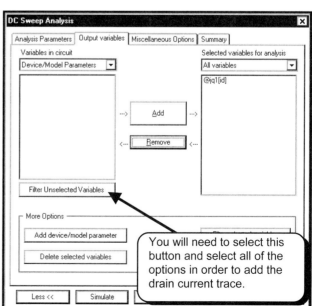

In order to display the drain current in the left pane of the right screen capture above, you will need to click on the **Filter Unselected Variables** button and select all of the options. Also, you will need to click the **Add device/model parameter** button and locate the model parameter that specifies the jFET drain current.

Simulate the circuit and modify the labels to obtain the plot below:

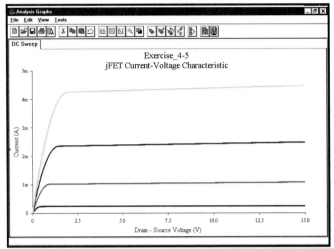

4.D.1. BJT I-V Characteristic Using the IV Plotter

We will now generate the same I-V plot of the 2N3904 BJT using the IV plotter instrument provided by Multisim 7. This is a new instrument added in version 7 of Multisim. To place the instrument, click the *LEFT* mouse button on the **IV-Analysis** button. The instrument will become attached to the mouse pointer and move with the mouse:

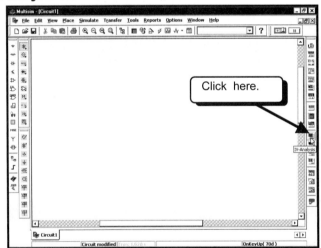
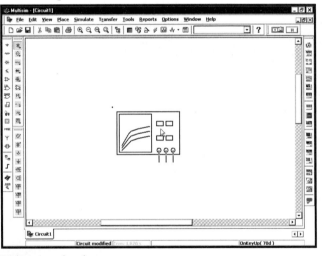

Place the instrument in your circuit and then place a 2N3904 BJT in your circuit:

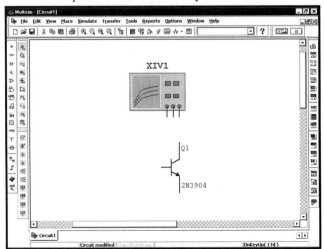

To understand how to connect the instrument to the device, we need to open the instrument. This instrument can be used to generate IV plots for two- and three-terminal devices. Double-click the *LEFT* mouse button on the instrument to open it:

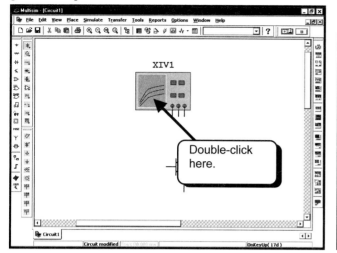 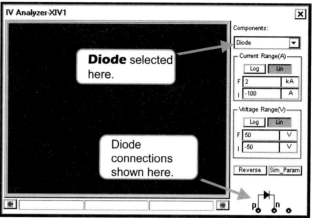

The bottom right portion of the screen shows us how to connect a diode to the instrument, and the **Components** field specifies a **Diode** as the type. We are using a BJT so we need to specify the component type as an NPN BJT. Click the *LEFT* mouse button on the down triangle ▼ as shown and specify the component type as a **BJT NPN**:

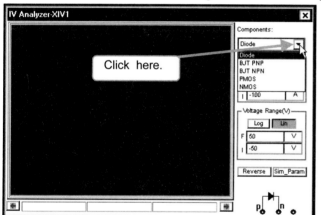 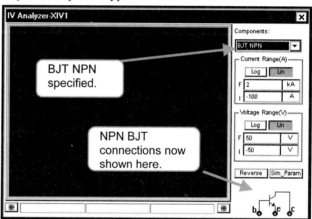

Once we specify the correct component type, the connections are specified at the bottom right of the dialog box. Now that we know how to connect the BJT, close the dialog box and connect the device to the instrument:

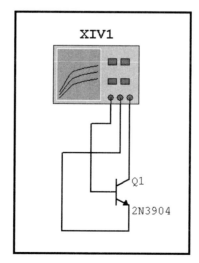

We are now ready to set up the parameters for the instrument. Double-click on the instrument to open it up again and then click on the **Sim_Param** button:

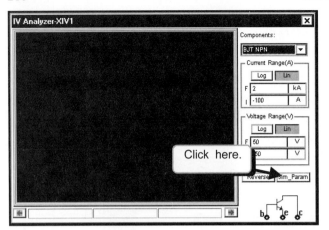
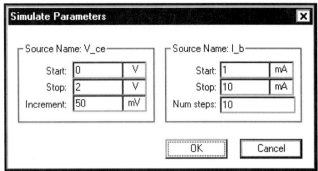

This dialog box specifies the parameters for the IV plot. In the plot, we sweep the collector-emitter voltage (V_{CE}) from 0 to 15 volts and step the base current from 0 to 100 µA in 10 µA steps just like we did for the DC sweep in the previous section. We will specify the same parameters here as we did for the DC sweep on page 212:

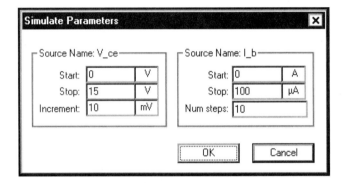

We will sweep V_{CE} from 0 to 15 volts, and step I_B from 0 to 100 µA in 10 µA steps. Click the **OK** button to accept the settings. Press the **F5** button to start the simulation and generate the IV plot:

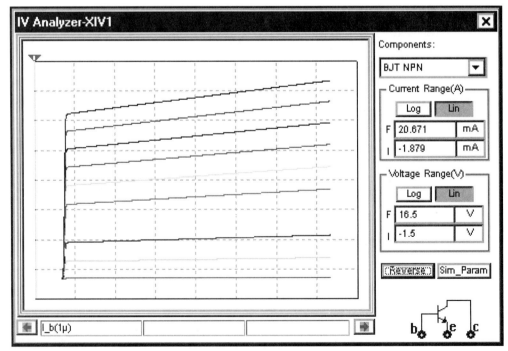

When the curve is complete, you should press the **F5** key again to stop the real-time simulation. You can also type **CTRL-G** to open the Grapher and view the IV plot with the Grapher:

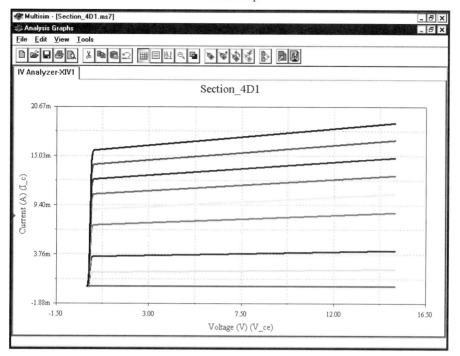

EXERCISE 4-5A: Use IV Plotter to find the characteristic curves of an IRF250 N-type power MOSFET.

SOLUTION:

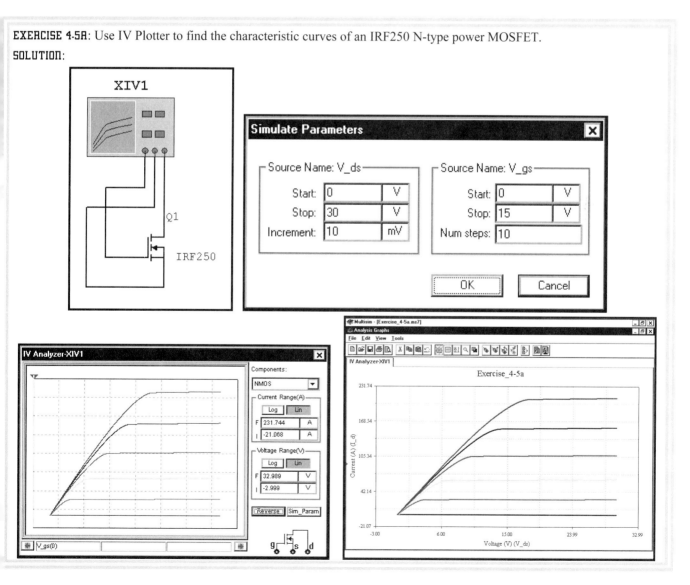

4.E. DC Current Gain of a BJT

In this section we will investigate how the DC current gain (H_{FE}) of a bipolar junction transistor varies with DC bias collector current I_{CQ} and DC bias collector-emitter voltage V_{CEQ}. We will use the basic circuit shown below for all simulations. Before running these simulations, you may want to shut down and then restart Multisim.

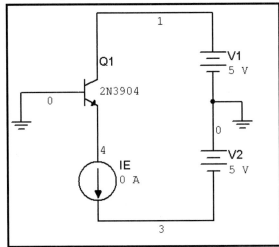

In this circuit we are making the approximations $I_C \cong I_E$ and $V_{CE} = V_1 + V_{BE} \approx V_1$. We make these approximations because we cannot directly sweep the collector current. The displayed curves will have the same shape as if we were sweeping the collector current.

4.E.1. H_{FE} Versus Emitter Current

Our first analysis will display how H_{FE} of a transistor varies with DC emitter current.[3] In many applications we want to use a transistor where it has its maximum current gain. This plot will tell us how to choose the emitter current for maximum current gain. We can then bias the transistor at this current. This plot is easily generated with a DC Sweep. We will generate this curve with V_{CE} constant at approximately 5.7 V (assuming that $V_{BE} \approx 0.7$ volts). Note in the circuit above that V_{CE} is held approximately constant by the DC source V1, and we will not change it during the simulation. V_{BE} will change slightly as the current changes, but we will assume that this change is small. Wire the circuit above and then select **Simulate**, **Analyses**, and then **DC Sweep**. Fill in the dialog box as shown below. You will need to select the **Change Filter** button for **Source 1** and enable all of the options in order to specify the **Source** as **i:xie**.

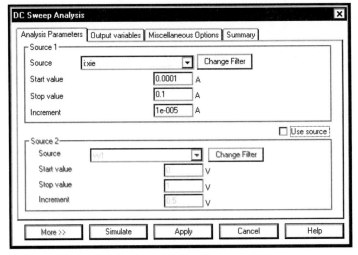

We will sweep the emitter current from 100 µA to 100 mA in 10 µA steps.

---

[3] Thanks to John D. Welkes of Arizona State University for this example.

To plot the DC current gain, H_{FE}, we need to plot the ratio of I_C/I_B using the Postprocessor. Since current data is not normally available to the Postprocessor, we need to specify the traces in the **Output variables** portion of the setup. Select the **Output variables** tab and then follow the procedure on pages 186-189 to add the base current and collector current as traces that will be plotted during the simulation:

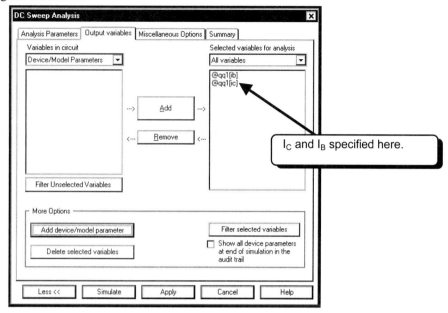

Click on the **Simulate** button to run the simulation. The **Analysis Graphs** window will open and display the base current and collector current:

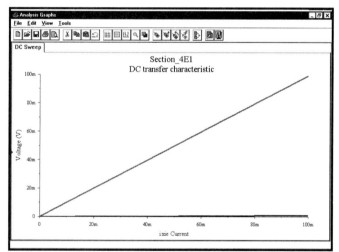

This graph shows that the simulation ran, but it is not the plot we are after. To generate a plot of I_C/I_B, we need to use the Postprocessor. Select **Simulate** and then **Postprocessor** from the Multisim menus:

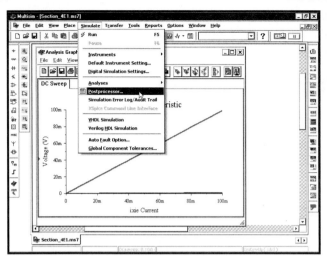
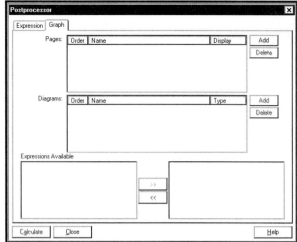

My Postprocessor shown above is empty. However, your Postprocessor may contain pages from a previous simulation. Delete any old pages listed in your Postprocessor. Select the **Expression** tab:

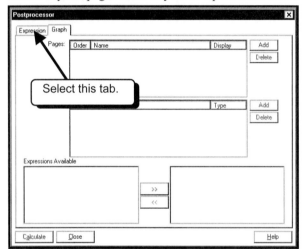
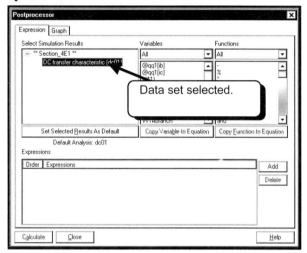

If you have run several simulations before this one, several data sets will be listed in this window. For this demonstration, I shut down and then restarted Multisim, so this is the only simulation I have run. Thus, only one data set is listed, and the proper data set is selected. Before you continue, you will need to make sure that the proper data set is selected.

Next, we need to create the expression for I_C/I_B. Click the **Add** button and then piece together the **Expression** **@qq1[ic]/@qq1[ib]**. Your **Expression** may be slightly different:

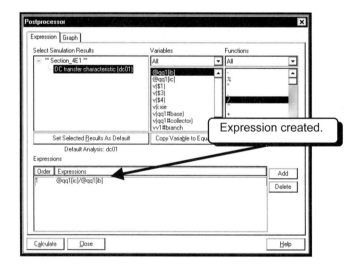

Select the **Graph** tab:

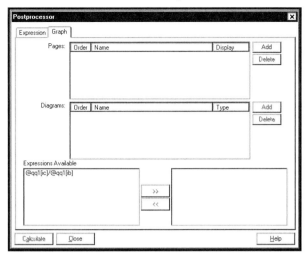

See sections 2.A to 2.C on creating new plots with the Postprocessor. **You should delete any old Pages before continuing. Multisim will try to plot those old pages, and if it cannot find the data for those pages, an error will occur.** Click the **Add** button to create a new **Page**, and then **Add** a new **Diagram** to that **Page** with the expression of I_C/I_B on it:

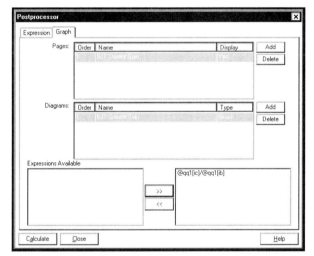

Click on the **Calculate** button to plot the trace. In the screen capture below, I have modified the labels slightly:

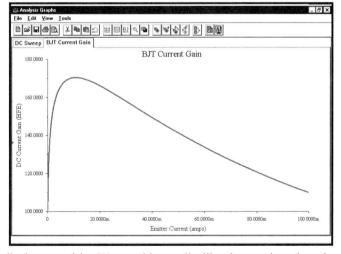

This is not the plot we are really interested in. We would actually like the x-axis to be a log scale. To change it, click the **RIGHT** mouse button on the graph:

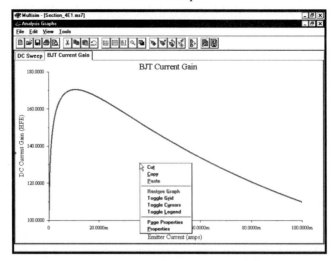

Select **Properties**:

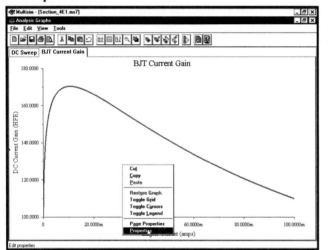
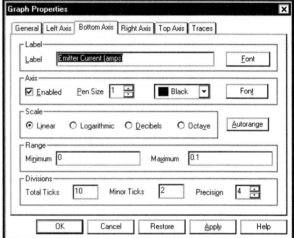

Select the **Bottom Axis** tab if it is not already selected:

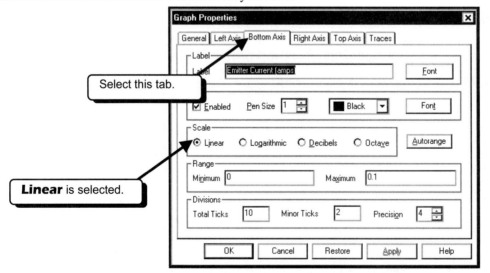

In the screen capture above, the x-axis (bottom axis) is specified as **Linear**. We need to change this axis to a logarithmic scale, so click the *LEFT* mouse button on the circle ○ next to the text **Logarithmic**. The circle should fill with a dot ⊙, indicating that the option is selected:

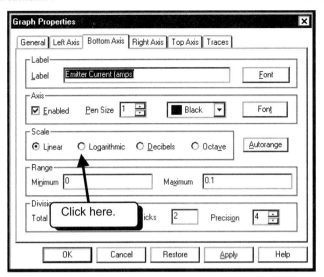
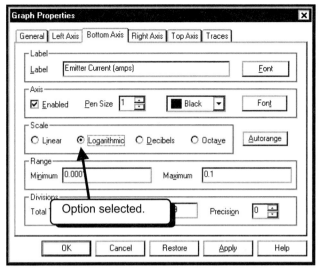

Click on the **OK** button to see the results. I also turned on the grid by clicking the right mouse button on the graph and selecting **Toggle Grid**.

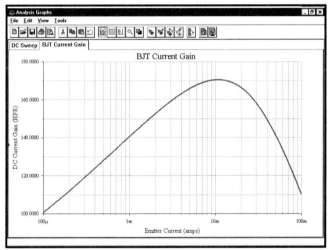

If we turn on the cursors, we see that the maximum current gain is **170.4555** and occurs at an emitter current of about 10.5 mA:

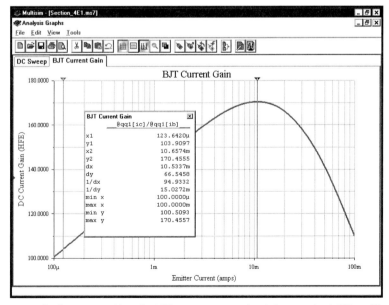

In our analysis, we made the assumption that V_{BE} was approximately constant. We will now plot the base-emitter voltage to see if this was a good approximation. V_{BE} is equal to zero minus the voltage at node **4** in our circuit:

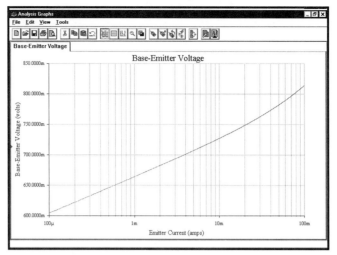

We see that V_{BE} varies by about 200 mV as the emitter current changes from 100 μA to 100 mA. We see that V_{BE} is roughly constant for the simulation, and therefore V_{CE} is roughly constant as well since V_{CE} is equal to **V1** minus V_{BE}.

EXERCISE 4-6: For the 2N4402 PNP bipolar junction transistor, find the collector current where H_{FE} is at its maximum value. Display the maximum value of H_{FE} and the emitter current where it occurs. Let V_{EC} be constant at about 5.7 V.

SOLUTION: Use a circuit similar to the one used in the previous example, but use the 2N4402 transistor:

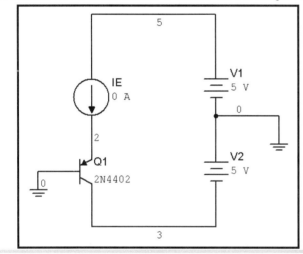
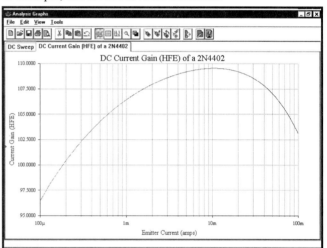

4.E.2. H_{FE} Versus I_C for Different Values of V_{CE}

The next thing we would like to do is to see how the H_{FE} versus I_E curve is affected for different values of DC collector-emitter voltage. The curve in the previous example was generated at $V_{CE} \cong 5.7$ V. We would now like to generate three curves at different values of V_{CE} and plot them all on the same graph. We will generate curves at $V_{CE} \cong 5$ V, 10 V, and 15 V. We will use the same circuit as in the previous section:

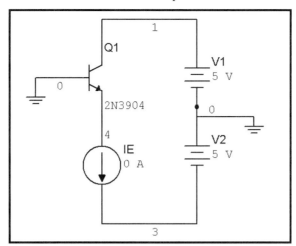

Select **Simulate**, **Analyses**, and then **DC Sweep** to bring up our previous sweep setup:

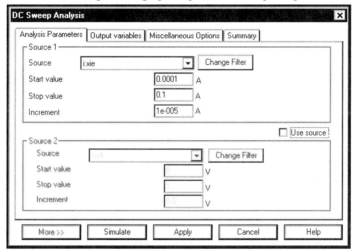

The dialog box should retain its settings from the previous simulation, so all we need to do is specify the nested sweep information. We need to specify **Source 2** as the source that controls VCE (in my case this is **V1**). Fill in the second source's information as shown:

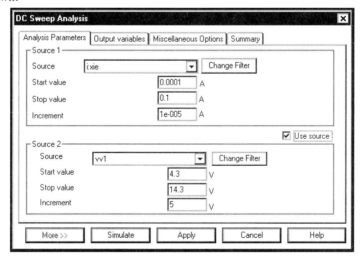

The rest of the setup, including the **Output variables** information, should be the same as in the previous run and should be retained from the last simulation. If not, set up the **Output variables** tab the same as you did in section 4.E.1. Run the simulation and then open the Postprocessor and select the **Expressions** tab:

My dialog box shows two sets of data because I have run two simulations since I started using Multisim in this session. For my most recent simulation I used the circuit named **Section_4E1**, but there are two data sets for this circuit. The most recent data set is the one listed at the bottom. If we expand the tree, we will see that the bottom data set contains three DC Sweeps:

When we set up our nested DC sweep, we ran multiple DC sweeps of the emitter current. Those multiple sweeps are listed here. It turns out that there are only three nested sweeps, one for VCE = 4.3 volts, one for VCE = 9.3 volts, and one for VCE = 14.4 volts. We can now generate plots for each individual sweep, or any combination of plots.

Click on one of the data sets to select it:

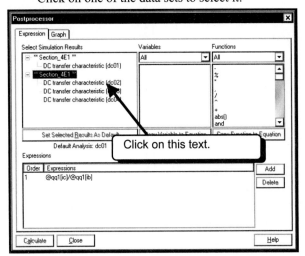 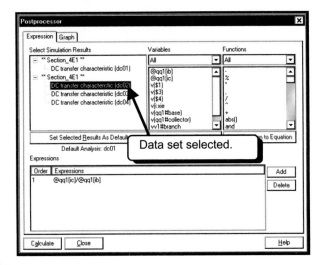

Click the **Add** button and then create the **Expression** of I_C/I_B:

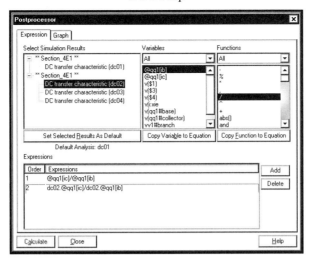

Notice that the text for the expression is **dc02.@qq1[ic]/dc02.@qq1[ib]**. The text **dc02** specifies data set dc sweep 02 and **qq1[ic]** specifies the collector current of the device called q1 in that data set.

The expression we just added was for a single value of V_{CE}. We would like to generate a plot with all three values of VCE, so we need to add two more expressions. Click the ***LEFT*** mouse button on the text as shown below to select the second value of V_{CE} for our simulation:

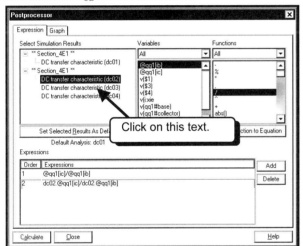
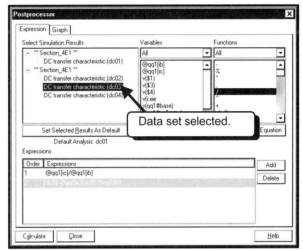

Next, create the expression `dc03.@qq1[ic]/dc03.@qq1[ib]`:

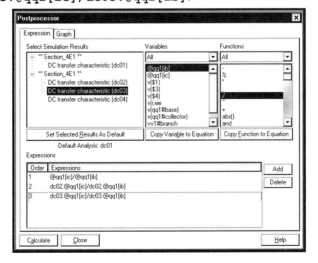

Note that this trace is the same as the previous one except that it contains the text **dc03**, which specifies the third DC sweep that I have run. This is actually the second DC sweep for this simulation, which has three, one at $V_{CE} \cong 5$ volts, one at $V_{CE} \cong 10$ volts, and one at $V_{CE} \cong 15$ volts.

Repeat the procedure to add the trace **Expression** for the third run:

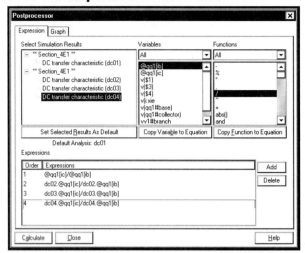

The third **Expression** is **dc04.@qq1[ic]/dc04.@qq1[ib]**.

We are now ready to generate the plot, so select the Graph tab. Add a new Page to the Postprocessor and add a new graph to the page. Add the three traces we just created to the new graph:

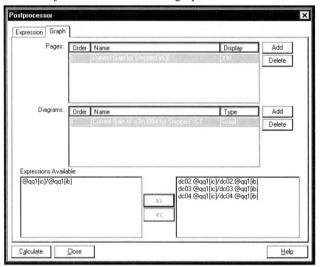

Click on the **Calculate** button to create the plot:

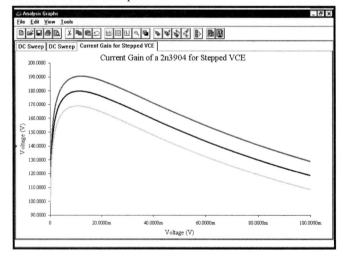

Change the labels to correctly annotate the plot, and change the x-axis to a log scale:

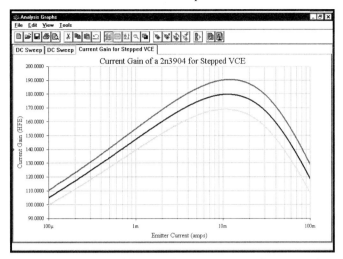

We see that the value of V_{CE} does affect the value of the transistor's current gain, H_{FE}. What we realize is that our bias point specifies the collector current (I_C) and collector-emitter voltage (V_{CE}), and these two parameters affect the performance of the device. If we wanted to bias this transistor for maximum H_{FE}, the curve above could be used as a guide for how to choose our bias values of V_{CE} and I_C.

4.F. Problems

Problem 4-1: For the circuit below, plot the voltage Vo versus Vin as Vin is varied from –10 volts to 15 volts.

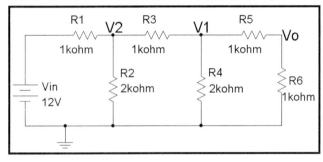

Problem 4-2: For the circuit of Problem 4-1, plot the voltage V1 versus Vin as Vin is varied from 0 volts to 15 volts.

Problem 4-3: For the circuit of Problem 4-1, plot the voltage V2 versus Vin as Vin is varied from 0 volts to 25 volts.

Problem 4-4: For the circuit of Problem 4-1, plot the sum of voltages V1 and V2 versus Vin as Vin is varied from 0 volts to 25 volts.

Problem 4-5: For the circuit of Problem 4-1, plot the voltage across resistor R3 versus Vin as Vin is varied from 0 volts to 25 volts.

Problem 4-6: For the circuit of Problem 4-1, plot the ratio Vo/Vin versus Vin as Vin is varied from 0 volts to 7 volts.

Problem 4-7: For the circuit of Problem 4-1, plot the current through resistor R1 versus Vin as Vin is varied from –8 volts to 16 volts.

Problem 4-8: For the circuit of Problem 4-1, plot the current through resistor R3 versus Vin as Vin is varied from –10 volts to 10 volts.

Problem 4-9: For the circuit of Problem 4-1, plot the power absorbed by resistor R1 versus Vin as Vin is varied from –25 volts to 25 volts.

Problem 4-10: For the circuit of Problem 4-1, plot the power supplied by voltage source Vin versus Vin as Vin is varied from –25 volts to 25 volts.

Problem 4-11: For the circuit below, plot the voltage Vo versus I1 as I1 is varied from –10 mA to 15 mA.

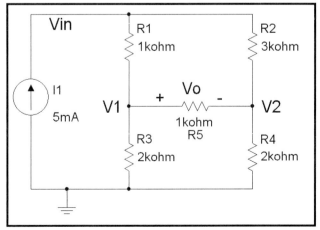

Problem 4-12: For the circuit of Problem 4-11, plot the voltage V1 versus I1 as I1 is varied from 0 mA to 15 mA.

Problem 4-13: For the circuit of Problem 4-11, plot the voltage V2 versus I1 as I1 is varied from 0 mA to 25 mA.

Problem 4-14: For the circuit of Problem 4-11, plot the sum of voltages V1 and V2 versus I1 as I1 is varied from 0 mA to 25 mA.

Problem 4-15: For the circuit of Problem 4-11, plot the voltage across resistor R3 versus I1 as I1 is varied from 0 mA to 25 mA.

Problem 4-16: For the circuit of Problem 4-11, plot the ratio Vo/Vin versus I1 as I1 is varied from 0 mA to 7 mA.

Problem 4-17: For the circuit of Problem 4-11, plot the current through resistor R1 versus I1 as I1 is varied from –8 mA to 16 mA.

Problem 4-18: For the circuit of Problem 4-11, plot the current through resistor R3 versus I1 as I1 is varied from –10 mA to 10 mA.

Problem 4-19: For the circuit of Problem 4-11, plot the power absorbed by resistor R1 versus I1 as I1 is varied from –25 mA to 25 mA.

Problem 4-20: For the circuit of Problem 4-11, plot the power supplied by current source I1 versus I1 as I1 is varied from –25 mA to 25 mA.

Problem 4-21: For the circuit below, plot the voltage Vo versus Vin as Vin is varied from –10 volts to 15 volts.

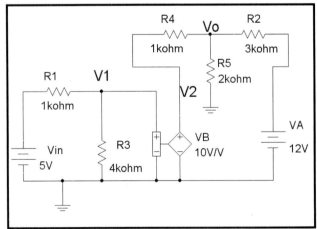

Problem 4-22: For the circuit of Problem 4-21, plot the voltage V1 versus Vin as Vin is varied from 0 volts to 15 volts.

Problem 4-23: For the circuit of Problem 4-21, plot the voltage V2 versus Vin as Vin is varied from 0 volts to 25 volts.

Problem 4-24: For the circuit of Problem 4-21, plot the sum of voltages V1 and V2 versus Vin as Vin is varied from 0 volts to 25 volts.

Problem 4-25: For the circuit of Problem 4-21, plot the voltage across resistor R4 versus Vin as Vin is varied from 0 volts to 25 volts.

Problem 4-26: For the circuit of Problem 4-21, plot the ratio Vo/Vin versus Vin as Vin is varied from 0 volts to 7 volts.

Problem 4-27: For the circuit of Problem 4-21, plot the current through resistor R2 versus Vin as Vin is varied from –8 volts to 16 volts.

Problem 4-28: For the circuit of Problem 4-21, plot the current through resistor R3 versus Vin as Vin is varied from –10 volts to 10 volts.

Problem 4-29: For the circuit of Problem 4-21, plot the current through resistor R4 versus VA as VA is varied from –10 volts to 12 volts.

Problem 4-30: For the circuit of Problem 4-21, plot the current through resistor R5 versus VA as VA is varied from –12 volts to 10 volts.

Problem 4-31: For the circuit of Problem 4-21, plot the power absorbed by resistor R1 versus Vin as Vin is varied from –25 volts to 25 volts.

Problem 4-32: For the circuit of Problem 4-21, plot the power supplied by voltage source Vin versus Vin as Vin is varied from –25 volts to 25 volts.

Problem 4-33: Plot the I-V characteristic of a 1N914 diode. From the plot, obtain a numerical value for the diode turn-on voltage.

Problem 4-34: Plot the I-V characteristic of a 1S1830 diode. From the plot, obtain a numerical value for the diode turn-on voltage.

Problem 4-35: Plot the I-V characteristic of a 1N4740A Zener diode. From the plot, obtain a numerical value for the Zener breakdown voltage and the turn-on voltage.

Problem 4-36: Plot the I-V characteristic of a 1N4756A Zener diode. From the plot, obtain a numerical value for the Zener breakdown voltage and the turn-on voltage.

Problem 4-37: Plot the I-V characteristic of a blue LED. From the plot, obtain a numerical value for the diode turn-on voltage.

Problem 4-38: Plot the I-V characteristic of a yellow LED. From the plot, obtain a numerical value for the diode turn-on voltage.

Problem 4-39: Plot the I-V characteristic of a 1 kΩ resistor.

Problem 4-40: Plot the I-V characteristic of a 500 Ω resistor.

Problem 4-41: Plot the I-V characteristic of a 15 Ω resistor.

Problem 4-42: Plot the I-V characteristic of a 15 MΩ resistor.

Problem 4-43: Plot the transfer curve Vo versus Vin for the circuit below. Let Vin vary from –20 volts to 20 volts.

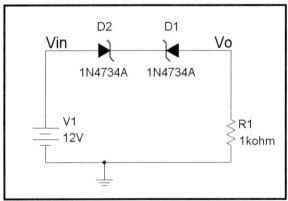

Problem 4-44: Plot the transfer curve Vo versus Vin for the circuit below. Let Vin vary from –20 volts to 20 volts.

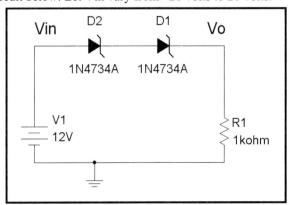

Problem 4-45: Plot the transfer curve Vo versus Vin for the circuit below. Let Vin vary from –20 volts to 20 volts.

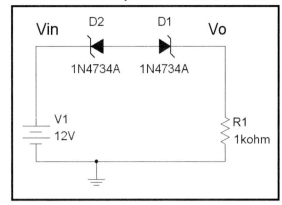

Problem 4-46: Plot the transfer curve Vo versus Vin for the circuit below. Let Vin vary from –20 volts to 20 volts.

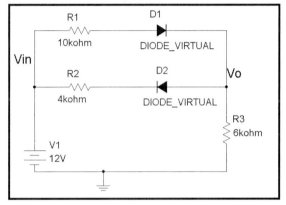

Problem 4-47: Plot the transfer curve Vo versus Vin for the circuit below. Let Vin vary from –20 volts to 20 volts.

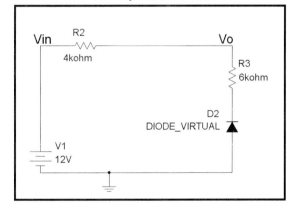

Problem 4-48: Plot the transfer curve Vo versus Vin for the circuit below. Let Vin vary from –20 volts to 20 volts.

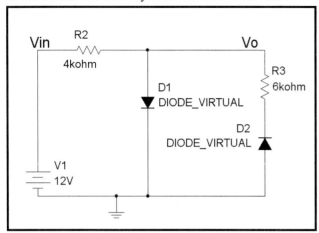

Problem 4-49: Plot the transfer curve Vo versus Vin for the circuit below. Let Vin vary from 0 volts to 5 volts. For the driver transistor, use $Kp = 60\ \mu A/V^2$ and $VTO = 1$ volt. For the load transistor, use $Kp = 30\ \mu A/V^2$ and $VTO = 0.5$ volt.

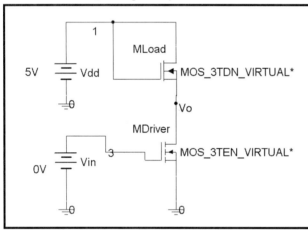

Problem 4-50: Plot the transfer curve Vo versus Vin for the circuit below. Let Vin vary from 0 volts to 5 volts. For the driver transistor, use $Kp = 120\ \mu A/V^2$ and $VTO = 1$ volt. For the load transistor, use $Kp = 30\ \mu A/V^2$ and $VTO = 0.5$ volt.

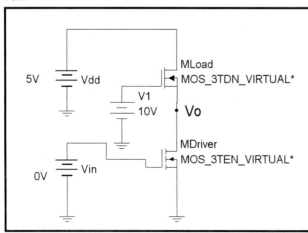

Problem 4-51: Redo the simulation of section 4.C.2 using the circuit below. Use the same parameters for the MOSFET as those used in the example. Plot Vo versus Vin.

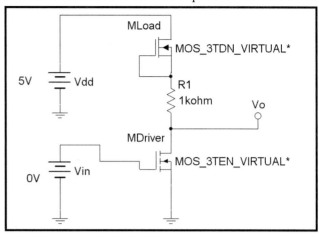

Problem 4-52: Plot the characteristic curves for a 2N2222A NPN BJT transistor.

Problem 4-53: Plot the characteristic curves for a 2N2222A NPN BJT transistor.

Problem 4-54: Plot the characteristic curves for a 2N4401 NPN BJT transistor.

Problem 4-55: Plot the characteristic curves for a 2N3906 PNP BJT transistor.

Problem 4-56: Plot the characteristic curves for a PN2907A PNP BJT transistor.

Problem 4-57: Plot the characteristic curves for a MOSFET with the following model parameters: $Kp = 120\ \mu A/V^2$, $VTO = 1$ V, and lambda $= 0.001$ V^{-1}.

Problem 4-58: Plot the characteristic curves for a MOSFET with the following model parameters: $Kp = 120\ \mu A/V^2$, $VTO = 1$ V, and lambda $= 0.000$ V^{-1}.

Problem 4-59: Plot the characteristic curves for a MOSFET with the following model parameters: $Kp = 120\ \mu A/V^2$, $VTO = 1$ V, and lambda $= 0.01$ V^{-1}.

Problem 4-60: Plot the characteristic curves for a 6C5 triode vacuum tube.

PART 5
Magnitude and Phase Simulations

Multisim provides several tools for sinusoidal steady-state magnitude and phase analysis, including AC voltmeters, power meters with power factor calculation, Bode plotters, and the SPICE AC Analysis. Circuits can be analyzed at a single frequency or at multiple frequencies.

It is important to realize the difference between magnitude and phase simulations and the time domain simulations discussed in Part 6. The simulations in this section are used to find the magnitude and phase of voltages and currents. The time domain analyses are used to look at waveforms versus time. Example waveforms versus time using the Transient Analysis and the Oscilloscope instrument are:

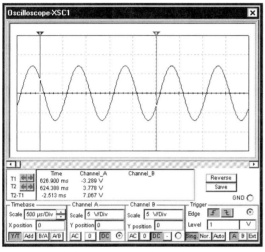 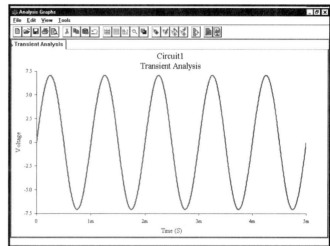

These graphs show us a voltage versus time, and it could be represented mathematically as $5\sin(2\pi 1000t + 0°)$. The magnitude of the above waveform is 5 V and the phase is zero degrees—in phasor notation, $5\angle 0°$. All of the magnitude and phase simulations will give a result such as $5\angle 0°$, while the time domain analyses allow us to view waveforms versus time such as the above screen captures.

Before we get started, we will note that there are two AC voltage sources that you can use. One is called the AC Power voltage source and the other is called the AC Signal voltage source. They both look about the same when you place them in your circuit. The schematic depiction and the properties for the AC Power source are:

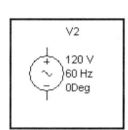 

Initially, the AC Power voltage source has an RMS amplitude of 120 V and a frequency of 60 Hz. The important things to note about the power source are that the amplitude is an RMS value and that we cannot tell from the schematic that the amplitude is an RMS value.

Next, we show the AC Signal voltage source:

233

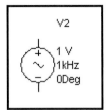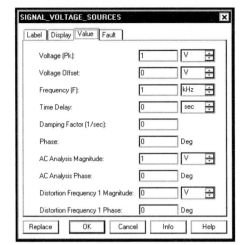

The schematic depiction of this source is basically identical to that for the power voltage source except for the initial numerical values. For the signal source, the peak amplitude is 1 V and the frequency is 1 Hz. The main difference between the two sources is in the properties dialog box. The signal voltage source specifies the peak values of the voltage source, while the power voltage source specifies the RMS value of the voltage source. This difference is not indicated on the schematic, and so we cannot tell from looking at the schematic which type of source we are using.

When you are editing a schematic, you can easily determine which type of source you are using because you can double-click on the source and look at the dialog box, which indicates RMS or peak. However, when you read this book, all you can do is look at the schematics, and the symbols for both types of sources are identical. To distinguish between the two sources, I will add the text "RMS" or "Peak" to all of my schematics.

5.A. Magnitude and Phase Measurements at a Single Frequency

Multisim provides two ways to measure AC waveforms. The magnitude of waveforms at a single frequency can be measured using the Multimeter or the current and voltage indicators. With these tools we can measure only the magnitude of currents and voltages; we cannot measure the phase. The AC Analysis is a more general analysis tool. It can be used to find both the magnitude and phase of voltages and currents at multiple frequencies, and the results can be viewed graphically or as text output.

5.A.1. Magnitude Measurements with Instruments

The Multisim voltage and current indicators and the Multimeter allow us to easily measure RMS voltages in a circuit at a single frequency. The use of these meters is the same as in section 3.A except we must change the meter to read AC voltages rather than DC. If you leave a meter on DC, it will read zero or close to zero for AC voltages. As we saw with the DC measurements, the current meter is modeled as a small series resistance, and the voltage meter is modeled as a large parallel resistance. We must keep these models in mind when making measurements because they could affect our results. Unlike real AC voltmeters, these meters do not have an upper frequency limitation. From simple tests, it appears that the meters give an accurate reading anywhere from 20 Hz up to 10^{12} Hz. Thus, except for the series and parallel resistances, the meters appear to be ideal. The one limitation of the meters and indicators is that we can measure only the magnitude of voltages and currents. For AC circuits, this is only half of what we need to know. In order to find the phase of a voltage or a current, we need to run the SPICE AC Analysis.

We will start by measuring the magnitude of the voltages and currents in the circuit below:

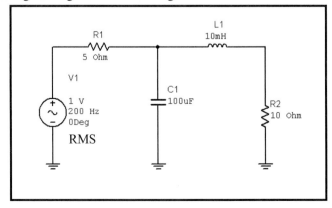

Notice that the size of the voltage source is displayed as **1** volt. This is the RMS amplitude of the source. We shall see that the indicators display the RMS value of voltages and currents. Remember that the RMS value is equal to the peak value divided by the square root of two: $V_{RMS} = V_{peak}/\sqrt{2}$. When we specify the value of a source, we specify its RMS value. To see the settings of this source, double-click the *LEFT* mouse button on the graphic of the voltage source:

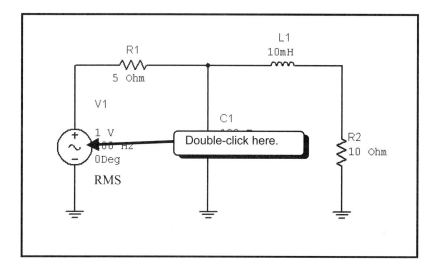
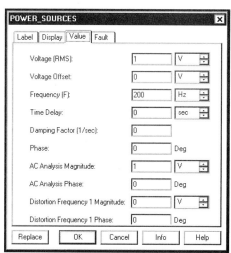

For this source, I specified the RMS value to be **1** volt. Also, note that the **Frequency** is **200 Hz** and the **Phase** is set to **0**. Make changes to your dialog box so that the settings match the ones shown and then click on the **OK** button to close the dialog box.

Next, we need to add some monitors to the circuit. As examples, we will measure the voltages at two nodes and the current through **L1**. Before doing this section, you may want to review section 3.A where we cover the use of the meters in DC applications. The use of the meters in AC circuit simulations is about the same.

To place an indicator, click the *LEFT* mouse button on the **Indicator** button as shown below:

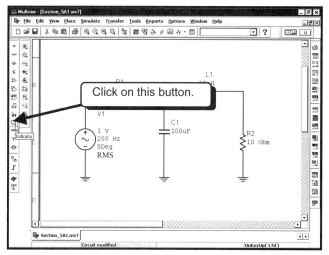
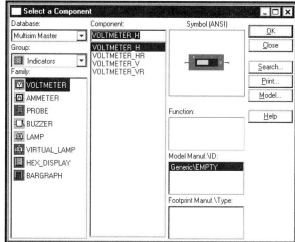

The meters all perform the same function. The only difference is the arrangement of the connection terminals. Select the **VOLTMETER_V** part and click on the **OK** button. Place the part in your circuit and wire the meter to your circuit as shown:

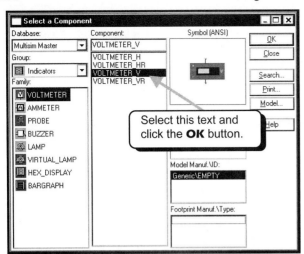
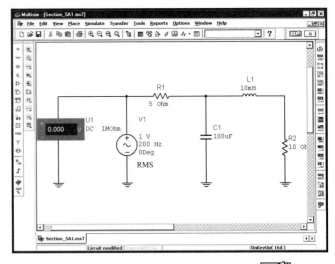

Before we add more meters, we will see how this meter works. Click on the **Run/stop simulation** button to start the simulation:

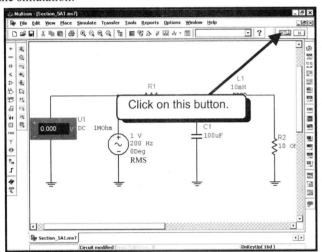
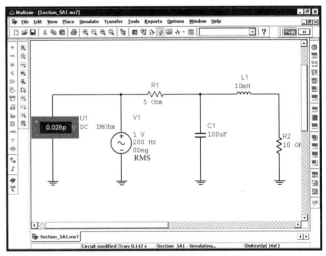

The indicator jumps around a little bit, but most of the time it reads zero. This is because the default setting for the indicator is DC. The DC component of our source is zero, so the indicator is reading the correct value. Stop the simulation by clicking on the **Run/stop simulation** button again.

To change the properties of the indicator, double-click the *LEFT* mouse button on the indicator:

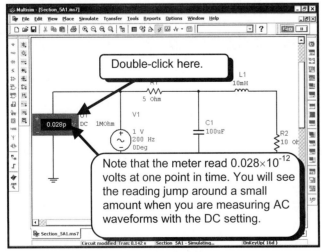
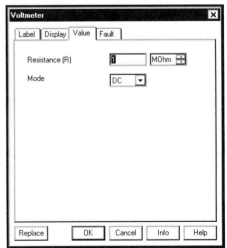

We see that the indicator is set to the **DC Mode**. Because we are measuring AC waveforms, we need to select the **AC Mode**. Change the **Mode** to **AC** and click on the **OK** button:

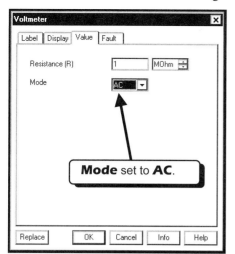
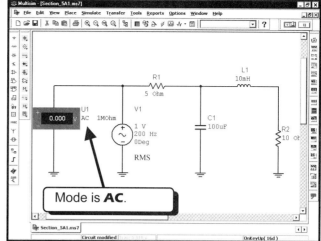

Note that the **Mode** of the indicator is displayed on the schematic as well. Click on the **Run/start simulation** button to restart the simulation. After a few moments the meter should display the RMS value of the voltage it is measuring:

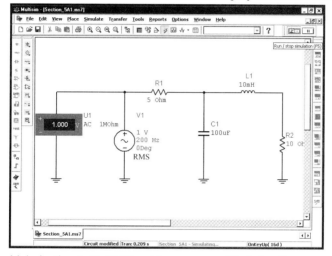

My meter reads 1.000 volt, which is the 1 V RMS setting we specified for the source. If you perform simple test measurements with the current indicator and the Multimeter, you will notice that all of the meters read the RMS value of voltages and currents rather than the peak values. Stop the simulation by clicking on the **Run/stop simulation** button again.

Next, we will measure the current through **L1** and the voltage across **R2** using Multimeters. To place a Multimeter, click the *LEFT* mouse button on the **Multimeter** icon in the Instruments toolbar:

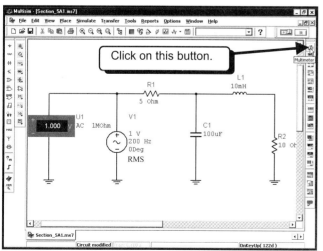
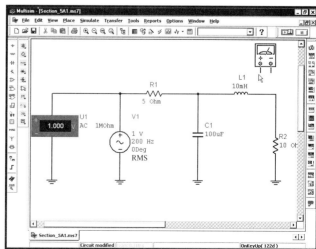

Place two Multimeters in your circuit as shown:

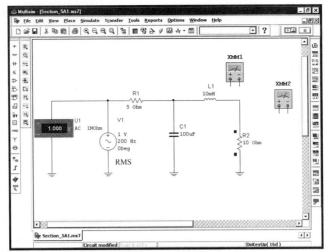

We will use **XMM2** to measure the voltage across **R2**, and use **XMM1** to measure the current through **L1**. Wire the meters as shown in the two screen captures below:

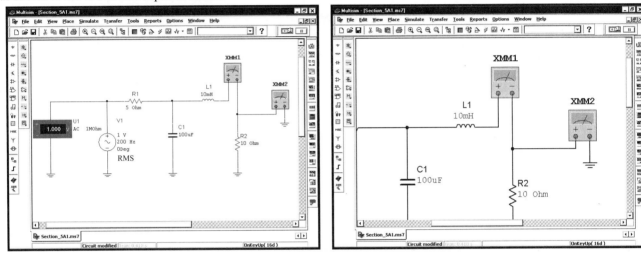

To set up and view meter results, we must open the meter windows by double-clicking on the meters. Double-click on each meter to open the windows. Arrange the windows as shown:

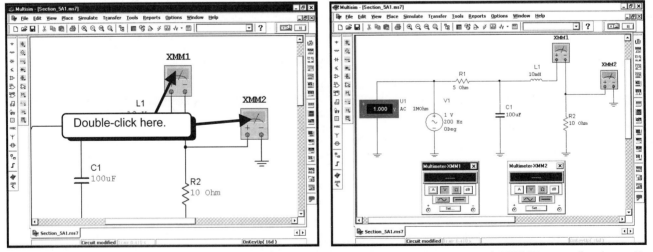

We must now set both meters to read AC waveforms, and set **XMM1** to read current and **XMM2** to read voltage. For **XMM1**, click on the [∼] button and the [A] button to make the meter read AC current. For **XMM2**, click on the [∼] button and the [V] button to make the meter read AC volts. When you have set up the meters correctly, click on the **Run/stop simulation** button [▭] to restart the simulation. After a few moments, the meter windows will display their results:

Multisim 7   Magnitude and Phase Simulations   239

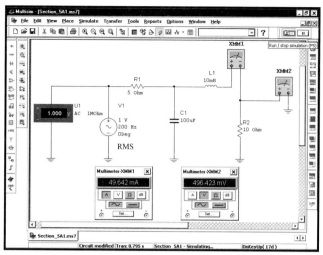

We see that the voltage across R2 is equal to about 496 mV and the current through L1 is approximately 49.6 mA. We can use this method to find the magnitude of any voltage or current in the circuit. Remember that the values displayed are RMS values. The input was 1 volt RMS. If you want peak values, multiply all of the values by the square root of two.

EXERCISE 5-1: Find the RMS value of the voltages at nodes 1 and 2, and the RMS current through L1 at a frequency of 1 kHz:

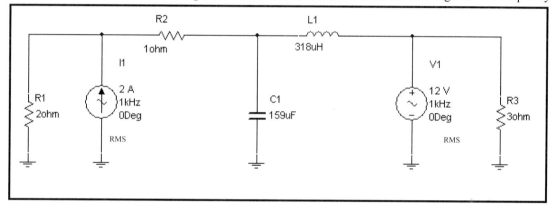

SOLUTION: Add three Multimeters or indicators and run the simulation:

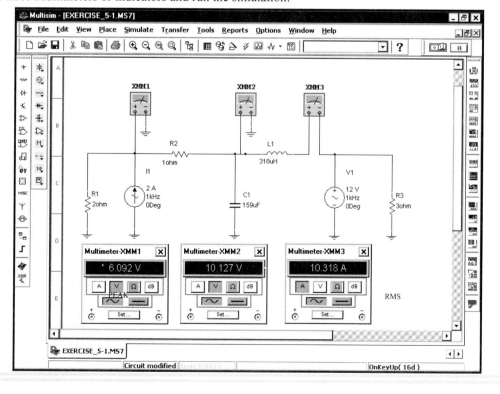

EXERCISE 5-2: Find the magnitude of the voltage at node 5 and the current through **R1** at a frequency of 60 Hz:

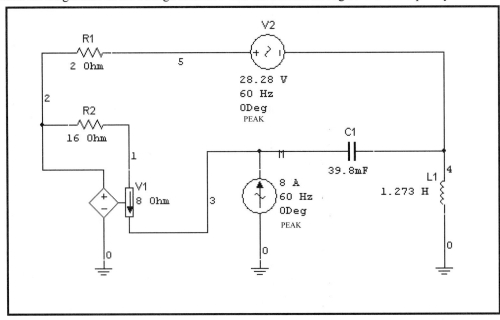

HINT: **V1** is a current-controlled voltage source. The voltage of **V1** is **8** times the current through **R2**.

SOLUTION: Add Multimeters or voltage and current indicators. Note that when I added the current indicator to the circuit above, the node that was labeled **5** in the circuit above was renamed to **6** in the circuit below.

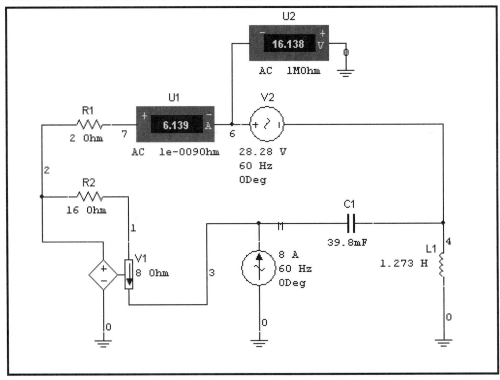

5.A.2. Magnitude and Phase Measurements – AC Analysis

We will now do the same analyses as in section 5.A.1 using the SPICE AC Analysis and we will also display the phase as well as the magnitude. You will need to use the Postprocessor for this section. If you are unfamiliar with the Postprocessor and Grapher, you may want to review Chapter 2 before continuing. We will start with the circuit below:

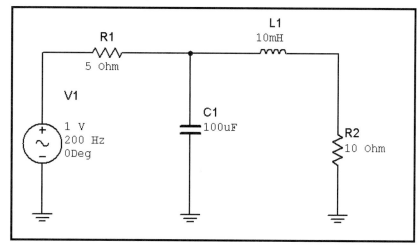

In order to use the AC Analysis, we must specify the properties of V1 for the sweep. The text displayed on the screen for V1 is used only with the virtual lab simulation and the instruments. There is another set of properties when the source is used for the AC Analysis. Double-click the **LEFT** mouse button on the V1 graphic:

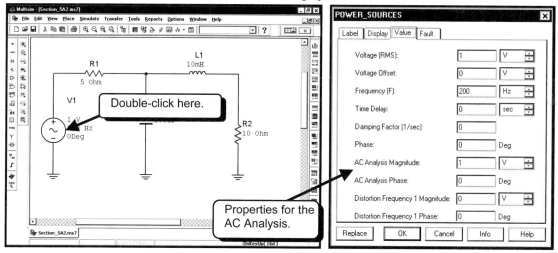

The last four lines of this dialog box allow us to specify parameters for the SPICE analyses. For the AC Analysis, the magnitude of the source is 1 volt, and the phase of the source is 0 degrees. The magnitude of the source can be interpreted as an RMS value or as a peak value. If we interpret the magnitude as an RMS value, then all voltages and currents we measure are RMS values. If we interpret the magnitude as a peak amplitude, then all measurements will be peak values. Our previous example used RMS values, so we shall interpret the 1 volt magnitude as 1 V RMS. The dialog box is already set up correctly, so click on the **OK** button to return to the schematic.

We would like to measure the current through **L1** and the voltage across **R2**. To measure current we will add a zero-volt voltage source and display the current through the source as shown in section 4.A. Wire the source as shown below:

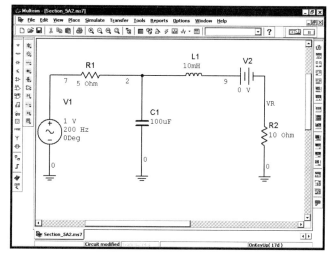

I have turned on the display of node names so that we can easily display node voltages. Note that the resistor voltage is the voltage at node **VR** relative to ground. To rename a node, double-click the *LEFT* mouse button on the wire.

We can now set up the AC Analysis. Select **Simulate**, **Analyses**, and then **AC Analysis** from the Multisim menus:

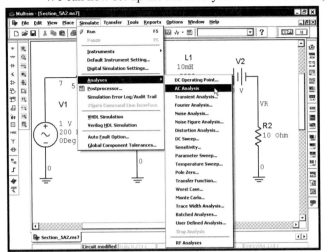
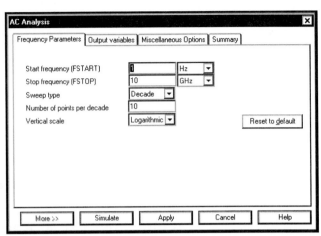

We want to analyze the circuit at a single frequency of 200 Hz. Fill in the dialog box as shown below:

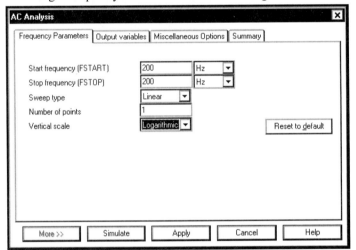

We have specified a **Linear Sweep type** with total **Number of points** equal to 1. The **Start frequency** and **Stop frequency** are set to **200 Hz**. Thus, this dialog box specifies a single frequency at 200 Hz. We do not need to specify an option for the **Vertical scale** because we will not be looking at the results graphically.

Next, we need to specify the output variables we would like to observe. Click the *LEFT* mouse button on the **Output variables** tab:

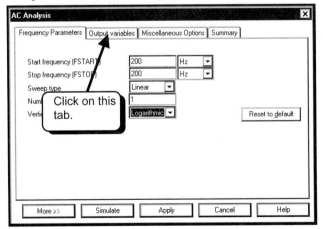
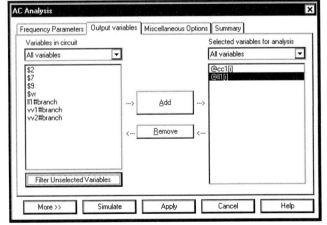

On my dialog box, two variables are selected, but these are not the variables we need so I will remove them. We are interested in the voltage at node VR and the current through L1. Click on the text **$vr** to select it:

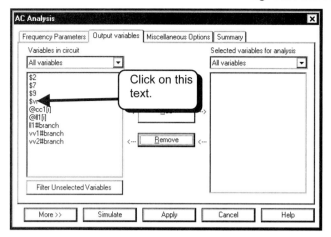
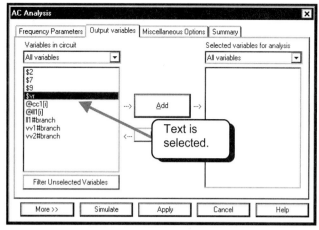

Click on the **Add** button:

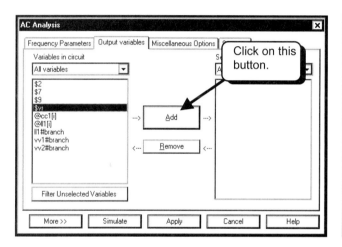
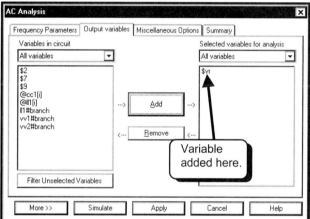

The voltage at node **vr** will now be plotted during the simulation and the data for the node will be available to the Postprocessor. Repeat the procedure to add the trace **vv2#branch** to the list, which is the current through source v2 and is also the current through L1:

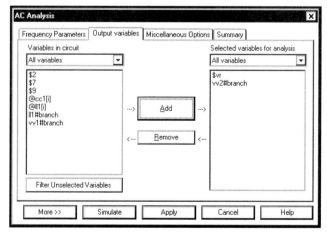

We are now ready to run the simulation, so click on the **Simulate** button:

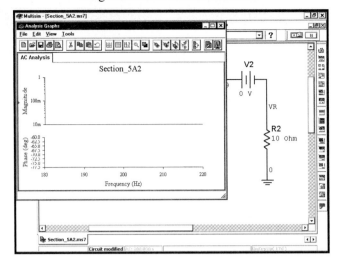

By default, a plot is displayed. This plot is not too useful because our simulation contained only a single datum point. To view the results, we need to use the Postprocessor. Select **Simulate** and then **Postprocessor** from the Multisim menus:

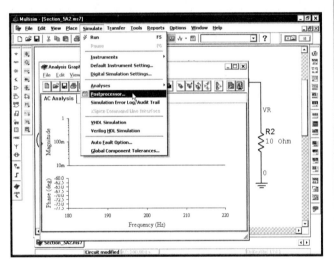 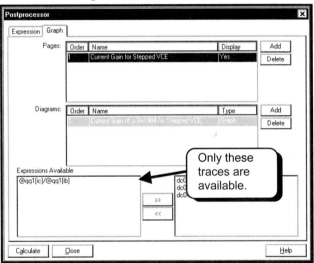

The Postprocessor contains pages and graphs from a previous simulation. If your Postprocessor contains old plots and graphs, as mine does, delete the older pages and graphs:

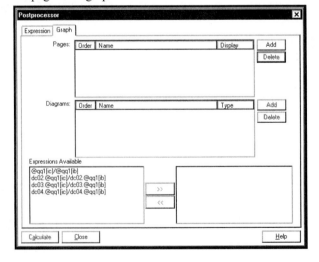

To view the data as text, we need to create a new Postprocessor page and add a chart to that page. First, click on the top **Add** button to create a new **Page**:

Multisim 7 — Magnitude and Phase Simulations — 245

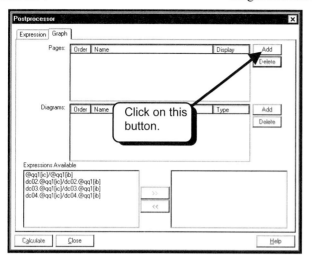
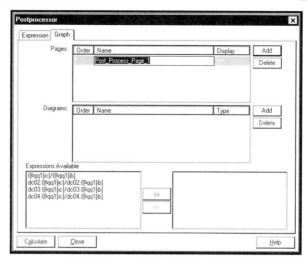

Enter a name for the new page, such as `Mag and Phase Data`, for example:

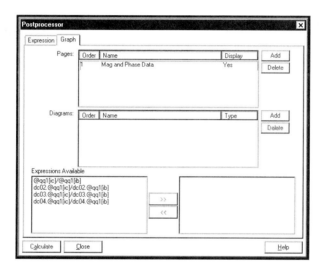

Next, click on the lower **Add** button to create a new Diagram on the Postprocessor page:

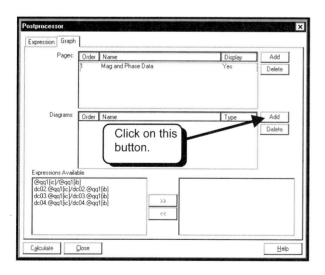
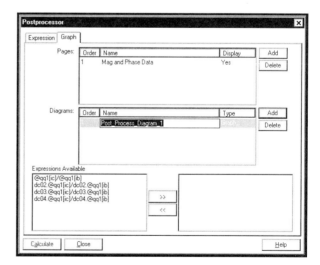

Enter a name for the chart, such as `AC Mag and Phase Text Output`:

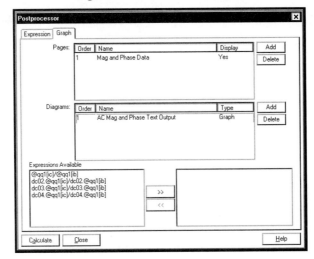

The last thing we need to do is specify this diagram as a chart. Click the **LEFT** mouse button on the text **Graph** to change the **Diagram Type**. A down triangle ▼ will appear:

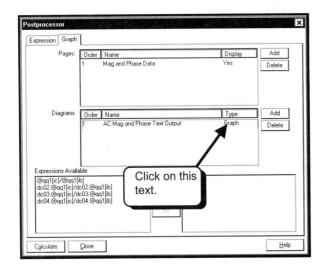
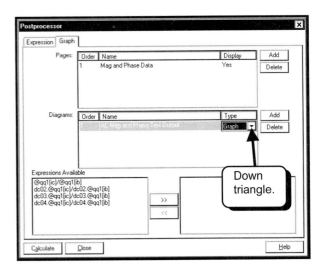

Click the **LEFT** mouse button on the down triangle ▼ to see the list of available **Diagram Type**s:

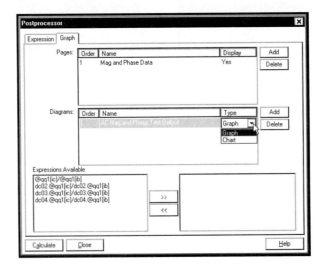

Select a **Chart** type.

The next thing we need to do is to specify the data to be displayed in the chart and create the expressions to be displayed. Select the **Expression** tab:

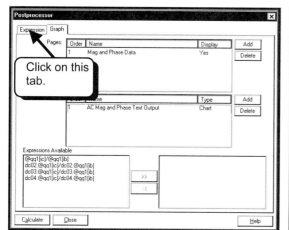 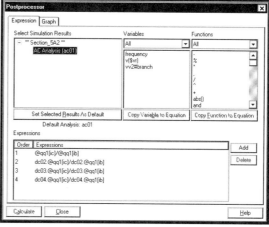

On my dialog box, there are a number of expressions from a previous simulation that I will delete. To delete an expression, select the text and then click the **Delete** button:

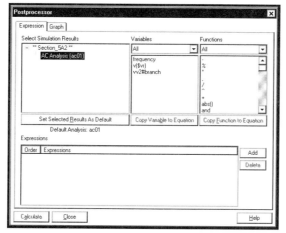

We see that the only data available in this simulation is the frequency information, the voltage at node **VR**, and the current through source v2. This is the data we specified when we initially set up our simulation. We would like to create the following expressions: `real(frequency)`, `vm($vr)`, `vp($vr)*180/pi`, `mag(vv2#branch)`, and `ph(vv2#branch)*180/pi`. By default, Multisim displays numbers with real and imaginary parts. To display frequency as a purely real number, we will use the real portion of the frequency. The frequency is purely real in all of our simulations anyway and always has a zero imaginary part. To save space, I have asked for only the real part. **vm($vr)** means the magnitude of the voltage at node **VR**. **vp($vr)*180/pi** is the phase of the voltage at node **VR**. By default, Multisim returns the angle in radians, so we convert the angle to degrees by multiplying by $180/\pi$. **mag(vv2#branch)** is the magnitude of the current through source v2, or the magnitude of the current through L1. **ph(vv2#branch)*180/pi** is the phase of the current through source v2. To add an expression, click the **Add** button and then either type the expression manually or use the **Copy Variable to Equation** button or the **Copy Function to Equation** button. For more details, see Section 2.C on using the Postprocessor.

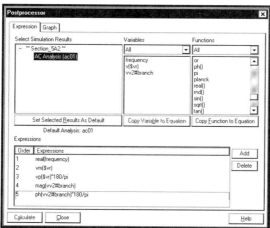

Click the **Graph** tab:

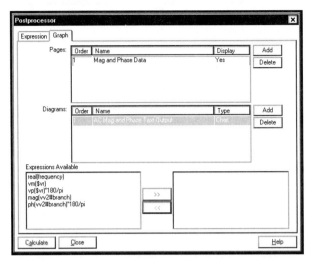

Add all of the **Expressions** to the chart:

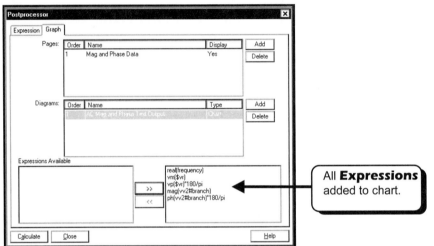

We are now done adding traces. Click on the **Calculate** button to create the chart:

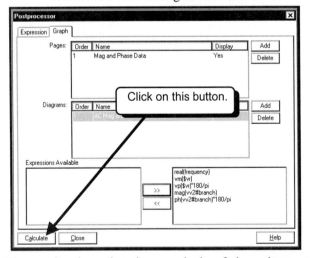
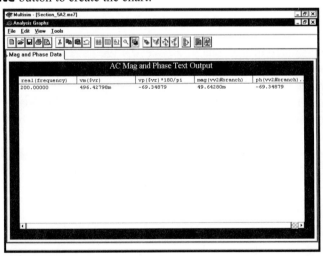

The results show that the magnitude of the voltage at node **vr** is **496** mV with a phase angle of **-69.3**5 degrees ($0.496\angle -69.35°$ volts). The current through L1 is **49.6** mA with a phase angle of **-69.3**5 degrees ($0.0496\angle -69.35°$ amps).

EXERCISE 5-3: Find the magnitude and phase of the voltages at nodes 1, 2, and 3 at a frequency of 1 kHz using the AC Analysis.

Magnitude and Phase Simulations

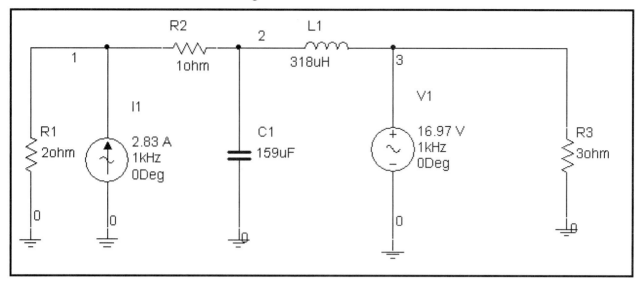

SOLUTION: Add a voltage source in series with the inductor to measure the current through the inductor. Set the value of the source to zero:

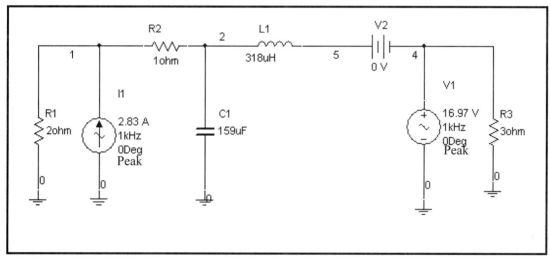

In the **Analysis Setup** for the two sources, specify **I1** as a 2 A source and **V1** as a 12-volt source. These values duplicate the RMS values used in **EXERCISE 5-1**.

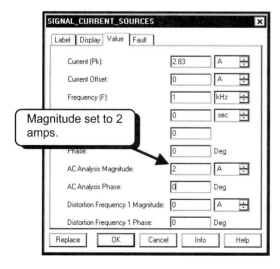
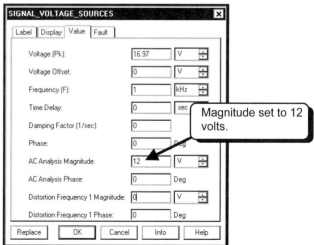

Set up the **AC Analysis** dialog boxes as shown:

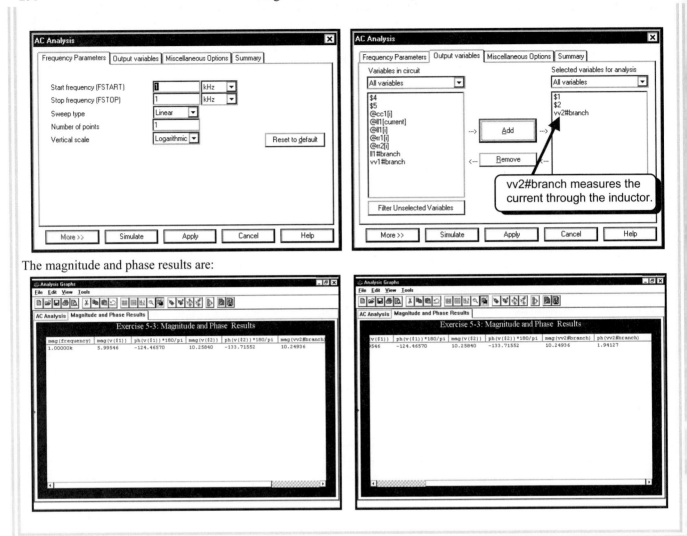

The magnitude and phase results are:

EXERCISE 5-4: Find the magnitude and phase of the voltage at node 5 and the current through **R1** at a frequency of 60 Hz using the AC Analysis:

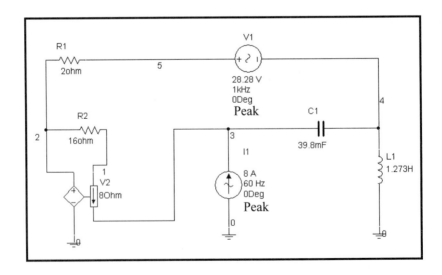

SOLUTION: To measure the current, we will add a zero-volt DC source to the circuit as shown below:

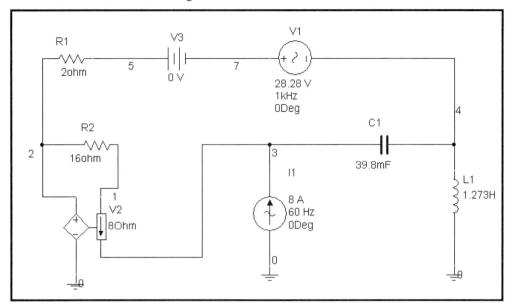

In the **Analysis Setup** for the two sources, specify **I1** as a 5.66 A source and **V1** as a 20-volt source. These values duplicate the RMS values used in EXERCISE 5-2.

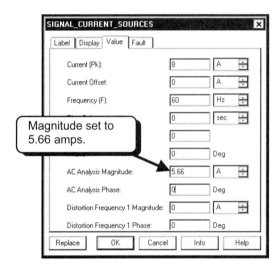
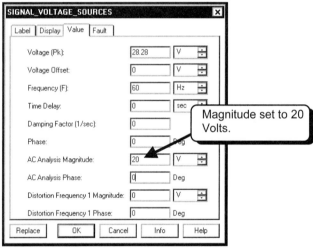

Set up the **AC Analysis** dialog boxes as shown:

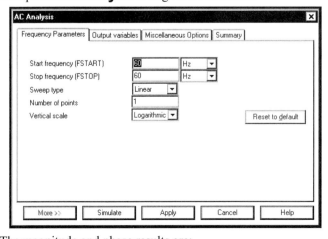
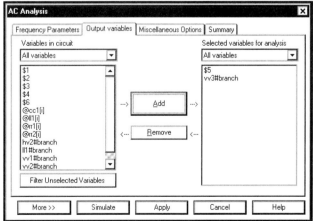

The magnitude and phase results are:

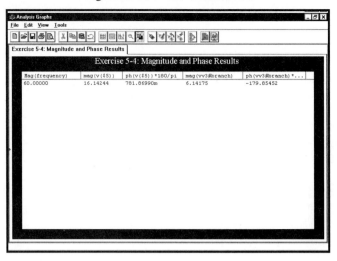

5.B. Bode Plots

Bode plots are plots of magnitude and phase versus frequency. Because we are usually interested in the magnitude of the gain, an AC 1-volt source will be used in the following sections. All AC analyses assume a linear network: If the output for a 1 V source is 3 V, the output for a 10 V source will be 30 V. Because gain is the ratio of an output to the source, and the networks are linear, the magnitude of the input does not matter. For convenience we will use a 1-volt magnitude source, and we will use the signal voltage source rather than the power voltage source. Thus, the magnitude of sources we discuss in this section refer to peak values. Bode phase plots will give us the phase of any voltage or current relative to the phase of the source. For simplicity we will set the phase of the source to zero.

As with most simulations in Multisim, we can simulate the circuit using a virtual laboratory with the Bode Plotter instrument, or we can use the standard SPICE AC Analysis. The Bode Plotter can only be used for Bode plots while the AC Analysis is a more general analysis and can be used for other applications in addition to Bode plots. We will show both methods in this section.

5.B.1. Bode Plots Using the Bode Plotter Instrument

We will first illustrate how to use the Bode Plotter instrument. This method gives us the quickest way of observing the frequency response of a network. But the Bode Plotter limits us in that we can only view a single node in the circuit. If you have a large circuit and want to look at the response of several nodes, you should use the AC Analysis.

Wire the low-pass filter shown:

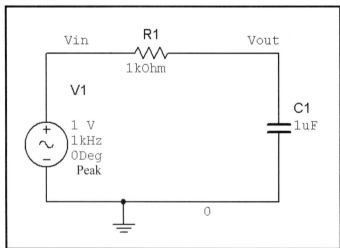

At low frequencies, the capacitor is an open circuit and **Vout** should equal **Vin**. At high frequencies the capacitor becomes a short, and the gain goes toward zero. The 3 dB frequency of the circuit is $\omega = 1/R_1C_1 = 1{,}000$ rad/s $\cong 159$ Hz.

Next, we need to place the Bode Plotter instrument in the circuit. The Bode Plotter button is located in the Instruments toolbar:

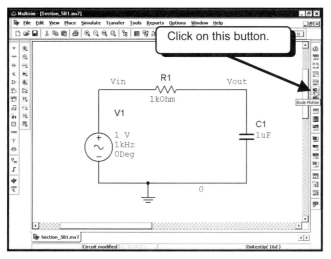

Place the Bode Plotter and connect it to your circuit as shown:

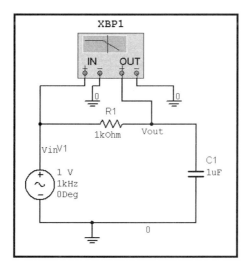

We can either run the simulation now, or modify the parameters for the Bode plot. To illustrate how to change the settings, we will modify the parameters. Double-click the **LEFT** mouse button on the Bode Plotter instrument to open the instrument window:

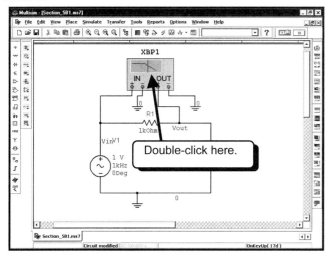
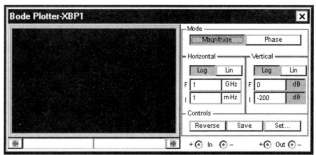

If we look at the instrument window a little closer, we see the following settings:

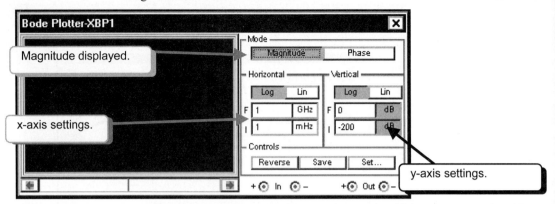

The settings specify that the instrument window will display the **Magnitude** plot, the y-axis will have a range of **0 dB** to **-200 dB**, and the x-axis will have a range of **1 mHz** (10^{-3} Hz) to **1 GHz** (10^9 Hz). The x-axis specifies the range of frequencies that the frequency will be swept. The y-axis is just an initial guess as to what the range of output gain might be. Because I know that the pole for this circuit is at 159 Hz, I will change the x-axis to sweep frequencies from 100 mHz to 10 MHz:

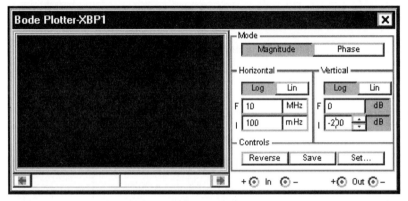

The Bode Plotter does not give us a very big window to view results, so we will not worry too much about the settings for the y-axis.

Click on the **Run/stop simulation** button to simulate the circuit. The Bode Plotter will trace out the Bode magnitude plot, which is the ratio of the magnitude of the voltage measured by the **out** terminals divided by the magnitude of the voltage measured by the **in** terminals:

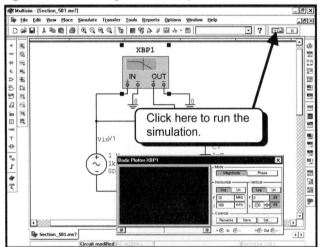
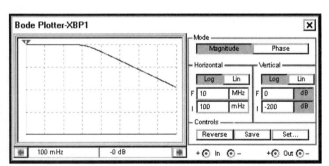

We see that the y-axis range was not a good guess because the plot traverses only about half the range of the y-axis. Also, the Bode Plotter does not number the axes. To get a better view of the results, we will use the Grapher.

Before we use the Grapher, we must close the Bode Plotter window. Click the *LEFT* mouse button on the ⊠ in the upper right corner of the Bode Plotter window to close the window:

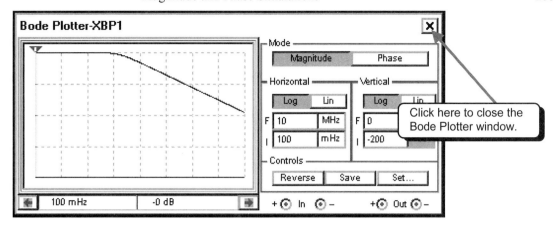

We can now view the results with the Grapher. Select **View** and then **Grapher** from the Multisim menus:

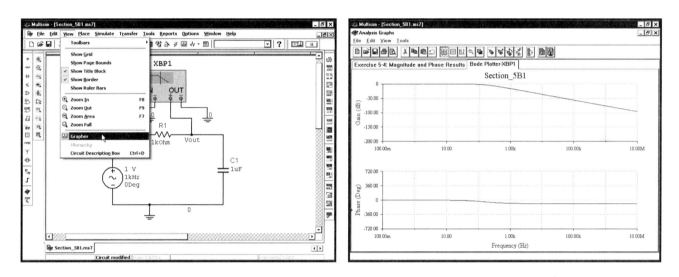

We can now modify the plot using the techniques covered in section 2.F. To change the properties of a plot, click the **RIGHT** mouse button on the plot you wish to change. For example, click the **RIGHT** mouse button on the lower plot and then select **Properties**:

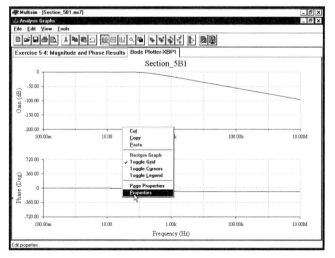
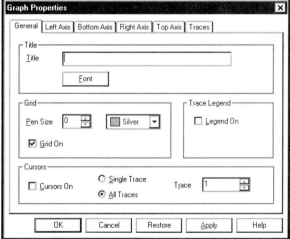

Use this dialog box to turn on the grid (if it is not already on), set the title to "Low Pass Filter – Bode Phase Plot," change the left axis range to –90 to 0 degrees, and specify a trace pen size of 2. When done, click on the **OK** button:

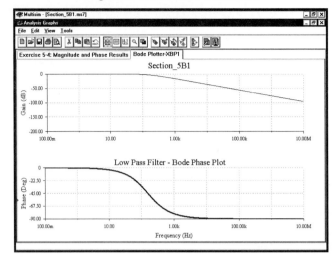

To modify the top plot, click the **RIGHT** mouse button on the top plot and select **Properties**. Make the same changes as we did for the phase plot except change the title to "Low Pass Filter – Bode Magnitude Plot," and change the y-axis range to –100 to 0 dB:

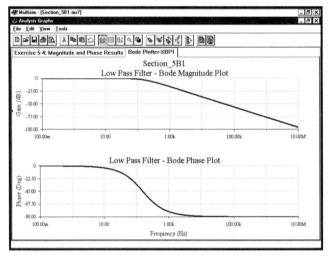

We can also use the cursors on the top plot to find the –3 dB frequency. To display the cursors, click the **RIGHT** mouse button on the top plot and select **Toggle Cursors**:

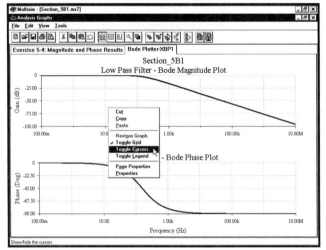
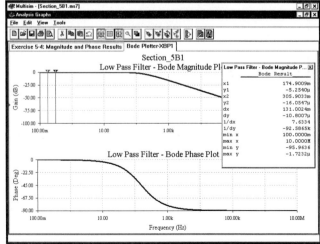

Place cursor 1 (red) so that it reads a **y1** value close to 0 dB. Place cursor 2 (blue) so that the cursor display box reads a **dy** value close to –3 dB:

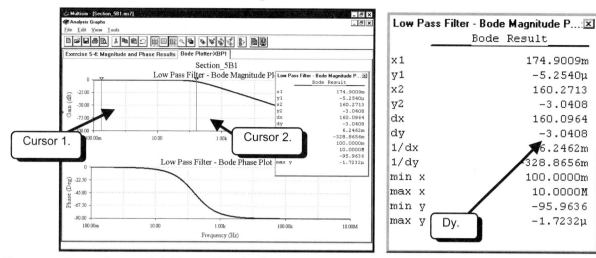

The x coordinate of cursor 2 (**x2**) specifies the frequency coordinate of cursor 2. This value is the –3 dB frequency and is approximately 160 Hz in this example. Our calculated value was 159 Hz. The values differ slightly because we could not hit the –3 dB point exactly with the cursors.

5.B.2. Bode Plots Using the AC Analysis

We will now simulate the same circuit using the SPICE AC Analysis. Wire the circuit below, or use the circuit used in the previous example. Note that we have removed the Bode Plotter instrument because it is unnecessary when running the AC Analysis.

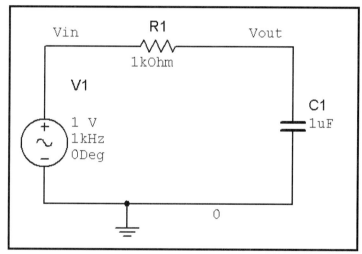

We first need to set up voltage source V1 for the sweep. To do this, double-click the **LEFT** mouse button on the graphic:

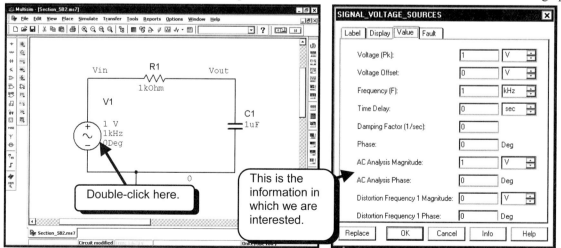

We are not running any of the distortion analyses, so we can ignore the distortion options of this dialog box. The **AC Analysis Magnitude** and **AC Analysis Phase** options specify the parameters of this source that are used when an AC analysis is run. For the AC Analysis, the magnitude of the source will be 1 volt and its phase will be zero degrees. In phasor notation V1 is 1∠0°.

Because the gain of this circuit is the ratio of the voltage at node Vout to the voltage at node Vin, a 1-volt source is as good a choice as any, and it also makes the ratio easier for humans to calculate. We do not need to change any of the parameters for this run. However, now that we have seen these properties, we know where to find them when we do need to change them. Click on the **OK** button to return to the schematic.

Next, we need to set up and then run the AC Analysis. Select **Simulate**, **Analyses**, and then **AC Analysis** from the Multisim menus:

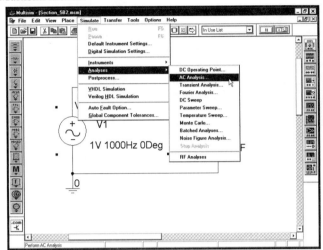
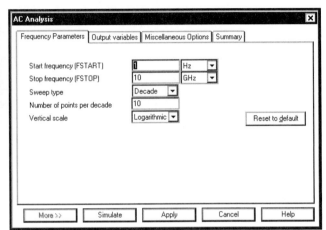

This first tab allows us to specify the range of frequencies for the sweep, the sweep type, and the number of points in the sweep. A decade is a factor of 10 in frequency. For example, 1 Hz to 10 Hz is a decade, 10 Hz to 100 Hz is a decade, 100 Hz to 1000 Hz is a decade, and so on. When you specify a sweep in **Decade**s, you also specify the number of points per decade. For example, with the current settings, there will be 10 simulation points between 1 and 10 Hz, 10 simulation points between 10 Hz and 100 Hz, and so on.

You can also specify sweep types of Octave and Linear. An octave is a factor of two in frequency, 1 Hz to 2 Hz, 2 Hz to 4 Hz, and so on. For an octave sweep type, you specify the number of points per octave. For a linear sweep type, you specify the total number of points in the sweep. The linear sweep is useful for frequency ranges where the starting frequency is of the same order of magnitude as the ending frequency, say 1 kHz to 5 kHz. A decade sweep is useful for wide frequency ranges that span several orders of magnitudes, say 1 Hz to 1 MHz.

We will set up this sweep with the parameters we specified in the Bode Plotter setup on page 254:

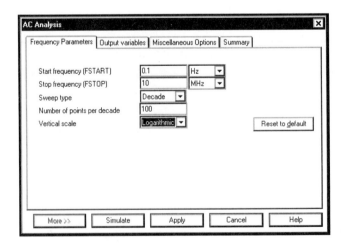

The dialog box is set to sweep frequencies from **0.1 Hz** (100 mHz) to **10 MHz** in decades, with **100** points per decade.

The last thing we need to do is to set the **Vertical scale** for the magnitude plot. Click the **LEFT** mouse button on the down triangle ▼ as shown:

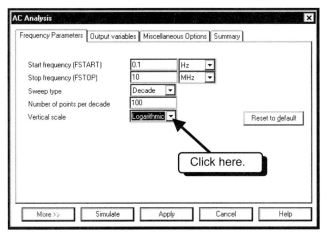 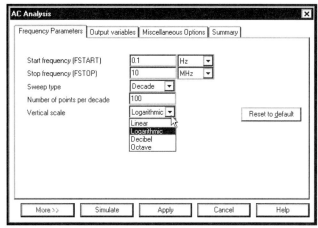

The **Linear** option specifies that the y-axis for the magnitude plot is to be a linear scale while the **Logarithmic** option specifies the y-axis as a log base 10 scale. With a log base 10 scale, factors of 10 are spaced the same distance apart. That is, the spacing between the numbers 1 and 10 is the same as the spacing between 10 and 100, which is the same as the spacing between 100 and 1000, and so on. The **Octave** option specifies that the y-axis is to be logarithmic in factors of 2. With a log base 2 scale, factors of 2 are spaced the same distance apart. That is, the spacing between the numbers 1 and 2 is the same as the spacing between 2 and 4, which is the same as the spacing between 4 and 8, and so on. The **Decibel** option specifies that the magnitude plot should be plotted in decibels ($20 \log 10 \left(v_{out} / v_{in} \right)$). In a standard Bode magnitude plot, the magnitude in decibels is plotted versus frequency, so we will select the **Decibel** option:

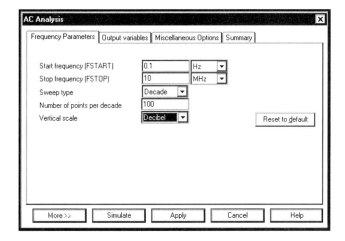

Next, we must specify the output variables for the simulation. Select the **Output variables** tab:

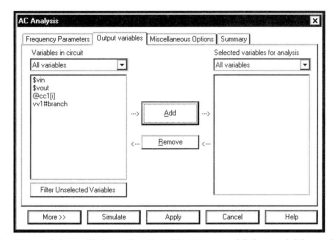

Click on the text **$vout** to select it, and then click on the **Add** button to add the variable to the right window pane:

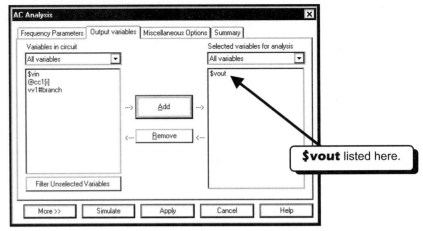

The data from node **Vout** will now be available after the simulation.

We are now ready to run the simulation, so click on the **Simulate** button:

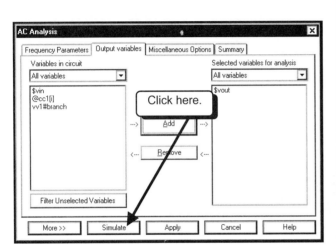
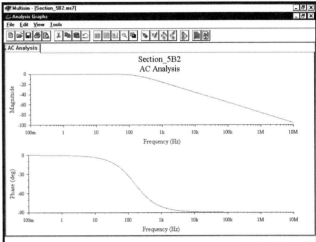

We see that the magnitude and phase plots of the voltage at node **Vout** are automatically displayed. These plots are nearly the same as the ones we obtained from the Bode Plotter shown on page 257. We can use the techniques shown on pages 255 to 257 to obtain the same plot as shown on page 257:

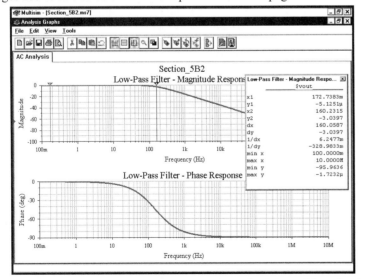

Low-Pass Filter - Magnitude Respo...	
	$vout
x1	172.7383m
y1	−5.1251μ
x2	160.2315
y2	−3.0397
dx	160.0587
dy	−3.0397
1/dx	6.2477m
1/dy	−328.9833m
min x	100.0000m
max x	10.0000M
min y	−95.9636
max y	−1.7232μ

The cursors show a −3 dB frequency of **160.2315** Hz. This value is nearly the same as that obtained from the previous section. Both −3 dB frequencies are slightly off the calculated value because we could not hit the −3 dB frequency exactly with the cursors. To come up with a closer value, we must specify more points per decade in our AC Analysis setup.

EXERCISE 5-5: Plot the Bode phase and magnitude plots for frequencies from 1 Hz to 1 MHz. Use the cursors to find the frequencies of the poles and zeros.

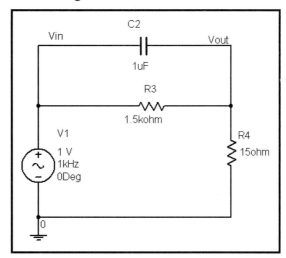

ANALYTIC SOLUTION: At low frequencies, the capacitor is an open circuit and the voltage at **Vout** can be obtained by the voltage divider of **R3** and **R4**:

$$\frac{V_{out}}{V_{in}} = \frac{R_4}{R_3 + R_4} = 0.0099 = -40 \text{ dB}$$

At high frequencies, the capacitor is a short circuit and **Vout** = **Vin** so

$$\frac{V_{out}}{V_{in}} = 1 = 0 \text{ dB}$$

Thus, we should expect the magnitude plot to start at −40 dB and finish at 0 dB. The next question is, where are the poles and zeros? Let

$$Z = Z_c \| R_3 = \frac{(1/j\omega C_2)R_3}{(1/j\omega C_2) + R_3} = \frac{R_3}{1 + j\omega R_3 C_2}$$

Substituting Z for $Z_c \| R_3$ yields the equivalent circuit:

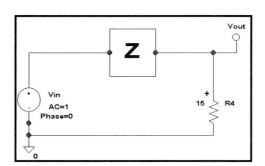

An expression for the gain can now be obtained from the voltage divider of **R4** and **Z**:

$$\frac{V_{out}}{V_{in}} = \frac{R_4}{Z + R_4} = \frac{R_4}{\left(\dfrac{R_3}{1 + j\omega R_3 C_2}\right) + R_4}$$

Multiplying the numerator and denominator by $1 + j\omega R_3 C_2$ gives

$$\frac{V_{out}}{V_{in}} = \frac{R_4(1 + j\omega R_3 C_2)}{R_3 + R_4 + j\omega C_2 R_3 R_4}$$

We see that the zero is at $1 = \omega R_3 C_2$, or

$$\omega_z = \frac{1}{R_3 C_2} = \frac{1}{(1500\ \Omega)(1\ \mu F)} = 666\ \text{rad/s} = 106\ \text{Hz}$$

The pole is at $R_3 + R_4 = \omega C_2 R_3 R_4$, or

$$\omega_p = \frac{R_3 + R_4}{C_2 R_3 R_4} = \frac{1}{C_2 \left(\dfrac{R_3 R_4}{R_3 + R_4}\right)} = \frac{1}{C_2 (R_3 \| R_4)}$$

$$= \frac{1}{(1\ \mu F)(1500\ \Omega \| 15\ \Omega)} = 66.6\ \text{krad/sec} = 10.6\ \text{kHz}$$

SOLUTION: Use the Bode Plotter and specify frequencies from 1 Hz to 1 MHz. Plot the results with the Grapher and use the cursors to locate the numerical values:

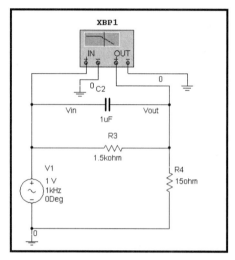

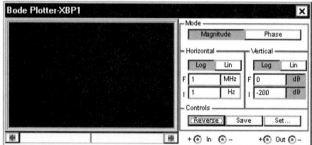

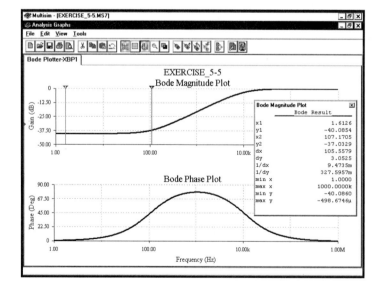

The cursors above show the location of the zero to be at approximately **107** Hz, close to the calculated value of 106 Hz. Next, we use the cursors to find the location of the pole:

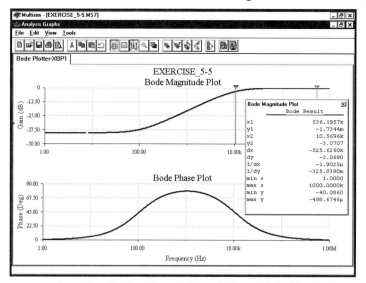

The cursors show that the location of the pole is at **10.57 kHz**, close to the calculated value of 10.6 kHz.

EXERCISE 5-6: Plot the Bode phase and magnitude plots for frequencies from 1 Hz to 1 MHz. Use the cursors to find the frequencies of the poles and zeros.

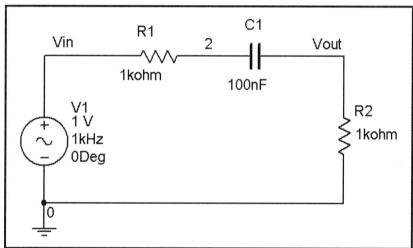

ANALYTIC SOLUTION: At low frequencies, the capacitor is an open circuit and the voltage at **Vout** is zero. As the frequency is increased, the impedance of the capacitor will decrease and the voltage at **Vout** will increase. Thus, at low frequencies we should expect the gain in decibels to be a large negative number and to increase at a rate of 20 dB/decade. At high frequencies, the capacitor is a short circuit and **Vout** can be obtained from the voltage divider of R1 and R2:

$$\frac{V_{out}}{V_{in}} = \frac{R_2}{R_2 + R_1} = 0.5 = -6 \text{ dB}$$

Thus, we should expect the magnitude plot to start at a negative value in decibels and finish at –6 dB. The next question is, where are the poles and zeros? The gain of the circuit can be obtained from the voltage divider of **R1**, **R2**, and **C1**:

$$\frac{V_{out}}{V_{in}} = \frac{R_2}{R_1 + R_2 + Z_c} = \frac{R_2}{R_1 + R_2 + \left(\frac{1}{j\omega C_1}\right)} = \frac{j\omega C_1 R_2}{1 + j\omega C_1 (R_1 + R_2)}$$

We see that there is a zero at ω = 0 rad/sec and a pole at 1 = ωC₁(R₁ + R₂), or

$$\omega_p = \frac{1}{C_1(R_1 + R_2)} = \frac{1}{(100 \text{ nF})(1000 \text{ }\Omega + 1000 \text{ }\Omega)} = 5000 \text{ rad/s} = 795.8 \text{ Hz}$$

SOLUTION: Set up an AC Analysis to sweep the frequency from 1 Hz to 1 MHz at 100 points per decade.

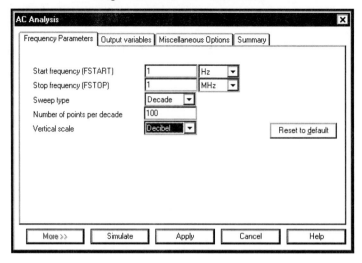

Plot **Vout** and use the cursors to locate the –3 dB point.

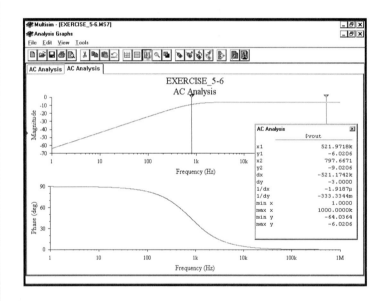

The cursors show the location of the pole to be at about **797.7** Hz. This is close to the calculated value of 795.8 Hz.

5.C. Amplifier Gain Analysis

One of the most important applications of the AC Analysis and the Bode Plotter instrument is to see the frequency response of an amplifier. If an AC Analysis or Bode Plotter is used on a circuit with a transistor, the DC bias point is calculated and the transistor is replaced by a small-signal model around the bias point. The frequency response is then measured using the linearized model of the transistor. The AC Analysis and the Bode Plotter instrument can only be used to find the small-signal gain and frequency response. Voltage swing, clipping, and saturation information must be obtained from the transient simulation, or by using the operating point information.

For this example, we could use either the AC Analysis or the Bode Plotter instrument. Both should yield the same results. For this example, we will use the Bode Plotter instrument. Wire the amplifier circuit as shown:

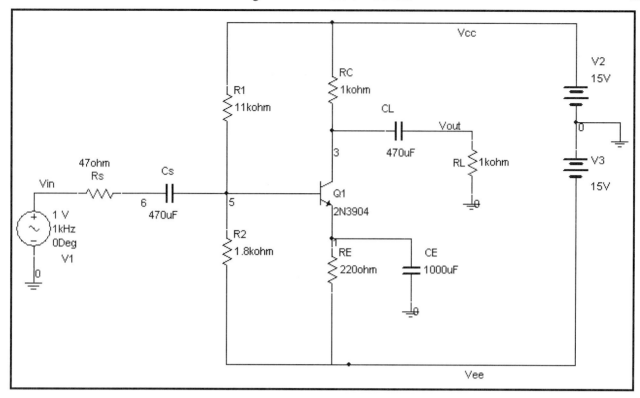

To place the 2N3904 NPN BJT transistor. Click on the **Transistor** button,

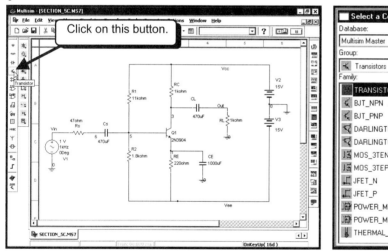 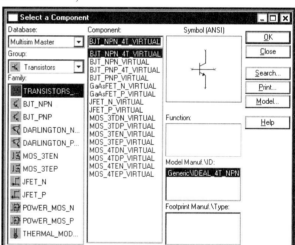

Click on the text **BJT_NPN** to view the list of NPN transistors available:

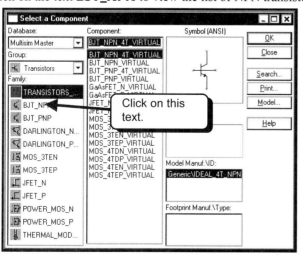 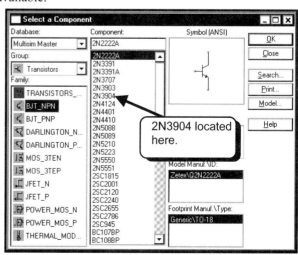

You can now select a 2N3904 and place it in your circuit.

We would like to see the gain of this circuit for frequencies from 1 Hz to 100 MHz. Place the Bode Plotter instrument in your schematic and wire it to nodes **Vin** and **Vout**:

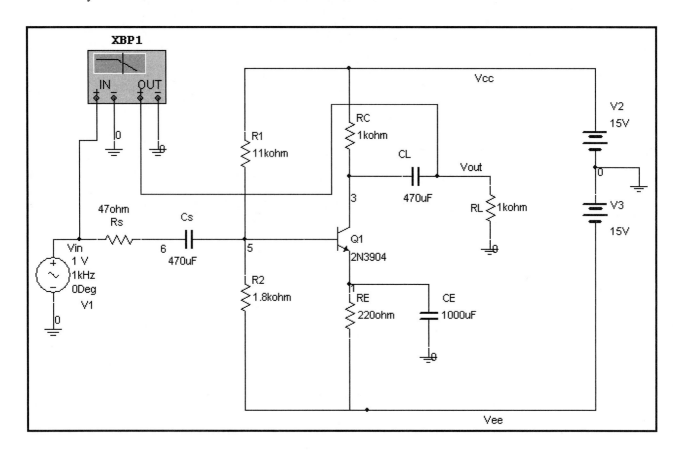

We must now change the parameters for the Bode plot. Double-click on the Bode Plotter instrument to bring up its window:

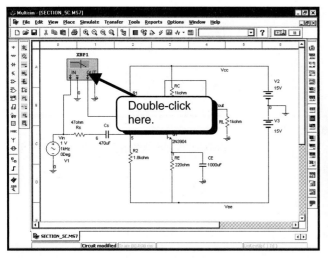
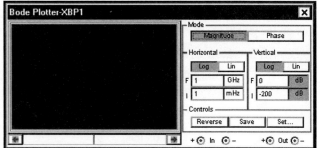

The instrument is set to sweep frequency from **1 mHz** to **1 GHz**. For our Bode plot, we would like to simulate the circuit from 1 Hz to 100 MHz, so modify the dialog box as shown:

Multisim 7 — Magnitude and Phase Simulations

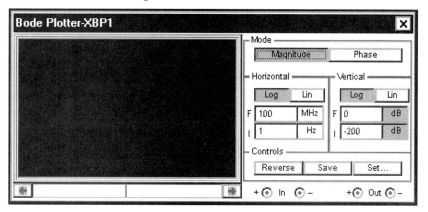

We are now ready to run the simulation, so click on the **Run/stop simulation** button to start the simulation. The Bode magnitude plot for this amplifier will be displayed on the Bode Plotter instrument screen:

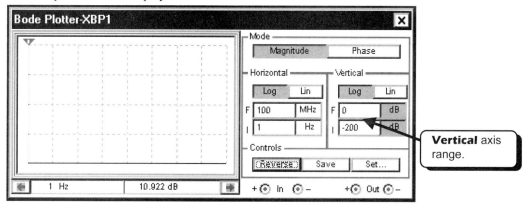

Where is the plot? It turns out that the default range for the **Vertical** axis (y-axis) is from –200 dB to 0 dB. The gain of this amplifier at midband is about 45 dB. This value is out of range for the **Vertical** scale. Change the **Vertical** scale to range from 20 dB to 55 dB. As you change the scale, the Bode plot will appear within the **Bode Plotter** instrument window:

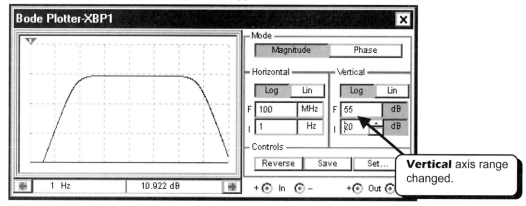

If we move the cursor to the center of the plot, we measure a gain of 45.715 dB, close to the expected value:

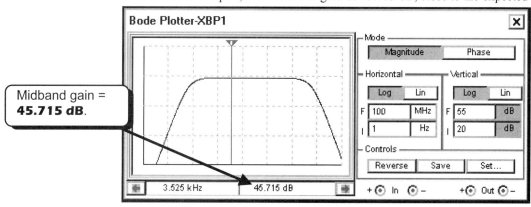

Midband gain = **45.715 dB**.

We can use the cursor to locate the upper and lower −3 dB frequencies of this plot. The midband gain is 45.7 dB, so the −3 dB frequencies will be the frequencies where the gain is 3 dB less than the midband gain, or 42.7 dB. We can locate these two points with the cursor:

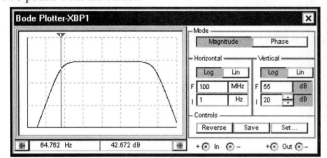 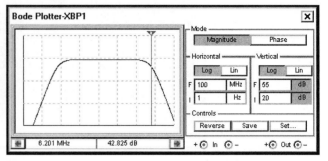

We see that the lower −3 dB frequency occurs at **64.762 Hz** and the upper −3 dB frequency occurs at **6.201 MHz.**

EXERCISE 5-7: Find the mid-band gain and the upper and lower −3 dB frequencies of the common-base amplifier shown below.

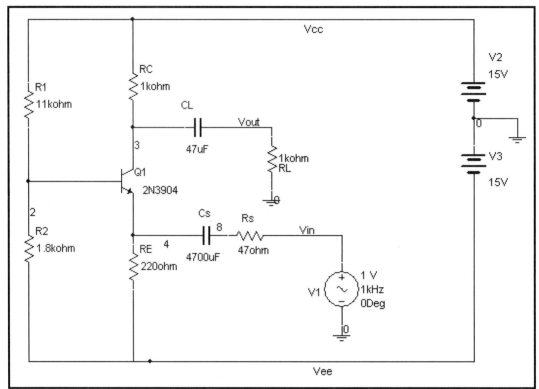

SOLUTION: Add the Bode Plotter instrument and simulate the circuit for frequencies from 10 mHz to 100 MHz:

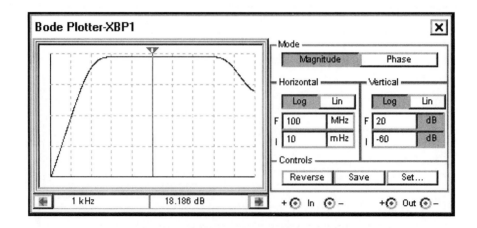

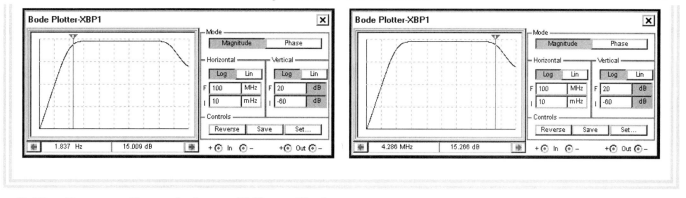

5.D. Operational Amplifier Gain

In this section we will use Multisim to determine the gain and bandwidth of an operational amplifier circuit with negative feedback. Since we used the Bode Plotter in the last section, we will use an AC Analysis for this example. We will find the gain and the upper –3 dB frequency for the circuit below:

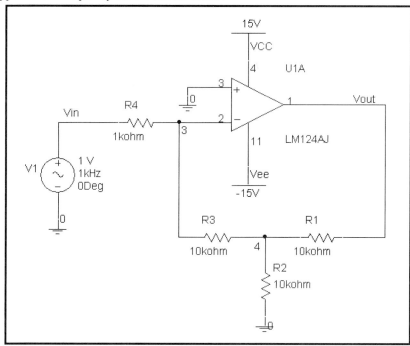

This circuit contains a few unfamiliar parts, so we will show how to draw the circuit here. First, we will place the LM124AJ OPAMP. To place this part, click the **LEFT** mouse button on the **Analog** button as shown:

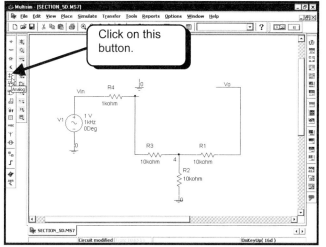
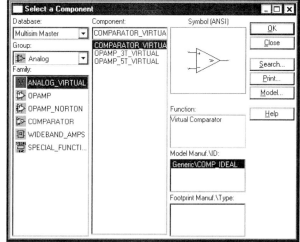

Select the **OPAMP Family** to see the list of available OPAMPs:

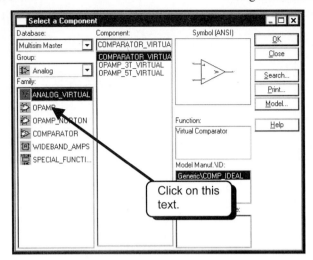
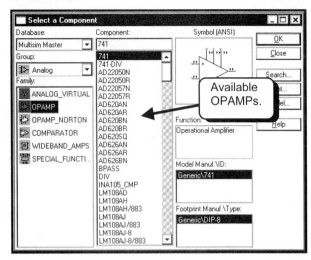

Locate and select the **LM124AJ** in the **Component** pane:

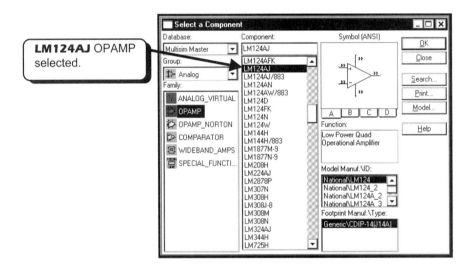

Click on the **OK** button to place the part:

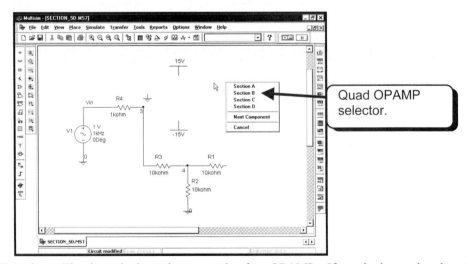

The LM124 is a quad OPAMP package. That is, a single package contains four OPAMPs. If you look at a datasheet for an LM124, you will see that each package contains four OPAMPs designated A, B, C, and D. This menu allows us to select a specific OPAMP in the quad package. For simulations, the OPAMP we select is not important, so select one of the OPAMPs and place the part in your circuit:

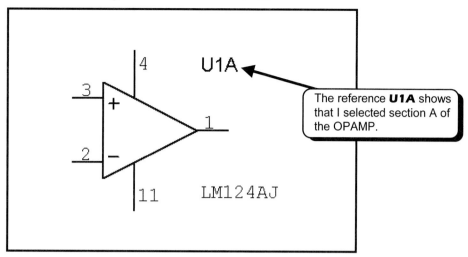

The next thing we will do is place the DC supplies in the circuit. We could do this using standard DC supplies as shown below (don't do this):

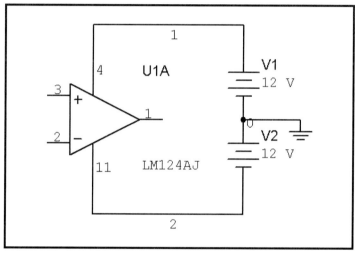

This method works fine, but it does clutter up the schematic a bit. Instead of using these DC supplies, we will use a part called Vcc. Close the analog parts toolbar and then click on the **Show Power Source Components Bar** button :

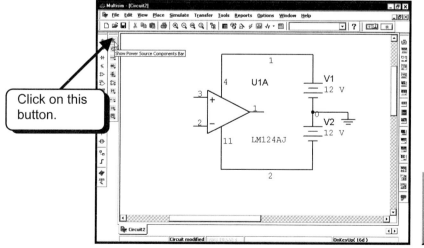

Click on the **Vcc** button  to select the part:

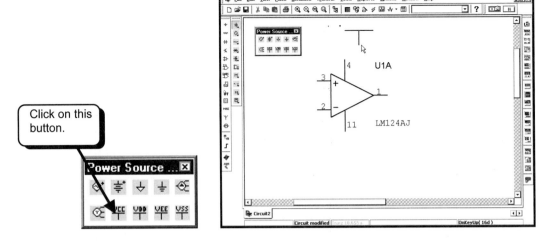

Place two copies of the part as shown:

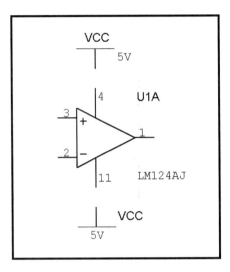

The first thing we will do is change the properties of the top **VCC** part. Double-click the *LEFT* mouse button on the graphic as shown below:

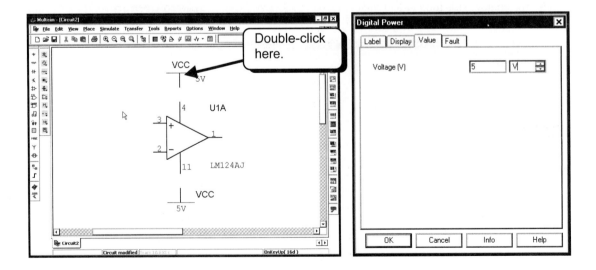

The **Value** tab sets the voltage of the node at **5** volts. Change the value to **15** volts:

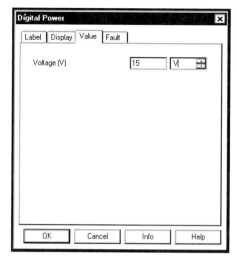

Next, we want to change how the part displays its information. Select the **Display** tab:

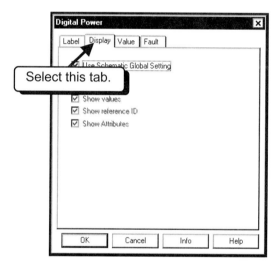

This part uses the Multisim defaults for specifying what information to display. Deselect the option as shown:

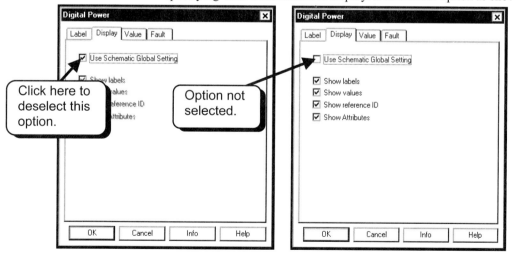

For our part, two values are displayed on the schematic. The **15V** is the value of the source and its display is enabled by selecting the **Show values** option, which is enabled. The text **VCC** is the reference ID for this part and its display is enabled by the **Show reference ID** option, which is also selected. We do not want to display the text VCC, so disable the **Show reference ID** option:

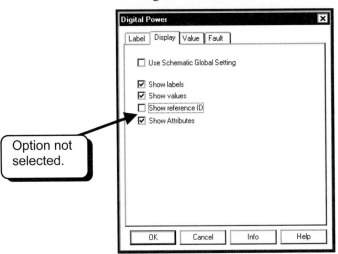

Click on the **OK** button to accept the changes and return to the schematic:

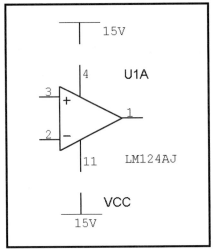

We see that the text VCC is no longer displayed and that the value of the part has changed to **15V**.

Next, we need to change the parameters of the lower **VCC** part. Change the value to **−15 V** and disable the **Show reference ID** option as shown below:

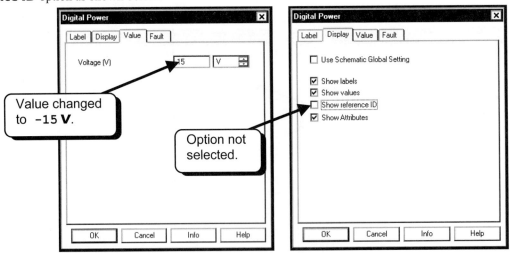

The last thing we need to change is the reference ID for this part. If we do not change the reference ID, both nodes will be labeled as VCC and will be electrically connected together. This would cause problems, so we will change the reference ID. Select the **Label** tab:

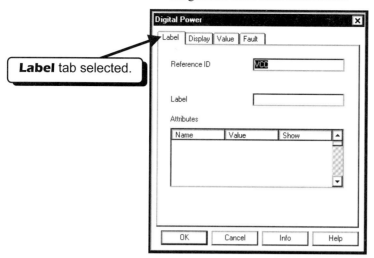

Change the **Reference ID** to **Vee**:

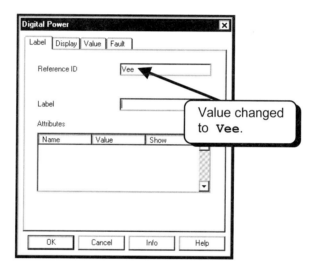

Click on the **OK** button to accept the changes:

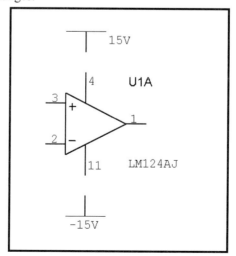

Connect the two parts to your circuit and move the text to make the schematic more readable:

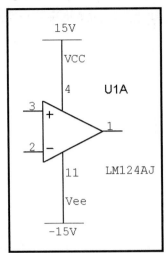

On my screen capture the text **VCC** and **Vee** is displayed. This is because I have selected the global option to display node names on my circuit so that we can easily label nodes and see their labels. You may not have this option selected, so you may not see the **VCC** and **Vee** labels. If you cannot see the labels, you can double-click on a wire to see its properties. For example, if I double-click on the wire as shown below, you will see information about the wire:

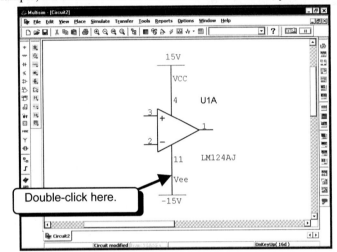
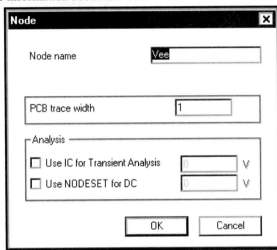

We see that the node is named **Vee**. If your nodes are not named differently (VCC and Vee) you need to double-click on the two parts again and change their reference IDs to make the parts different. Click on the **OK** button to close the window. The remainder of the circuit is similar to ones we have done before, so complete the circuit as shown:

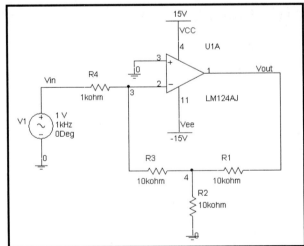

We would like to find the gain of this amplifier, V_{out}/V_{in}. Because the magnitude of V_{in} is one volt, the magnitude of V_{out} is the gain. If the gain of the amplifier is 200, then the magnitude of the output will be 200 V. This may seem a little unreasonable since the DC supplies are ±15 V. When an AC Analysis is performed, the operating points of all nonlinear parts are found, and the parts are replaced by their linear models. Because the models are linear, the voltages and currents in the circuit are not limited by the DC supplies. For a linear circuit, if an input of 1 V produces an output of 10 V, then an input of 1,000,000 V will produce an output of 10,000,000 V. Thus, for gain purposes the magnitude of the input does not matter and a 1 V input is chosen for convenience. For an AC simulation a magnitude of 200 V is not unreasonable. A magnitude of 200 V is not physically possible for our OPAMP, but it is a valid number when using the AC Sweep to calculate the gain. If you wish to observe the maximum voltage output of a circuit, you must run one of the Time Domain Analyses. An example of the voltage swing of a transistor amplifier is shown in Section 6.H on page 392. The same techniques could be used to find the voltage swing limits of this OPAMP circuit.

The next thing we need to do is set up the AC source to be used in the AC Analysis. Double-click the **LEFT** mouse button on the sinusoidal source. You should set the **AC Analysis Magnitude** to 1 V and the **AC Analysis Phase** to 0 degrees. Your dialog box should match the one shown below:

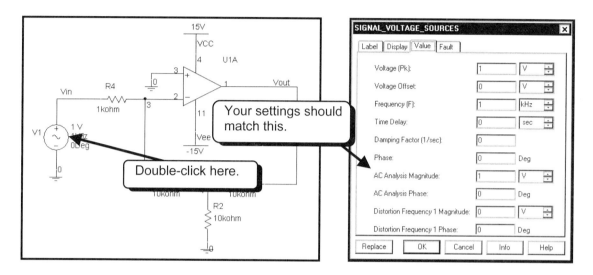

These settings specify that the source we selected will be used in the AC Analysis and that it will have a magnitude of 1 V and a phase of zero degrees. These are the default settings, but it is a good idea to check these settings before running an AC Analysis. We are not going to be running a distortion analysis, so the other settings under this tab are not relevant. Click on the **OK** button to return to the schematic.

Next, we need to set up the simulation. Select **Simulate**, **Analyses**, and then **AC Analysis** from the Multisim menus:

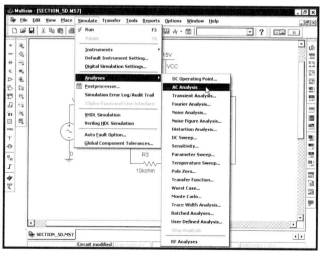
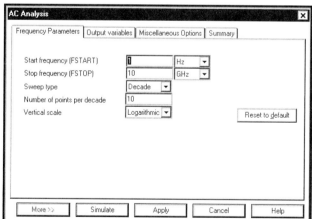

We would like to simulate this circuit from 1 Hz to 10 MHz. Because our frequency range spans several orders of magnitude in frequency (several decades of frequency), we will use the **Decade Sweep type**. To get a detailed plot, we will choose 100 **points per decade**. For a Bode plot, we would like the y-axis to be plotted in decibels, so we will select **Decibel** for the **Vertical scale**:

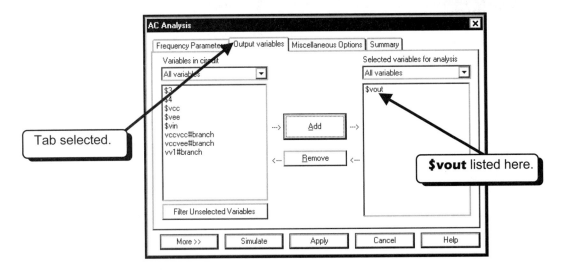

The last thing we must do is specify the variable we would like to plot. Select the **Output variables** tab and specify **$vout** as the output variable to plot during the simulation:

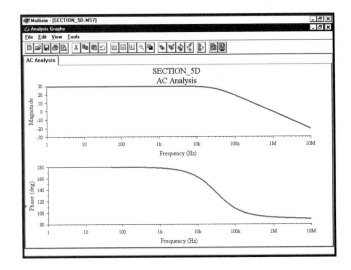

We are now ready to run the simulation, so click on the **Simulate** button. The following screen appears:

We can now use the cursors to find the midband gain and the upper −3 dB frequency. Click the **RIGHT** mouse button on the upper plot and then select **Toggle Cursors**:

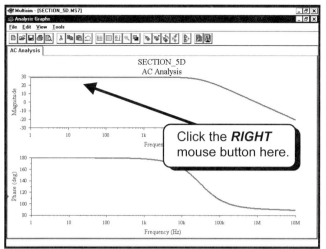
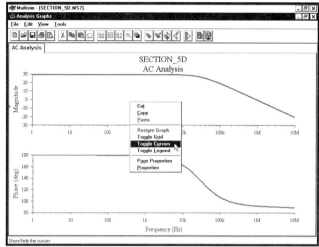

Use cursor 1 to find the low-frequency gain and use cursor 2 to find the frequency where the gain is down by 3 dB from the low-frequency gain:

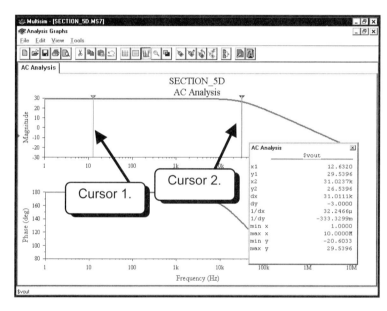

Coordinate **y1** shows us that the gain at cursor 1 is **29.5396** dB, or about 30. **dy** is the difference between the y-coordinates of cursor 1 and cursor 2. Thus, when **dy** = –3, we know that the gain at cursor 2 is 3 dB less than the gain at cursor 1 and that cursor 2 is placed at the –3 dB frequency. Coordinate **x2** shows us that the 3 dB frequency is **31.0237 k**Hz.

As a side note, we see that the midband gain of this amplifier is 30, and the –3 dB frequency is 31.0237 kHz. The gain-bandwidth product of this amplifier is:

$$GBW = (30)(31.024 \text{ kHz}) = 930{,}720 \text{ Hz}$$

This agrees well with the GBW specification of the LM124, which is specified as 1 MHz.

EXERCISE 5-8: Find the midband gain and the upper –3 dB frequency for the amplifier shown below, and find the gain-bandwidth product (GBW) from the simulated results.

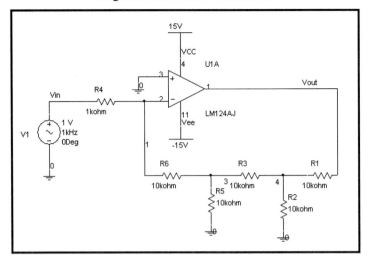

SOLUTION:

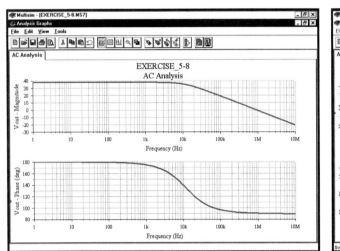
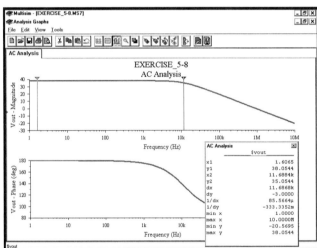

AC Analysis	$vout
x1	1.6065
y1	38.0544
x2	11.6884k
y2	35.0544
dx	11.6868k
dy	−3.0000
1/dx	85.5664µ
1/dy	−333.3352m
min x	1.0000
max x	10.0000M
min y	−20.5695
max y	38.0544

The midband gain is **38.0544** dB, which corresponds to a gain of 79.932. The upper −3 dB frequency is **11.6884 kHz**. The gain-bandwidth product is

$$GBW = (79.932)(11.6884 \text{ kHz}) = 934{,}277 \text{ Hz}$$

This value is very close to the value from the previous example. This is expected since the GBW product should be approximately constant for all circuits made from the same OPAMP.

5.E. Parameter Sweep — OPAMP Gain Bandwidth

In this section we will use Multisim to determine the bandwidth of an OPAMP circuit with varying amounts of negative feedback. For an OPAMP circuit, the product of the closed-loop gain times the bandwidth is approximately constant. To observe this property, we will run a simulation that creates a Bode plot for several different closed-loop gains. We will use the circuit below:

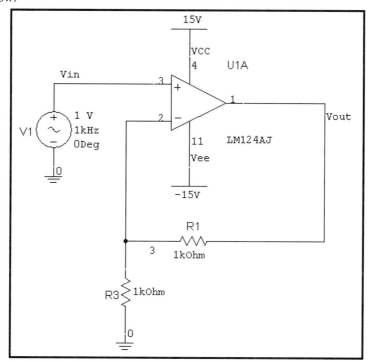

This circuit uses an **LM124AJ** OPAMP macro model. All of the real OPAMP models include bias currents, offset voltages, slew rate limitations, and frequency limitations. Also note that real OPAMP models require DC supplies. The models will not work without the supplies.

We would like to see the frequency response of the OPAMP circuit for different values of the gain, so we must set up a Parameter Sweep and then specify the analysis type to be an AC Analysis. Select **Simulate**, **Analyses**, and then **Parameter Sweep**:

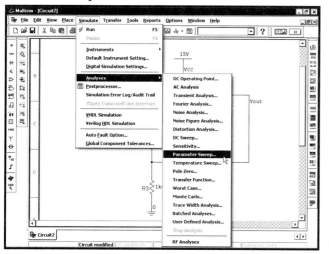
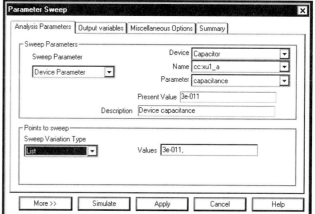

The **Sweep Parameter** field can be either a **Device Parameter** or a model parameter. A **Device Parameter** is something like the resistance of a resistor, the inductance of an inductor, the temperature coefficient of a capacitor, and so on. A model parameter would be a parameter in a diode or transistor model. Examples would be the breakdown voltage of a Zener diode (model parameter BV) or the forward current gain of a bipolar junction transistor (model parameter BF).

We would like to observe the frequency response of this OPAMP circuit for different gains. The gain of this circuit is $1 + \dfrac{R_1}{R_3}$, so we can change the gain by changing the value of R_1 or R_3. To change the gain, we will change the value of R_1, so we will use the Parameter Sweep to change the value of the resistance of R_1. Resistance of a resistor is a **Device Parameter**, so the **Sweep Parameter** should be set to **Device Parameter** as shown above. The **Device** field lists all of the available devices we can specify such as diodes, transistors, capacitors, and resistors. Select the **Resistor Device**.

Once you select the **Resistor Device**, all available resistors in your circuit will be listed in the **Name** field:

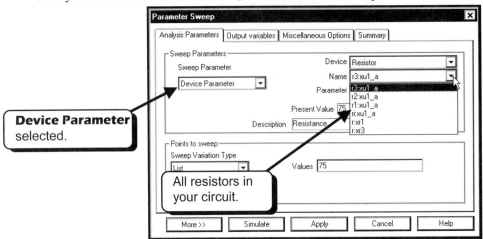

The pull-down menu lists many more resistors than we see on the schematic. This is because this list includes all of the resistors used in the OPAMP subcircuit. Resistors with the text **xu1_a** are inside the subcircuit for the OPAMP because this OPAMP was labeled U1A in the circuit. Thus, the only two resistors that we actually placed in our circuit, R1 and R3, are labeled as **r:xr1** and **r:xr3** in this list. We want to vary R1 to select **r:xr1**:

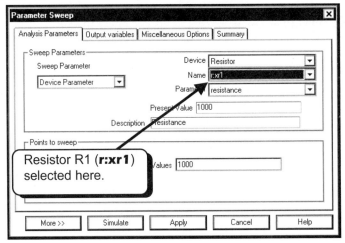

Next, we need to specify what parameter of the resistor we want to sweep. The **Parameter** field lists all of the items we can sweep for a resistor:

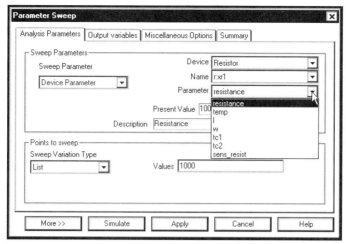

We see that we can sweep the temperature at which this resistor operates (**temp**), the length and width of the resistor (**l** and **w** – used for integrated circuits and film resistors), and other properties as well. We want to change the resistance, so select **resistance**. Note that the **Description** field describes the function of the selected **Parameter**:

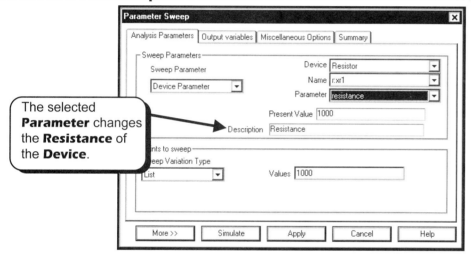

Next, we need to specify the values we want R1 to have. We will use the **List Sweep Variation Type** and specify the values for R1 to be 1000, 10000, 100000, 1000000. Values are separated by commas:

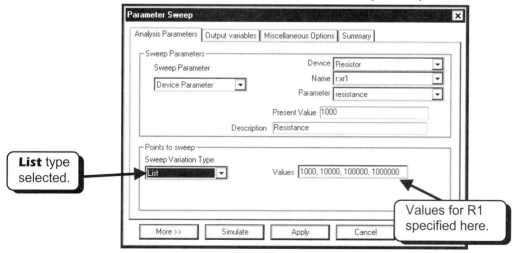

Note that we could have specified the same values for R1 using a Decade **Sweep Variation Type**.

We have now finished the setup for the **Parameter Sweep**. The next thing we must do is specify that this **Parameter Sweep** should run in conjunction with an AC Analysis. Click on the **More** button:

284 Magnitude and Phase Simulations Multisim 7

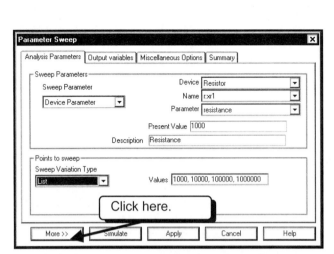
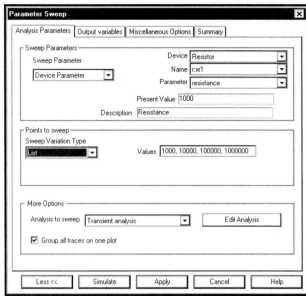

The default **Analysis to sweep** is a **Transient analysis**. To find the frequency response we need to run an AC Analysis. Select the **AC analysis** and also specify that we want to **Group all traces on one plot**. Selecting this option will allow us to view all of the results from the **Parameter Sweep** on a single plot. Select options as shown below. Make sure that your dialog box matches all of the options selected below:

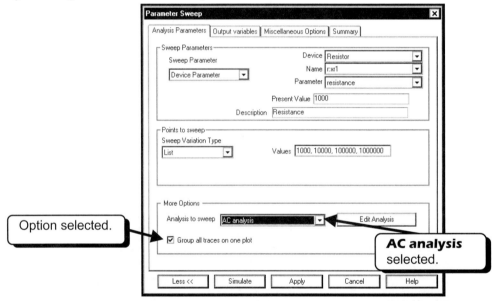

Next, we must set up the parameters for the AC Analysis. Select the **Edit Analysis** button as shown below and set up the parameters for the **AC analysis** as shown. This setup is the same as we have done for analyses in section 5.D:

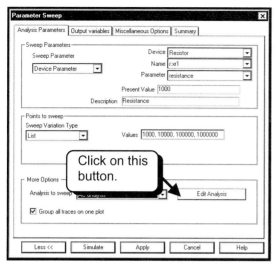
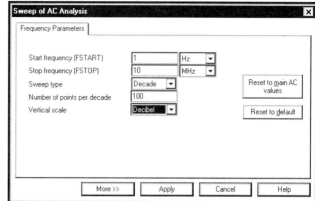

Click on the **Apply** button to return to the **Parameter Sweep** dialog box.

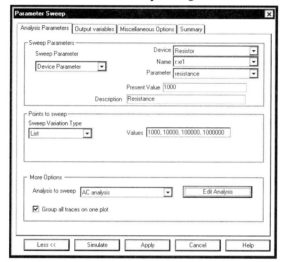

We have specified the following simulation. First, R1 will be set to 1,000 Ω and then the AC Analysis will sweep from 1 Hz to 10 MHz. Next, R1 will be set to 10,000 Ω and then the AC Analysis will sweep from 1 Hz to 10 MHz. Next, R1 will be set to 100,000 Ω and then the AC Analysis will sweep from 1 Hz to 10 MHz. Finally, R1 will be set to 1,000,000 Ω and then the AC Analysis will sweep from 1 Hz to 10 MHz. We will look at the results from all of the AC Analyses on a single plot.

The last thing we need to do is to specify the data we wish to plot. Select the **Output variables** tab and specify that **$vout** should be plotted during the simulation:

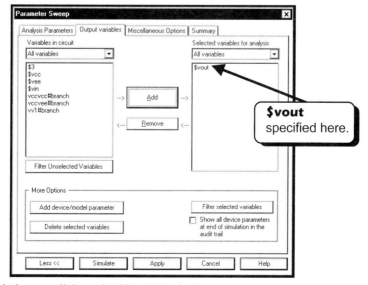

We are now ready to run the simulation, so click on the **Simulate** button:

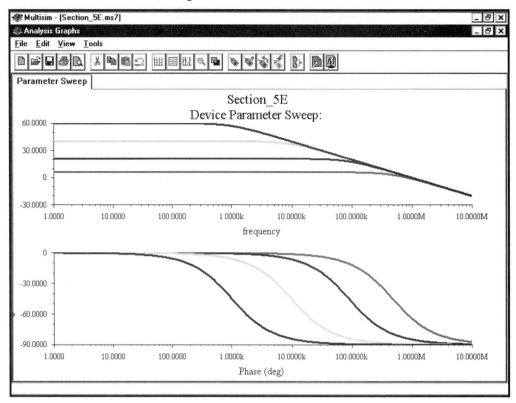

The top plot shows us the gain in decibels versus frequency. We see that higher gain rolls off at lower frequency. Using the cursors we find that for a gain of 60 dB (1000) the upper –3 dB frequency is 1 kHz. This corresponds to a gain bandwidth of 1 MHz (1000 times 1000). For a gain of 40 dB (100) the upper –3 dB frequency is 10 kHz. This also corresponds to a gain bandwidth of 1 MHz (100 times 10,000). For a gain of 20 dB (10) the upper –3 dB frequency is 91.2 kHz, corresponding to a gain bandwidth of 912 kHz (10 times 91,200). For a gain of 3 dB (2) the upper –3 dB frequency is 490 kHz, corresponding to a gain bandwidth of 980 kHz (2 times 490,000).

We see that for all of these gains, the gain-bandwidth product is relatively constant. This means that for high-gain OPAMP circuits the gain will roll-off at lower frequencies, and vice versa. This also means that we can trade gain for bandwidth for any particular OPAMP circuit. If we need a larger or smaller bandwidth, we can change the gain to achieve that bandwidth.

EXERCISE 5-9: For the inverting OPAMP below, find the upper –3 dB frequency for mid-band gains of –1, –10, –100, and –1000. Show that, except for the gain of –1, the magnitude of the product of the gain times the –3 dB bandwidth is relatively constant for all of the plots. This circuit with a gain of –1 is one of the few exceptions where the product of gain times bandwidth is not constant.

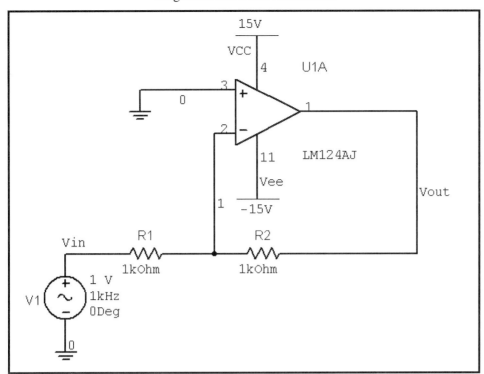

SOLUTION: Run the Parameter Sweep with the AC Analysis.

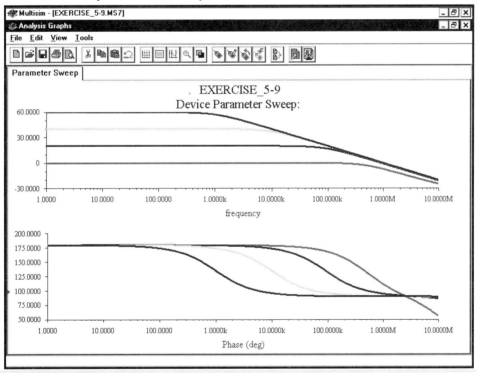

5.F. AC Power and Power Factor Correction

The AC power absorbed by a load and the power factor of the load can be easily measured using the Multisim Wattmeter. The Wattmeter can be used to measure both DC and AC power. In section 3.A.3, we demonstrated its use with the DC power absorbed by a resistor in a circuit with a DC source. Here we will demonstrate its use by finding the power absorbed by a complex load and the power factor of the load.

All linear loads can be reduced to an equivalent single impedance Z_{Load}, where Z_{Load} can be expressed as a magnitude and a phase, $Z_{Load} = Z_L \angle \theta_L$, or in rectangular coordinates as $Z_{Load} = A + jB$. In an AC circuit, the power absorbed by this load is

$$P_{Load} = \frac{V_M I_M}{2}\cos(\theta_L) = V_{RMS} I_{RMS} \cos(\theta_L)$$

where V_M and I_M are the peak values of the load voltage and current, and V_{RMS} and I_{RMS} are the RMS values of the load voltage and current. θ_L is the angle of the load and is also the angle between the load voltage and the load current. $\cos(\theta_L)$ is called the power factor (PF) and is a measurement of how inductive or capacitive a load appears. The Wattmeter displays the AC power absorbed by a load and the power factor.

As an example, we will calculate the power absorbed by the load below:

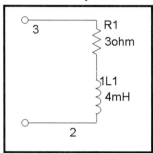

We will assume that this load is connected to the AC line where the voltage is 115 VRMS at 60 Hz. The impedance of this load at 60 Hz is

$$Z_{Load} = R_1 + j\omega L_1 = 3\Omega + j(2\pi 60\,\text{Hz})(0.004\,\text{H}) = (3 + j1.508)\Omega = (3.358\angle 26.687°)\Omega$$

If this load is connected to the 115 VRMS line, the power absorbed by this load is

$$P_{Load} = V_{RMS} I_{RMS} \cos(\theta_L) = V_{RMS}\left(\frac{V_{RMS}}{|Z_{Load}|}\right)\cos(\theta_L) = \frac{(115\,\text{V})^2}{3.358\Omega}\cos(26.687°) = 3.519\,\text{kW}$$

The power factor is

$$PF = \cos(\theta_L) = \cos(26.687°) = 0.893$$

We have now made a few simple calculations and we can compare them to results of Multisim simulations. We would like to measure the power absorbed by this load when connected to a 115 VRMS source as shown below:

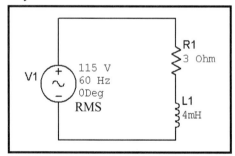

The voltage source used is the power voltage source, so the magnitude displayed is the RMS magnitude. We would like to measure the power absorbed by the load when connected to this source using the Wattmeter. This meter measures the load current and voltage and displays the power absorbed and the power factor.

To place the Wattmeter in your circuit, click the **LEFT** mouse button on the Wattmeter button in the Instruments toolbar. The Wattmeter will become attached to your mouse pointer:

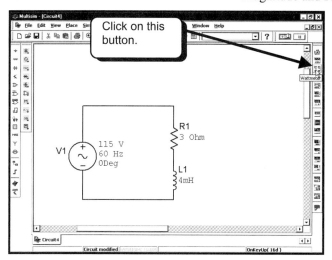 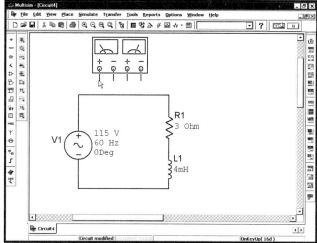

Place the Wattmeter in a convenient location in your circuit.

To measure the power absorbed by the load, we must measure the current through the load and the voltage across the load. Measuring current with the Wattmeter is the same as measuring current with a Multimeter or with a current indicator; if current enters the positive terminal and leaves the negative terminal, the meter will interpret it as a positive current. Wire the meter as shown:

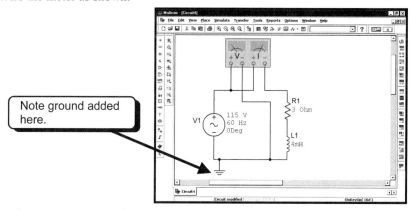

As shown, the meter measures the top terminal of the load as the positive voltage reference and the bottom terminal of the load as the negative reference. Current is measured as positive entering the top terminal of the load. Thus, the positive direction of current is entering the positive voltage terminal. Note that our equation for power, $P_{Load} = V_{RMS} I_{RMS} \cos(\theta_L)$, applies only if you follow the convention that the positive direction of current enters the positive voltage reference terminal.

One last thing to note in the circuit above is that we added a ground to the circuit. Although this is not necessary for the real circuit, Multisim and SPICE require all circuits to be grounded. If you do not ground your circuit, you will receive an error message.

Double-click the *LEFT* mouse button on the power meter to open its window:

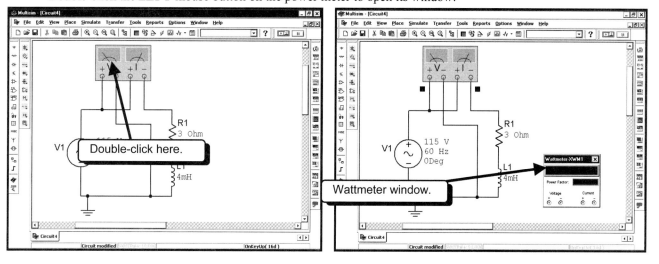

We are now ready to run the simulation. Select **Simulate** and then **Run** from the Multisim menus, press the **F5** key, or click on the **Run/stop simulation** button to start the simulation.

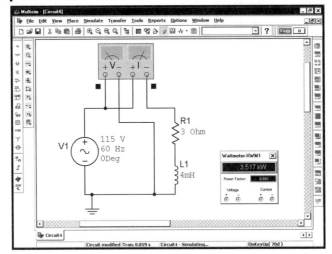

We see that the load absorbs **3.517 kW** of power and that the power factor is **0.893**. Both of these values agree with our calculated values.

5.F.1. Power Factor Correction

Now that we have found the power absorbed by the load and the power factor using calculations and simulation, we can look at correcting the power factor to unity. The basic idea behind power factor correction is that we can add an element in series or parallel with the load to cancel out the reactive portion of the load and make the load look pure real. A pure real load has a unity power factor. For the inductive load of the previous example, the easiest way to correct the power factor and make the load look pure real is to add a capacitor in parallel with the load.

Remember that admittances add in parallel. Because we are placing a capacitor in parallel with our load, the admittance of the capacitor will add to the admittance of the load. We must first calculate the admittance of our load:

$$Y_{Load} = \frac{1}{Z_{Load}} = \frac{1}{(3+j1.508)\Omega} = \frac{1}{(3.358\angle 26.687°)\Omega} = (0.298\angle -26.687)\text{S} = (0.266 - j0.134)\text{S}$$

The admittance of a capacitor is $Y_C = j\omega C$. When we place the capacitor in parallel with our load, the total admittance is

$$Y_{Total} = Y_{Load} + Y_C = (0.266 - j0.134) + j\omega C = 0.266 + j(\omega C - 0.134)$$

To make the load admittance pure real we need $\omega C - 0.134 = 0$, or $C = \frac{0.134}{\omega}$. Our example is at a frequency of 60 Hz, or $\omega = 2\pi 60$. For this frequency, the needed capacitor is 354.8 µF.

To check our calculations and method, we will simulate the circuit of the previous example with the addition of this parallel capacitor. Add the capacitor as shown:

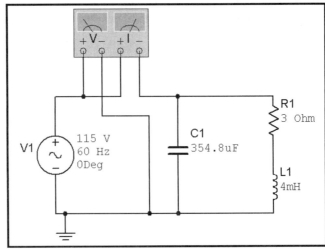

Select **Simulate** and then **Run** from the Multisim menus, press the **F5** key, or click on the **Run/stop simulation** button to start the simulation.

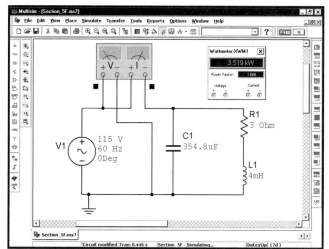
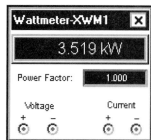

We see that the load absorbs the same amount of power as the previous simulation. This is the expected result because we placed the capacitor in parallel with the load, so the voltage across the load is the same for both simulations. If the voltage across the load is the same in both simulations, the power absorbed by the load must be the same. Capacitors and inductors do not absorb power; they only store energy, so placing a capacitor in parallel with our load will not change the power. We also see that the power factor is 1, so we accomplished our goal of creating a new load with a unity power factor.

EXERCISE 5-10: Calculate the power absorbed by the load and the power factor of the load in the circuit below. Assume that the frequency is 50 Hz and the voltage is 230 V RMS.

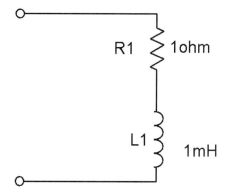

SOLUTION: Add an AC voltage power source to the circuit and set the voltage to 230 V RMS and the frequency to 50 Hz. Add the Wattmeter and wire it as shown. The results are displayed on the meter window:

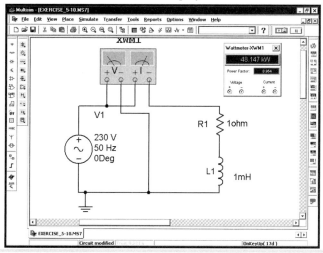
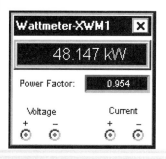

For power factor correction, we need to add a 910.2 μF capacitor in parallel with the load:

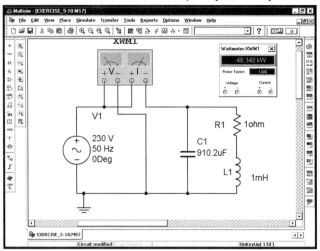
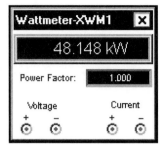

5.G. Measuring Impedance

The technique presented here can be used to measure the impedance or resistance between any two nodes. For purely resistive circuits, we could use the Multimeter. However, the Multimeter cannot measure the impedance of a circuit with capacitances or inductances, nor can it measure the input or output impedance of a circuit that contains active components such as transistors or OPAMPs. We will find the AC impedance between two nodes. We will illustrate using two examples. The first will be a passive circuit with resistors only. The second will be a jFET source follower. For the passive circuit, we will first measure the resistance using the Multimeter so that we can compare the result to our new measurement technique.

5.G.1. Resistive Measurement with the Multimeter

We will find the impedance between nodes **1** and **2** in the circuit below:

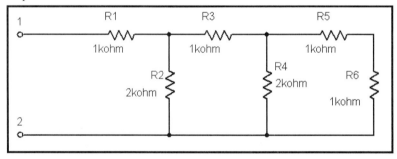

Because this is a resistive circuit, we can add the Multimeter instrument and measure the impedance. Click on the Multimeter button as shown to place the instrument, and wire it to your circuit as shown:

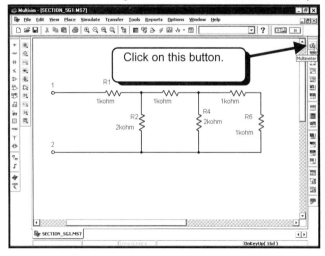
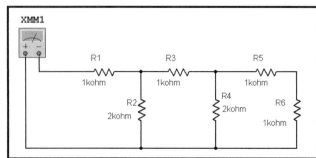

Double-click the **LEFT** mouse button on the instrument to open its window:

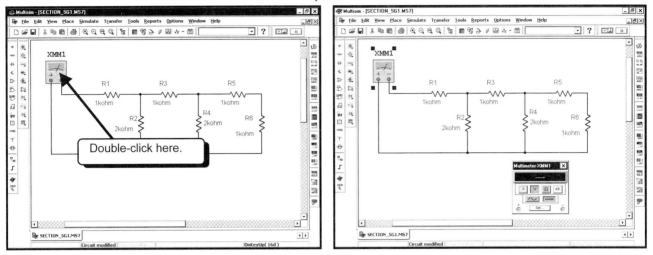

The last thing we must do is ground our circuit. Although it would not be necessary in the lab to ground a circuit before using the Multimeter to measure a resistance, it is necessary with MultiSim because Multisim must have a ground reference in order to calculate voltages in a circuit. Add a ground symbol as shown:

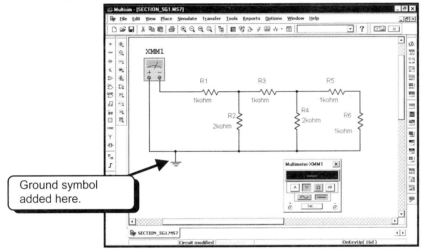

We can now run the simulation. Click the **LEFT** mouse button on the **Run/stop simulation** button to start the simulation:

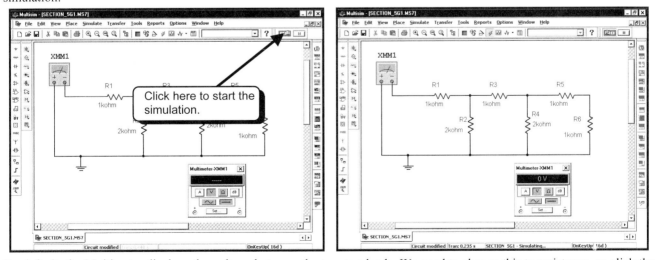

By default, the Multimeter displays the voltage between the two test leads. We need to change this to resistance, so click the **LEFT** mouse button on the Ω button to measure the resistance:

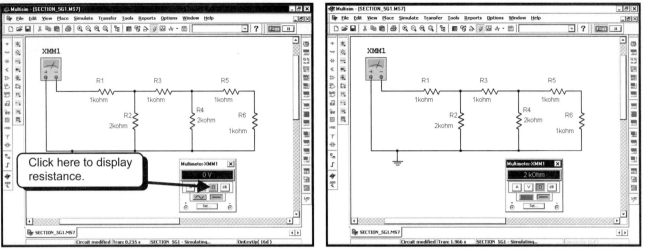

After a few moments the meter indicates that the resistance is 2 kΩ, which is the expected result. Click on the **Run/stop simulation** button again to stop the simulation.

Using the Multimeter is a very easy way to measure resistance, but it can only measure the resistance of a network of resistors. To measure anything more complicated, we will need to use the technique presented in the next section.

5.G.2. Impedance Measurement of a Passive Circuit Using SPICE

We will now use a new measurement technique to find the resistance of the resistor network used in the previous section. We use the same circuit here so that we can show that the two methods yield the same answer:

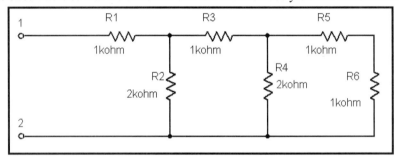

Add an AC current source as shown below. The test current source is added between nodes 1 and 2 because we are interested in the impedance between those two nodes. Either an AC current source or an AC voltage source can be used. Wire the circuit as shown. We have grounded one of the nodes because it is required for the simulation.

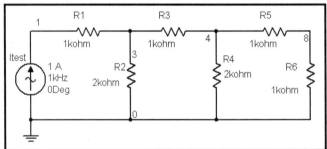

Although this circuit has no dependent sources, inductors, or capacitors, the circuit could contain any circuit element and we could still use this method to find the impedance. This circuit was chosen because the impedance is easily calculated and can be compared to the SPICE result or the result obtained with the Multimeter instrument. For this circuit, the calculated impedance is 2 kΩ.

Next, we need to set up and run the AC Analysis. Select **Simulate**, **Analyses**, and then **AC Analysis** from the menus:

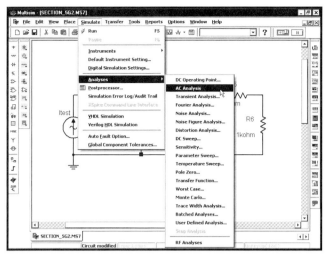
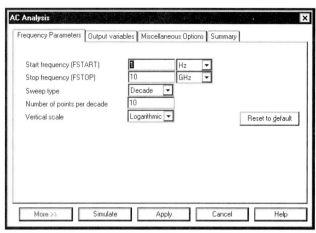

Because this is a resistive circuit, the frequency settings are not important and we will leave them set to their default values. The one parameter we will change is the **Vertical scale**. This is a passive circuit and the resistance will not change with frequency, so we will change the vertical scale to **Linear**:

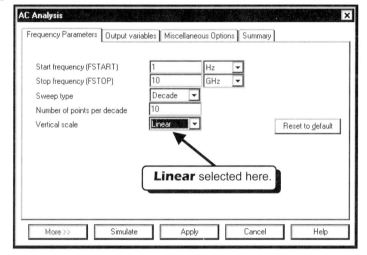

The resistance of our circuit is the voltage between node 1 and ground divided by the magnitude of the test current source, V(1)/I(I_test). Because we did not specify a value for the AC current source, its value will be 1 A. With this value of current, the resistance of the network is numerically equal to the voltage at node 1. Thus, we only need to plot the voltage at node 1.

The last thing we need to do is specify the output variables. We are only interested in the voltage at node **$1**, so select the **Output variables** tab and then specify to plot node **$1** during the simulation:

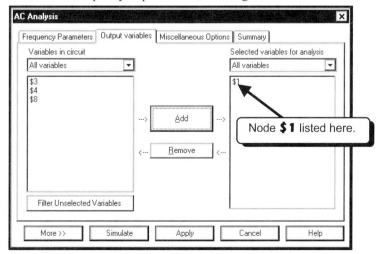

Click on the **Simulate** button to run the simulation. You will see this trace:

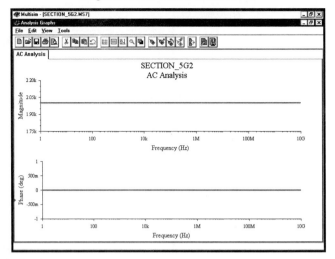

Use the cursors to find a numerical value:

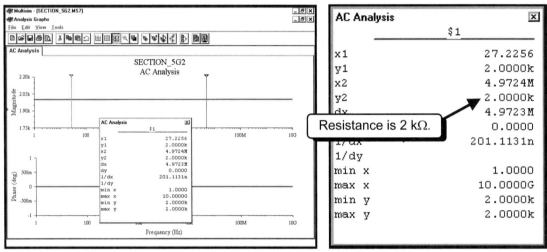

This trace shows that the impedance is constant with frequency, and the y-coordinates of the cursors show that the resistance is at the expected value of 2 kΩ. If we had inductors or capacitors in the circuit, the impedance would have changed with frequency.

EXERCISE 5-11: Find the equivalent impedance of the following resistor network:

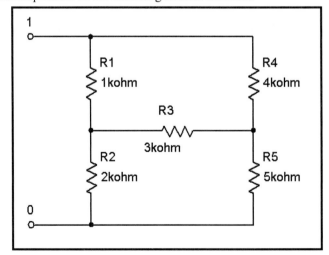

SOLUTION: Add an independent current source between the two terminals where you wish to measure the impedance:

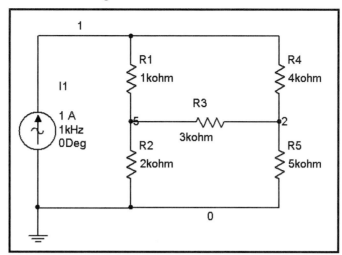

The equivalent impedance is V(**1**)/I(**Itest**). Because **Itest** is a 1-amp source, we can just plot the voltage at node 1. Set up and run an AC Analysis and plot the voltage at node 1 during the simulation:

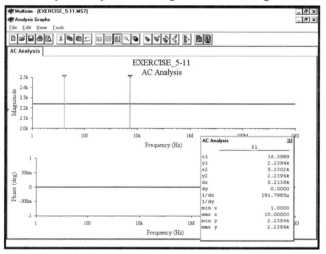

The impedance is **2.2394** kΩ.

5.G.3. Impedance Measurement of an Active Circuit Using SPICE

We would now like to find the output impedance of the jFET source follower circuit shown below. The output is taken at the source terminal of the jFET, so the impedance we are interested in is between nodes **Vout** and **0** (ground). Although this is a jFET circuit, we can use this technique with any circuit.

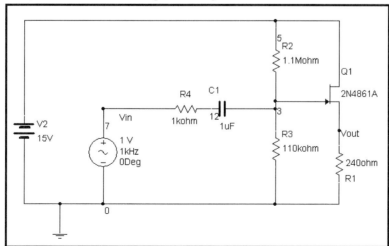

When you are finding output impedance you must set the input AC source to zero. Because the input is a voltage source, we replace **Vin** with a short:

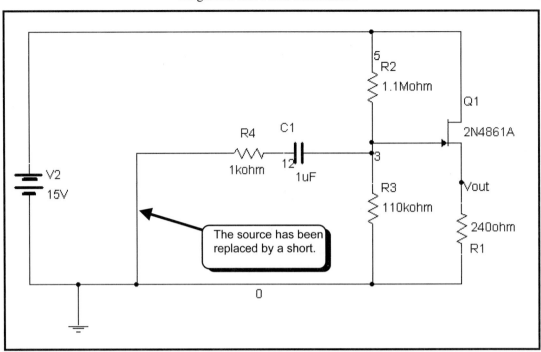

Next we need to add an AC source at the output to measure the impedance. We could use either an AC voltage source or an AC current source. If you use a voltage source, wire the circuit shown below:

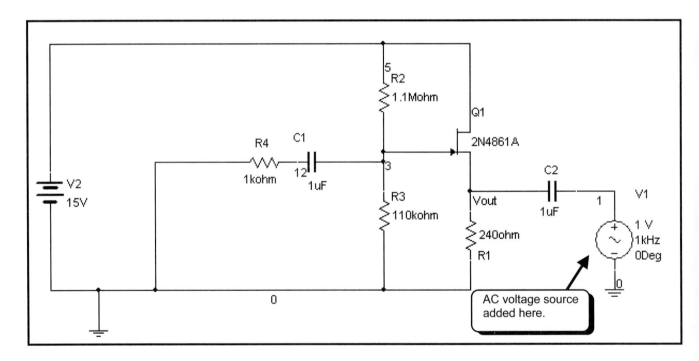

Capacitor **C2** is necessary to preserve the bias of the amplifier. Remember that **V1** is an AC source. For biasing, all AC sources are set to zero. Since **V1** is a voltage source, it would be replaced by a short. Without **C2**, the source terminal of the jFET would be grounded when calculating the bias. This would destroy the bias and render the impedance measurement invalid.

The output impedance of the source follower can be observed by plotting the ratio V(Vout)/I(V1) in the circuit above. It is important to note that the ratio V(V1)/I(V1) is not the output impedance of the source follower, but the output impedance plus the impedance of the capacitor.

The second method to measure the impedance of a circuit uses an AC current source:

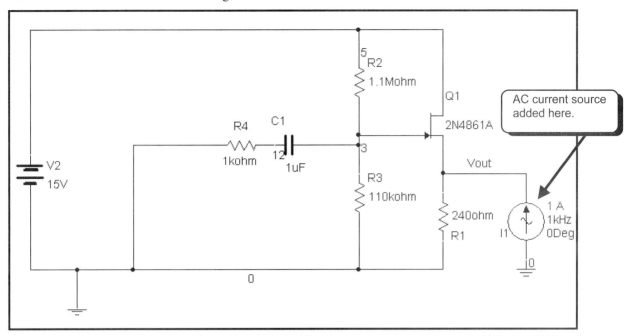

I1 is an AC current source. Note that there is no blocking capacitor between the current source and the amplifier. When the bias is calculated, all AC sources are set to zero. When a current source is set to zero, it is replaced by an open circuit. An open circuit at **Vout** is equivalent to the original circuit for bias calculations, and the bias is preserved. The output impedance of the circuit can be calculated by the ratio V(Vout)/I(I1). Also note that because the magnitude of I1 is 1 amp, the impedance is numerically equal to the voltage at node **Vout**.

We next need to set up an AC Analysis and then plot the voltage at node **Vout**. Select **Simulate**, **Analyses**, and then **AC Analysis** from the menus, and then fill in the dialog box as shown:

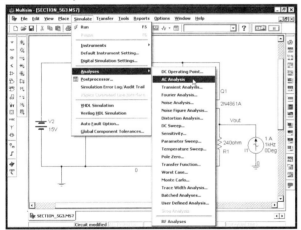 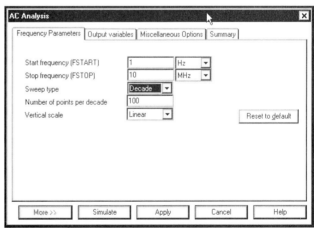

Select the **Output variables** tab and specify to plot **$vout** during the simulation:

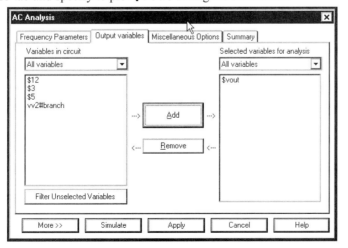

We are now ready to run the simulation. Click on the **Simulate** button and then use the cursors to obtain a numerical value of the output resistance:

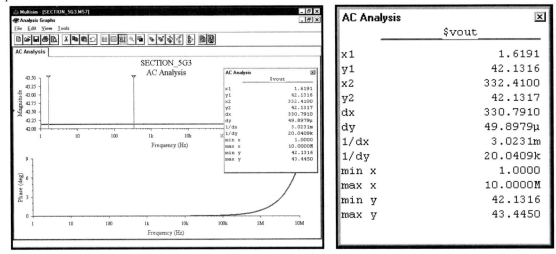

The trace shows that the impedance is about 42 Ω for frequencies up to 1 MHz.

EXERCISE 5-12: Find the output impedance of the emitter-follower:

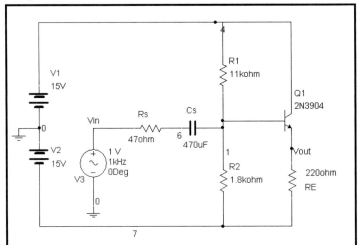

SOLUTION: Replace the independent voltage source with a short circuit and add an independent current source between the two terminals where you wish to measure the impedance:

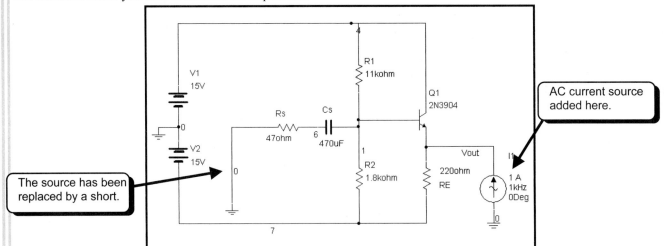

The output impedance is V(Vout)/I(I1). Because I1 has a magnitude of 1, the output impedance is numerically equal to the voltage at **Vout**:

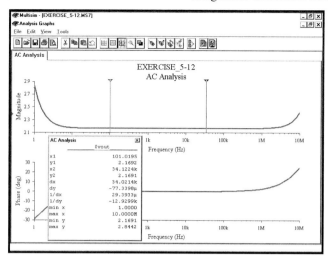

At mid-band, the impedance is about 2.2 Ω. The change in impedance at low frequencies is due to **Cs**. The change in impedance at high frequencies is due to capacitances internal to the transistor.

5.H. Problems

Problem 5-1: Find the RMS magnitude of the voltage at node V1 using indicators in the circuit below:

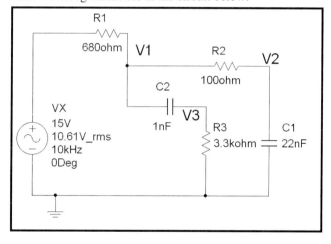

Problem 5-2: For the circuit of Problem 5-1, find the RMS magnitude of the voltage at node V2 and the current through R1 using indicators.

Problem 5-3: For the circuit of Problem 5-1, find the RMS magnitude of the voltage at node V3 and the current through C2 using indicators.

Problem 5-4: For the circuit of Problem 5-1, find the RMS magnitude of the voltage across R1 and the current through C2 using indicators.

Problem 5-5: For the circuit of Problem 5-1, find the amplitude (peak value) of the voltage at node V1 and the current through R2 using indicators.

Problem 5-6: For the circuit of Problem 5-1, find the amplitude (peak value) of the voltage at node V2 and the current through C1 using indicators.

Problem 5-7: For the circuit of Problem 5-1, find the amplitude (peak value) of the voltage at node V3 and the current through R3 using indicators.

Problem 5-8: For the circuit of Problem 5-1, find the RMS magnitude of the voltage across C2 and the current through C2 using indicators.

Problem 5-9: For the circuit of Problem 5-1, find the RMS magnitude of the voltage at node V1 and the current through R1 using Multimeters.

Problem 5-10: For the circuit of Problem 5-1, find the RMS magnitude of the voltage at node V2 and the current through R2 using Multimeters.

Problem 5-11: For the circuit of Problem 5-1, find the RMS magnitude of the voltage at node V3 and the current through R3 using Multimeters.

Problem 5-12: For the circuit of Problem 5-1, find the RMS magnitude of the voltage across C1 and the current through R2 using Multimeters.

Problem 5-13: For the circuit of Problem 5-1, find the amplitude (peak value) of the voltage at node V1 and the current through C2 using Multimeters.

Problem 5-14: For the circuit of Problem 5-1, find the amplitude (peak value) of the voltage at node V2 and the current through C1 using Multimeters.

Problem 5-15: For the circuit of Problem 5-1, find the amplitude (peak value) of the voltage at node V3 and the current through C2 using Multimeters.

Problem 5-16: For the circuit of Problem 5-1, find the amplitude (peak value) of the voltages at nodes V1, V2, and V3 using the SPICE AC Sweep analysis.

Problem 5-17: For the circuit of Problem 5-1, find the phase of the voltages at nodes V1, V2, and V3 using the SPICE AC Sweep analysis.

Problem 5-18: Find the RMS magnitude of the voltage at node V1 using indicators in the circuit below:

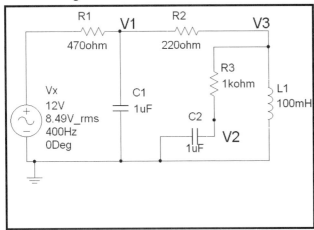

Problem 5-19: For the circuit of Problem 5-18, find the RMS magnitude of the voltage at node V2 and the current through R1 using indicators.

Problem 5-20: For the circuit of Problem 5-18, find the RMS magnitude of the voltage at node V3 and the current through C2 using indicators.

Problem 5-21: For the circuit of Problem 5-18, find the amplitude (peak value) of the voltage at node V1 and the current through R2 using indicators.

Problem 5-22: For the circuit of Problem 5-18, find the amplitude (peak value) of the voltage at node V2 and the current through C1 using indicators.

Problem 5-23: For the circuit of Problem 5-18, find the amplitude (peak value) of the voltage across C2 and the current through C1 using indicators.

Problem 5-24: For the circuit of Problem 5-18, find the amplitude (peak value) of the voltage at node V3 and the current through L1 using indicators.

Problem 5-25: For the circuit of Problem 5-18, find the RMS magnitude of the voltage at node V1 and the current through R1 using Multimeters.

Problem 5-26: For the circuit of Problem 5-18, find the RMS magnitude of the voltage at node V2 and the current through R2 using Multimeters.

Problem 5-27: For the circuit of Problem 5-18, find the RMS magnitude of the voltage at node V3 and the current through L1 using Multimeters.

Problem 5-28: For the circuit of Problem 5-18, find the RMS magnitude of the voltage across R3 and the current through L1 using Multimeters.

Problem 5-29: For the circuit of Problem 5-18, find the amplitude (peak value) of the voltage at node V1 and the current through C2 using Multimeters.

Problem 5-30: For the circuit of Problem 5-18, find the amplitude (peak value) of the voltage at node V2 and the current through C1 using Multimeters.

Problem 5-31: For the circuit of Problem 5-18, find the amplitude (peak value) of the voltage at node V3 and the current through L1 using Multimeters.

Problem 5-32: For the circuit of Problem 5-18, find the amplitude (peak value) of the voltage across C1 and the current through L1 using Multimeters.

Problem 5-33: For the circuit of Problem 5-18, find the amplitude (peak value) of the voltages at nodes V1, V2, and V3 using the SPICE AC Sweep analysis.

Problem 5-34: For the circuit of Problem 5-18, find the phase of the voltages at nodes V1, V2, and V3 using the SPICE AC Sweep analysis.

Problem 5-35: Find the RMS magnitude of the voltage at node V1 using indicators in the circuit below:

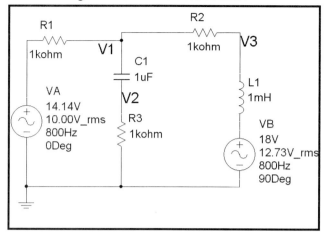

Problem 5-36: For the circuit of Problem 5-35, find the RMS magnitude of the voltage at node V2 and the current through R1 using indicators.

Problem 5-37: For the circuit of Problem 5-35, find the RMS magnitude of the voltage at node V3 and the current through C1 using indicators.

Problem 5-38: For the circuit of Problem 5-35, find the amplitude (peak value) of the voltage at node V1 and the current through R2 using indicators.

Problem 5-39: For the circuit of Problem 5-35, find the amplitude (peak value) of the voltage across L1 and the current through R2 using indicators.

Problem 5-40: For the circuit of Problem 5-35, find the amplitude (peak value) of the voltage at node V2 and the current through C1 using indicators.

Problem 5-41: For the circuit of Problem 5-35, find the amplitude (peak value) of the voltage at node V3 and the current through L1 using indicators.

Problem 5-42: For the circuit of Problem 5-35, find the RMS magnitude of the voltage at node V1 and the current through R1 using Multimeters.

Problem 5-43: For the circuit of Problem 5-35, find the RMS magnitude of the voltage at node V2 and the current through R2 using Multimeters.

Problem 5-44: For the circuit of Problem 5-35, find the RMS magnitude of the voltage across C1 and the current through R2 using Multimeters.

Problem 5-45: For the circuit of Problem 5-35, find the RMS magnitude of the voltage at node V3 and the current through L1 using Multimeters.

Problem 5-46: For the circuit of Problem 5-35, find the amplitude (peak value) of the voltage at node V1 and the current through C1 using Multimeters.

Problem 5-47: For the circuit of Problem 5-35, find the amplitude (peak value) of the voltage at node V2 and the current through C1 using Multimeters.

Problem 5-48: For the circuit of Problem 5-35, find the amplitude (peak value) of the voltage at node V3 and the current through L1 using Multimeters.

Problem 5-49: For the circuit of Problem 5-35, find the amplitude (peak value) of the voltages at nodes V1, V2, and V3 using the SPICE AC Sweep analysis.

Problem 5-50: For the circuit of Problem 5-35, find the phase of the voltages at nodes V1, V2, and V3 using the SPICE AC Sweep analysis.

Problem 5-51: For the circuit below, generate the Bode magnitude and phase plots of Vo/Vin using the Bode Plotter instrument. Use the cursors to find the frequency where the gain is down by –3 dB.

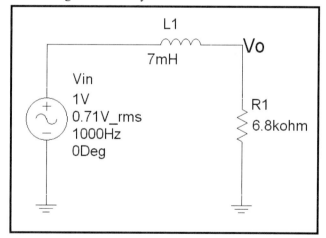

Problem 5-52: For the circuit of Problem 5-51, generate the Bode magnitude and phase plots of Vo/Vin using the SPICE AC Sweep analysis. Use the cursors to find the frequency where the gain is down by –3 dB. What are the gain and phase at this point?

Problem 5-53: For the circuit of Problem 5-51, generate the Bode magnitude and phase plots of Vo/Vin using the SPICE AC Sweep analysis. Use the cursors to find the frequency where the gain is down by –20 dB. What are the gain and phase at this point?

Problem 5-54: For the circuit of Problem 5-51, generate the Bode magnitude and phase plots of Vo/Vin using the SPICE AC Sweep analysis. Use the cursors to find the frequency where the gain is down by –40 dB. What are the gain and phase at this point?

Problem 5-55: For the circuit of Problem 5-51, generate the Bode magnitude and phase plots of Vo/Vin using the SPICE AC Sweep analysis. Use the cursors to find the frequency where the gain is down by –60 dB. What are the gain and phase at this point?

Problem 5-56: For the circuit of Problem 5-51, generate the Bode magnitude and phase plots of Vo/Vin using the SPICE AC Sweep analysis. Use the cursors to find the frequency where the phase is –45°. What are the gain and phase at this point?

Problem 5-57: For the circuit of Problem 5-51, generate the Bode magnitude and phase plots of Vo/Vin using the SPICE AC Sweep analysis. Use the cursors to find the frequency where the phase is –22°. What are the gain and phase at this point?

Problem 5-58: For the circuit of Problem 5-51, generate the Bode magnitude and phase plots of Vo/Vin using the SPICE AC Sweep analysis. Use the cursors to find the frequency where the phase is –75°. What are the gain and phase at this point?

Problem 5-59: For the circuit of Problem 5-51, generate the Bode magnitude and phase plots of Vo/Vin using the Bode Plotter instrument. Use the cursors to find the frequency where the gain is down by –20 dB. What are the gain and phase at this point?

Problem 5-60: For the circuit of Problem 5-51, generate the Bode magnitude and phase plots of Vo/Vin using the Bode Plotter instrument. Use the cursors to find the frequency where the gain is down by –40 dB. What are the gain and phase at this point?

Problem 5-61: For the circuit of Problem 5-51, generate the Bode magnitude and phase plots of Vo/Vin using the Bode Plotter instrument. Use the cursors to find the frequency where the gain is down by –60 dB. What are the gain and phase at this point?

Problem 5-62: For the circuit of Problem 5-51, generate the Bode magnitude and phase plots of Vo/Vin using the Bode Plotter instrument. Use the cursors to find the frequency where the phase is –45°. What are the gain and phase at this point?

Problem 5-63: For the circuit of Problem 5-51, generate the Bode magnitude and phase plots of Vo/Vin using the Bode Plotter instrument. Use the cursors to find the frequency where the phase is –22°. What are the gain and phase at this point?

Problem 5-64: For the circuit of Problem 5-51, generate the Bode magnitude and phase plots of Vo/Vin using the Bode Plotter instrument. Use the cursors to find the frequency where the phase is –75°. What are the gain and phase at this point?

Problem 5-65: For the circuit below, generate the Bode magnitude and phase plots of Vo/Vin using the Bode Plotter instrument. Use the cursors to find the frequencies where the gain is down by –3 dB from the midband gain. Simulate the circuit for frequencies from 10 mHz to 10 GHz.

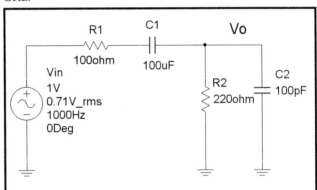

Problem 5-66: For the circuit of Problem 5-65, generate the Bode magnitude and phase plots of Vo/Vin using the SPICE AC Sweep analysis. Use the cursors to find the frequencies where the gain is down by –3 dB from the midband gain. What are the gain and phase at this point? Simulate the circuit for frequencies from 10 mHz to 10 GHz.

Problem 5-67: For the circuit of Problem 5-65, generate the Bode magnitude and phase plots of Vo/Vin using the SPICE AC Sweep analysis. Use the cursors to find the frequencies where the gain is down by –20 dB from the midband gain. What are the gain and phase at this point? Simulate the circuit for frequencies from 10 mHz to 10 GHz.

Problem 5-68: For the circuit of Problem 5-65, generate the Bode magnitude and phase plots of Vo/Vin using the SPICE AC Sweep analysis. Use the cursors to find the frequencies where the gain is down by –40 dB from the midband gain. What are the gain and phase at this point? Simulate the circuit for frequencies from 10 mHz to 10 GHz.

Problem 5-69: For the circuit of Problem 5-65, generate the Bode magnitude and phase plots of Vo/Vin using the SPICE AC Sweep analysis. Use the cursors to find the frequencies where the gain is down by –60 dB from the midband gain. What are the gain and phase at this point? Simulate the circuit for frequencies from 10 mHz to 10 GHz.

Problem 5-70: For the circuit of Problem 5-65, generate the Bode magnitude and phase plots of Vo/Vin using the SPICE AC Sweep analysis. Use the cursors to find the frequency where the phase is –45°. What are the gain and phase at this point? Simulate the circuit for frequencies from 10 mHz to 10 GHz.

Problem 5-71: For the circuit of Problem 5-65, generate the Bode magnitude and phase plots of Vo/Vin using the SPICE AC Sweep analysis. Use the cursors to find the frequency where the phase is +22°. What are the gain and phase at this point? Simulate the circuit for frequencies from 10 mHz to 10 GHz.

Problem 5-72: For the circuit of Problem 5-65, generate the Bode magnitude and phase plots of Vo/Vin using the SPICE AC Sweep analysis. Use the cursors to find the frequency where the phase is +75°. What are the gain and phase at this point? Simulate the circuit for frequencies from 10 mHz to 10 GHz.

Problem 5-73: For the circuit of Problem 5-65, generate the Bode magnitude and phase plots of Vo/Vin using the Bode Plotter instrument. Use the cursors to find the frequencies where the gain is down by –20 dB from the midband gain. What are the gain and phase at this point? Simulate the circuit for frequencies from 10 mHz to 10 GHz.

Problem 5-74: For the circuit of Problem 5-65, generate the Bode magnitude and phase plots of Vo/Vin using the Bode Plotter instrument. Use the cursors to find the frequencies where the gain is down by –40 dB from the midband gain. What are the gain and phase at this point? Simulate the circuit for frequencies from 10 mHz to 10 GHz.

Problem 5-75: For the circuit of Problem 5-65, generate the Bode magnitude and phase plots of Vo/Vin using the Bode Plotter instrument. Use the cursors to find the frequencies where the gain is down by –60 dB from the midband gain. What are the gain and phase at this point? Simulate the circuit for frequencies from 10 mHz to 10 GHz.

Problem 5-76: For the circuit of Problem 5-65, generate the Bode magnitude and phase plots of Vo/Vin using the Bode Plotter instrument. Use the cursors to find the frequency where the phase is +45°. What are the gain and phase at this point? Simulate the circuit for frequencies from 10 mHz to 10 GHz.

Problem 5-77: For the circuit of Problem 5-65, generate the Bode magnitude and phase plots of Vo/Vin using the Bode Plotter instrument. Use the cursors to find the frequency where the phase is –22°. What are the gain and phase at this point? Simulate the circuit for frequencies from 10 mHz to 10 GHz.

Problem 5-78: For the circuit of Problem 5-65, generate the Bode magnitude and phase plots of Vo/Vin using the Bode Plotter instrument. Use the cursors to find the frequency where the phase is –75°. What are the gain and phase at this point? Simulate the circuit for frequencies from 10 mHz to 10 GHz.

Problem 5-79: For the circuit below, use the Bode Plotter to generate a magnitude plot and find the midband gain (Vout/Vin) and the upper and lower –3 dB frequencies for this amplifier. (Note that this is nearly the same circuit as the one used in section 5.C. The values of the components are the same.)

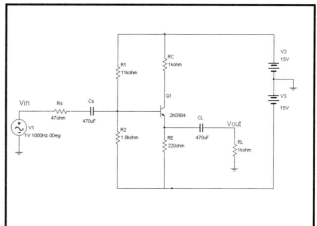

Problem 5-80: Repeat Problem 5-79 using the SPICE AC Sweep analysis.

Problem 5-81: Repeat the simulations of Problem 5-79 with RL = 10 kΩ, RL = 100 Ω, and RL = 10 Ω.

Problem 5-82: Repeat the simulations of Problem 5-79 with RC = 0 Ω. What is the effect of RC on this circuit?

Problem 5-83: For the circuit below, use the Bode Plotter to generate a magnitude plot and find the midband gain

(Vout/Vin) and the upper and lower −3 dB frequencies for this amplifier.

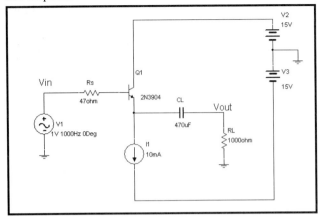

Problem 5-84: Repeat Problem 5-83 using the SPICE AC Sweep analysis.

Problem 5-85: Repeat the simulations of Problem 5-83 with RL = 10 kΩ, RL = 100 Ω, and RL = 10 Ω.

Problem 5-86: For the circuit below, use the Bode Plotter to generate a magnitude plot and find the midband gain (Vout/Vin) and the upper and lower −3 dB frequencies for this amplifier. (Note that this circuit is similar to the one used in section 5.C.)

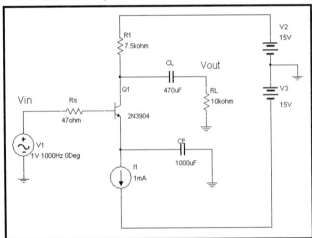

Problem 5-87: Repeat Problem 5-86 using the SPICE AC Sweep analysis.

Problem 5-88: Repeat the simulations of Problem 5-86 with RL = 1 kΩ, RL = 100 Ω, and RL = 10 Ω.

Problem 5-89: For the circuit below, use the Bode Plotter to generate a magnitude plot and find the midband gain (Vout/Vin) and the upper and lower −3 dB frequencies for this amplifier. (Note that this is nearly the same circuit used in section 5.C. The values of the components are the same, and CE has been removed.)

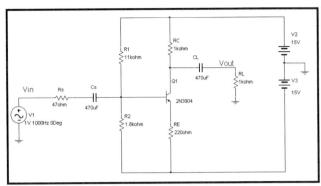

Problem 5-90: Repeat Problem 5-89 using the SPICE AC Sweep analysis.

Problem 5-91: Repeat the simulations of Problem 5-89 with RL = 10 kΩ, RL = 100 Ω, and RL = 10 Ω.

Problem 5-92:

(a) Design a non-inverting OPAMP circuit with a gain of 1 using a LM741CH. Create a Bode magnitude plot, verify the midband gain, and find the upper −3 dB frequency.

(b) Design an inverting OPAMP circuit with a gain of −1 using an LM741CH. Create a Bode magnitude plot, verify the midband gain, and find the upper −3 dB frequency. Use a minimum resistor size of 10 kΩ.

(c) Although the circuits from parts (a) and (b) have the same magnitude of gain, the upper −3 dB frequency is different. Why?

Problem 5-93:

(a) Design a non-inverting OPAMP circuit with a gain of 1 using a TL072ACD. Create a Bode magnitude plot, verify the midband gain, and find the upper −3 dB frequency.

(b) Design an inverting OPAMP circuit with a gain of −1 using a TL072ACD. Create a Bode magnitude plot, verify the midband gain, and find the upper −3 dB frequency. Use a minimum resistor size of 10 kΩ.

(c) Although the circuits from parts (a) and (b) have the same magnitude of gain, the upper −3 dB frequency is different. Why?

Problem 5-94:

(a) Design a non-inverting OPAMP circuit with a gain of 1 using an OPA602BM. Create a Bode magnitude plot, verify the midband gain, and find the upper −3 dB frequency.

(b) Design an inverting OPAMP circuit with a gain of −1 using an OPA602BM. Create a Bode magnitude plot, verify the midband gain, and find the upper −3 dB frequency. Use a minimum resistor size of 10 kΩ.

(c) Although the circuits from parts (a) and (b) have the same magnitude of gain, the upper −3 dB frequency is different. Why?

Problem 5-95:

(a) Design a non-inverting OPAMP circuit with a gain of 100 using an LM741CH. Create a Bode magnitude plot, verify the midband gain, and find the upper −3 dB frequency.

(b) Design an inverting OPAMP circuit with a gain of −100 using an LM741CH. Create a Bode magnitude plot, verify the midband gain, and find the upper −3 dB frequency.

(c) The simulations from parts (a) and (b) show the same gain and upper −3 dB frequency. This is different than in Problem 5-92. Why?

Problem 5-96:

(a) Design a non-inverting OPAMP circuit with a gain of 100 using a TL072ACD. Create a Bode magnitude plot, verify the midband gain, and find the upper −3 dB frequency.

(b) Design an inverting OPAMP circuit with a gain of −100 using a TL072ACD. Create a Bode magnitude plot, verify the midband gain, and find the upper −3 dB frequency.

(c) The simulations from parts (a) and (b) show the same gain and upper −3 dB frequency. This is different than in Problem 5-93. Why?

Problem 5-97:

(a) Design a non-inverting OPAMP circuit with a gain of 100 using an OPA602BM. Create a Bode magnitude plot, verify the midband gain, and find the upper −3 dB frequency.

(b) Design an inverting OPAMP circuit with a gain of −100 using an OPA602BM. Create a Bode magnitude plot, verify the midband gain, and find the upper −3 dB frequency.

(c) The simulations from parts (a) and (b) show the same gain and upper −3 dB frequency. This is different than in Problem 5-94. Why?

Problem 5-98: Design an inverting summing amplifier that implements the equation $V_{OUT} = -V_A - 10V_B - 100V_C$ using your favorite OPAMP. (Do not use the virtual OPAMPs.)

(a) With the inputs to V_B and V_C set to zero (grounded) obtain a Bode plot of V_{OUT}/V_A. Verify the midband gain and find the upper −3 dB frequency.

(b) With the inputs to V_A and V_C set to zero (grounded) obtain a Bode plot of V_{OUT}/V_B. Verify the midband gain and find the upper −3 dB frequency.

(c) With the inputs to V_A and V_B set to zero (grounded) obtain a Bode plot of V_{OUT}/V_C. Verify the midband gain and find the upper −3 dB frequency.

(d) Use the Postprocessor to generate Bode magnitude plots of V_{OUT}/V_A, V_{OUT}/V_B, and V_{OUT}/V_C on a single graph.

(e) Although the gain from each input is different, the upper −3 dB frequencies for each plot generated above are the same. Why?

Problem 5-99: For the circuit below, use the Parameter Sweep to plot the gain V_{OUT}/I_{IN} in decibels for values of RF equal to 1 kΩ, 10 kΩ, 100 kΩ, and 1 MΩ.

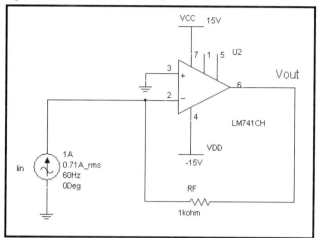

Problem 5-100: For the circuit below, use the Parameter Sweep to plot the gain V_{OUT}/V_{IN} in decibels for values of R4 equal to 1 kΩ, 10 kΩ, 100 kΩ, and 1 MΩ.

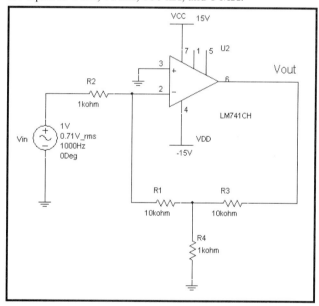

Problem 5-101: A complex load has an impedance of 15∠−37° at 60 Hz.

(a) Create a circuit of inductors, capacitors, and/or resistors that implements this impedance. You can use as many or as few components as you like.

(b) By hand, calculate the power factor. Does this circuit have a leading or lagging power factor?

(c) Calculate by hand the amount of power this load will absorb in watts if it is connected to a 115 V_{RMS} 60 Hz source.

(d) Verify with Multisim the power factor and amount of power this load absorbs.

(e) Specify a single component that can be added to the load to make the compensated load have a unity power factor.

(f) Calculate by hand the amount of power the compensated load will absorb in watts if it is connected to a 115 V_{RMS} 60 Hz source.

(g) Verify with Multisim the power factor and amount of power the compensated load absorbs.

Problem 5-102: A complex load has an impedance of $15\angle 37°$ at 60 Hz.

(a) Create a circuit of inductors, capacitors, and/or resistors that implements this impedance. You can use as many or as few components as you like.

(b) By hand, calculate the power factor. Does this circuit have a leading or lagging power factor?

(c) Calculate by hand the amount of power this load will absorb in watts if it is connected to a 115 V_{RMS} 60 Hz source.

(d) Verify with Multisim the power factor and amount of power this load absorbs.

(e) Specify a single component that can be added to the load to make the compensated load have a unity power factor.

(f) Calculate by hand the amount of power the compensated load will absorb in watts if it is connected to a 115 V_{RMS} 60 Hz source.

(g) Verify with Multisim the power factor and amount of power the compensated load absorbs.

Problem 5-103: A complex load has an impedance of $25\angle -25°$ at 400 Hz.

(a) Create a circuit of inductors, capacitors, and/or resistors that implements this impedance. You can use as many or as few components as you like.

(b) By hand, calculate the power factor. Does this circuit have a leading or lagging power factor?

(c) Calculate by hand the amount of power this load will absorb in watts if it is connected to a 220 V_{RMS} 400 Hz source.

(d) Verify with Multisim the power factor and amount of power this load absorbs.

(e) Specify a single component that can be added to the load to make the compensated load have a unity power factor.

(f) Calculate by hand the amount of power the compensated load will absorb in watts if it is connected to a 220 V_{RMS} 400 Hz source.

(g) Verify with Multisim the power factor and amount of power the compensated load absorbs.

Problem 5-104: A complex load has an impedance of $25\angle 25°$ at 400 Hz.

(a) Create a circuit of inductors, capacitors, and/or resistors that implements this impedance. You can use as many or as few components as you like.

(b) By hand, calculate the power factor. Does this circuit have a leading or lagging power factor?

(c) Calculate by hand the amount of power this load will absorb in watts if it is connected to a 220 V_{RMS} 400 Hz source.

(d) Verify with Multisim the power factor and amount of power this load absorbs.

(e) Specify a single component that can be added to the load to make the compensated load have a unity power factor.

(f) Calculate by hand the amount of power the compensated load will absorb in watts if it is connected to a 220 V_{RMS} 400 Hz source.

(g) Verify with Multisim the power factor and amount of power the compensated load absorbs.

Problem 5-105: A complex load has an impedance of $1\angle -30°$ at 60 Hz.

(a) Create a circuit of inductors, capacitors, and/or resistors that implements this impedance. You can use as many or as few components as you like.

(b) By hand, calculate the power factor. Does this circuit have a leading or lagging power factor?

(c) Calculate by hand the amount of power this load will absorb in watts if it is connected to a 115 V_{RMS} 60 Hz source.

(d) Verify with Multisim the power factor and amount of power this load absorbs.

(e) Specify a single component that can be added to the load to make the compensated load have a unity power factor.

(f) Calculate by hand the amount of power the compensated load will absorb in watts if it is connected to a 115 V_{RMS} 60 Hz source.

(g) Verify with Multisim the power factor and amount of power the compensated load absorbs.

Problem 5-106: A complex load has an impedance of $1\angle 30°$ at 60 Hz.

(a) Create a circuit of inductors, capacitors, and/or resistors that implements this impedance. You can use as many or as few components as you like.

(b) By hand, calculate the power factor. Does this circuit have a leading or lagging power factor?

(c) Calculate by hand the amount of power this load will absorb in watts if it is connected to a 115 V_{RMS} 60 Hz source.

(d) Verify with Multisim the power factor and amount of power this load absorbs.

(e) Specify a single component that can be added to the load to make the compensated load have a unity power factor.

(f) Calculate by hand the amount of power the compensated load will absorb in watts if it is connected to a 115 V_{RMS} 60 Hz source.

(g) Verify with Multisim the power factor and amount of power the compensated load absorbs.

Problem 5-107: A complex load has an impedance of 0.5∠−20° at 400 Hz.

(a) Create a circuit of inductors, capacitors, and/or resistors that implements this impedance. You can use as many or as few components as you like.

(b) By hand, calculate the power factor. Does this circuit have a leading or lagging power factor?

(c) Calculate by hand the amount of power this load will absorb in watts if it is connected to a 220 V_{RMS} 400 Hz source.

(d) Verify with Multisim the power factor and amount of power this load absorbs.

(e) Specify a single component that can be added to the load to make the compensated load have a unity power factor.

(f) Calculate by hand the amount of power the compensated load will absorb in watts if it is connected to a 220 V_{RMS} 400 Hz source.

(g) Verify with Multisim the power factor and amount of power the compensated load absorbs.

Problem 5-108: A complex load has an impedance of 0.5∠20° at 400 Hz.

(a) Create a circuit of inductors, capacitors, and/or resistors that implements this impedance. You can use as many or as few components as you like.

(b) By hand, calculate the power factor. Does this circuit have a leading or lagging power factor?

(c) Calculate by hand the amount of power this load will absorb in watts if it is connected to a 220 V_{RMS} 400 Hz source.

(d) Verify with Multisim the power factor and amount of power this load absorbs.

(e) Specify a single component that can be added to the load to make the compensated load have a unity power factor.

(f) Calculate by hand the amount of power the compensated load will absorb in watts if it is connected to a 220 V_{RMS} 400 Hz source.

(g) Verify with Multisim the power factor and amount of power the compensated load absorbs.

Problem 5-109: A load that is connected to a 120 V_{RMS} 60 Hz source has a power factor of 0.93 lagging and absorbs 360 watts of power.

(a) What is the impedance of the load?

(b) Create a circuit of inductors, capacitors, and/or resistors that implements this impedance. You can use as many or as few components as you like.

(c) By hand, calculate the power factor. Does this circuit have a leading or lagging power factor?

(d) Calculate by hand the amount of power this load will absorb in watts if it is connected to a 220 V_{RMS} 400 Hz source.

(e) Verify with Multisim the power factor and amount of power this load absorbs.

(f) Specify a single component that can be added to the load to make the compensated load have a unity power factor.

(g) Calculate by hand the amount of power the compensated load will absorb in watts if it is connected to a 220 V_{RMS} 400 Hz source.

(h) Verify with Multisim the power factor and amount of power the compensated load absorbs.

Problem 5-110: A load that is connected to a 220 V_{RMS} 60 Hz source has a power factor of 0.93 leading and absorbs 1200 watts of power.

(a) What is the impedance of the load?

(b) Create a circuit of inductors, capacitors, and/or resistors that implements this impedance. You can use as many or as few components as you like.

(c) By hand, calculate the power factor. Does this circuit have a leading or lagging power factor?

(d) Calculate by hand the amount of power this load will absorb in watts if it is connected to a 220 V_{RMS} 400 Hz source.

(e) Verify with Multisim the power factor and amount of power this load absorbs.

(f) Specify a single component that can be added to the load to make the compensated load have a unity power factor.

(g) Calculate by hand the amount of power the compensated load will absorb in watts if it is connected to a 220 V_{RMS} 400 Hz source.

(h) Verify with Multisim the power factor and amount of power the compensated load absorbs.

Problem 5-111: A load that is connected to a 208 V_{RMS} 400 Hz source has a power factor of 0.89 leading and absorbs 10 kW of power.

(a) What is the impedance of the load?

(b) Create a circuit of inductors, capacitors, and/or resistors that implements this impedance. You can use as many or as few components as you like.

(c) By hand, calculate the power factor. Does this circuit have a leading or lagging power factor?

(d) Calculate by hand the amount of power this load will absorb in watts if it is connected to a 220 V_{RMS} 400 Hz source.

(e) Verify with Multisim the power factor and amount of power this load absorbs.

(f) Specify a single component that can be added to the load to make the compensated load have a unity power factor.

(g) Calculate by hand the amount of power the compensated load will absorb in watts if it is connected to a 220 V_{RMS} 400 Hz source.

(h) Verify with Multisim the power factor and amount of power the compensated load absorbs.

Problem 5-112: A load that is connected to a 208 V_{RMS} 400 Hz source has a power factor of 0.85 lagging and absorbs 5 kW of power.

(a) What is the impedance of the load?

(b) Create a circuit of inductors, capacitors, and/or resistors that implements this impedance. You can use as many or as few components as you like.

(c) By hand, calculate the power factor. Does this circuit have a leading or lagging power factor?

(d) Calculate by hand the amount of power this load will absorb in watts if it is connected to a 220 V_{RMS} 400 Hz source.

(e) Verify with Multisim the power factor and amount of power this load absorbs.

(f) Specify a single component that can be added to the load to make the compensated load have a unity power factor.

(g) Calculate by hand the amount of power the compensated load will absorb in watts if it is connected to a 220 V_{RMS} 400 Hz source.

(h) Verify with Multisim the power factor and amount of power the compensated load absorbs.

Problem 5-113: Find the impedance of the circuit below using the Multimeter.

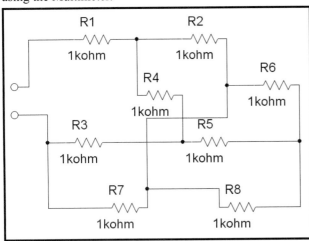

Problem 5-114: Find the impedance of the circuit below using the Multimeter.

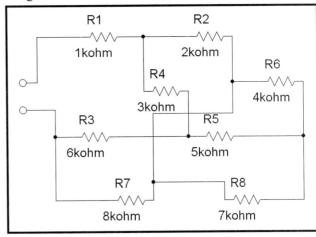

Problem 5-115: Find the impedance of the circuit below using the Multimeter.

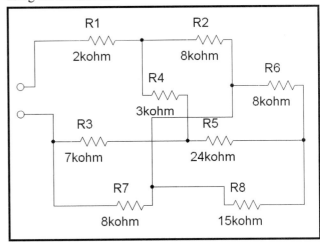

Problem 5-116: Find the impedance of the circuit below using the Multimeter.

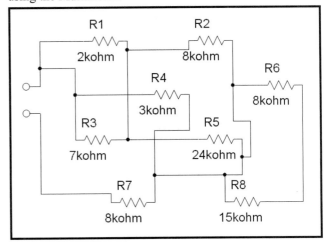

Problem 5-117: Find the impedance of the circuit below using the Multimeter.

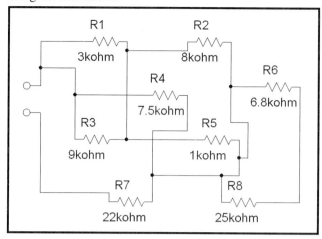

Problem 5-118: Find the impedance of the circuit below using the Multimeter.

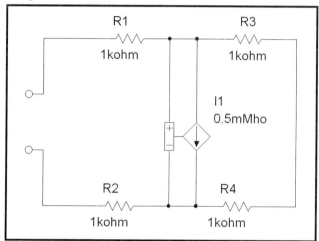

Problem 5-119: Find the impedance of the circuit below using the Multimeter.

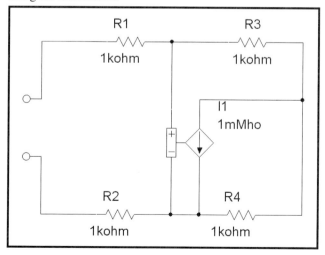

Problem 5-120: Find the impedance of the circuit below using the Multimeter.

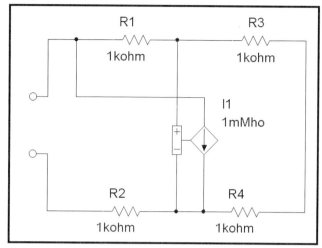

Problem 5-121: Find the impedance of the circuit below using the Multimeter.

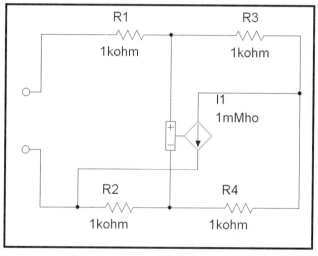

Problem 5-122: Find the impedance of the circuit below using the AC Sweep analysis.

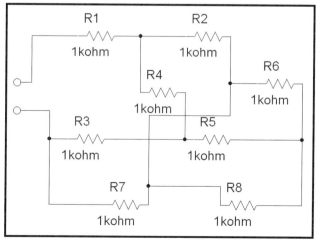

Problem 5-123: Find the impedance of the circuit below using the AC Sweep analysis.

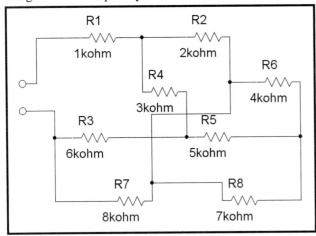

Problem 5-124: Find the impedance of the circuit below using the AC Sweep analysis.

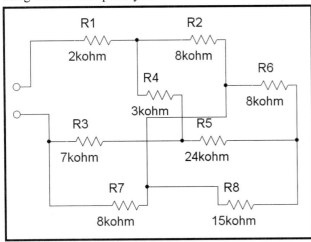

Problem 5-125: Find the impedance of the circuit below using the AC Sweep analysis.

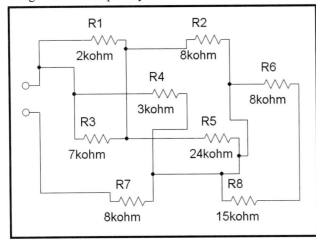

Problem 5-126: Find the impedance of the circuit below using the AC Sweep analysis.

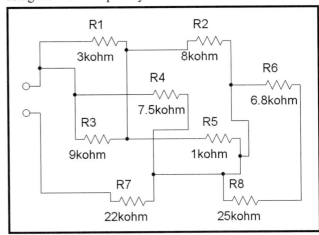

Problem 5-127: Find the impedance of the circuit below using the AC Sweep analysis.

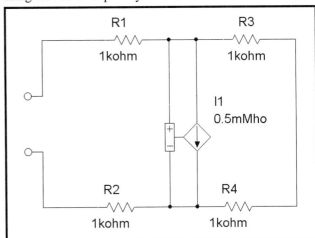

Problem 5-128: Find the impedance of the circuit below using the AC Sweep analysis.

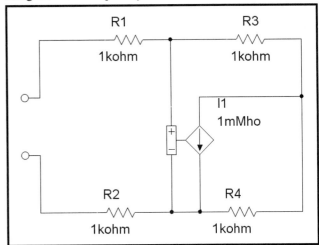

Problem 5-129: Find the impedance of the circuit below using the AC Sweep analysis.

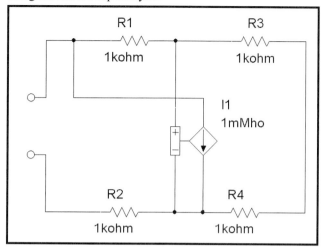

Problem 5-130: Find the impedance of the circuit below using the AC Sweep analysis.

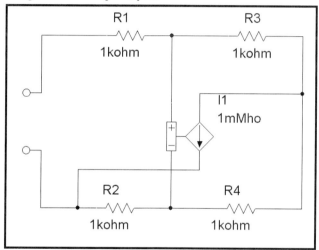

Problem 5-131: For the circuit below, find the AC output impedance between node V_{OUT} and ground using the SPICE AC Analysis.

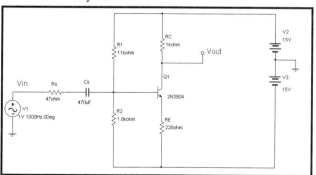

Problem 5-132: For the circuit below, find the AC input impedance between node V_{IN} and ground using the SPICE AC Analysis.

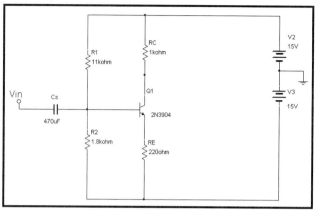

Problem 5-133: For the circuit below, find the AC input impedance between node V_{IN} and ground using the SPICE AC Analysis.

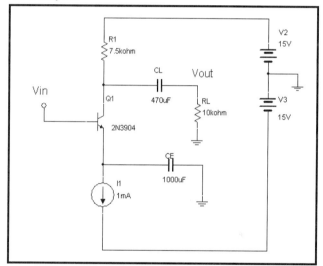

Problem 5-134: For the circuit below, find the AC output impedance between node V_{OUT} and ground using the SPICE AC Analysis.

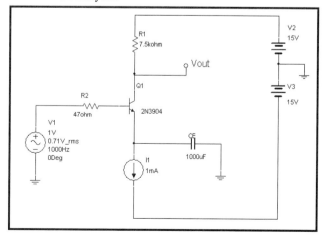

Problem 5-135: For the circuit below, find the AC output impedance between node V_{OUT} and ground using the SPICE AC Analysis.

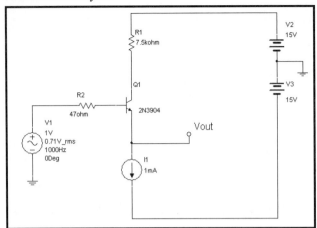

Problem 5-136: For the circuit below, find the AC output impedance between node V_{OUT} and ground using the SPICE AC Analysis.

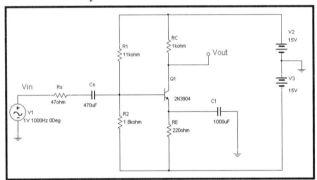

Problem 5-137: For the circuit below, find the AC input impedance between node V_{IN} and ground using the SPICE AC Analysis.

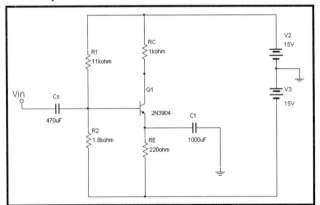

PART 6
Time Domain Analyses

The time domain analyses are used to look at waveforms versus time. Multisim provides two methods of running time domain analyses, the virtual lab simulation using an Oscilloscope instrument, and the standard SPICE Transient Analysis. With these analyses, waveforms are displayed similar to how we would see them on an oscilloscope screen in the laboratory. With the Oscilloscope instrument, the waveforms are displayed almost exactly as you would see them in the lab. You have to set the oscilloscope settings properly to see the oscilloscope traces, and the waveform will dance around the oscilloscope screen if you do not properly trigger the scope. With the Oscilloscope instrument, it is easy to see a few cycles of a periodic waveform on the screen, and you usually do not view the complete waveform from time equals zero. With the SPICE Transient Analysis, you can look at the entire waveform or zoom in to a portion of the waveform. You do not need to set any parameters analogous to the oscilloscope settings to view the results of a Transient Analysis. Example waveforms versus time using the Transient Analysis and the Oscilloscope instrument are:

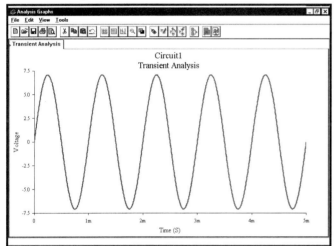 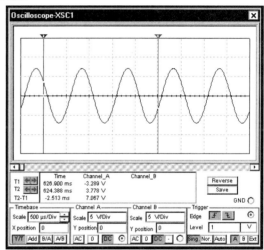

The graphs show us a voltage versus time. Use the Transient Analysis to obtain a waveform versus time. The time domain equation for this waveform is $v_x(t) = 5\sin(2\pi \bullet 1000t + 0°)$. This waveform has an amplitude of 5 V and a frequency of 1000 Hz. If you want to look at the magnitude and phase of voltages and currents, use the AC Sweep or Bode Plotter instrument. The magnitude of the above waveform is 5 V and the phase is zero degrees—in phasor notation, $5\angle 0°$. The AC Sweep or Bode Plotter instrument will give us the result $5\angle 0°$. The Transient Analysis and Oscilloscope instrument will give us the graphs shown above.

6.A. Using the Oscilloscope Instrument

In Multisim 7 three different oscilloscopes are available. The basic two-channel oscilloscope:

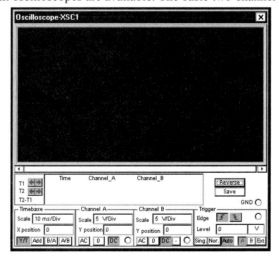

a four-channel oscilloscope:

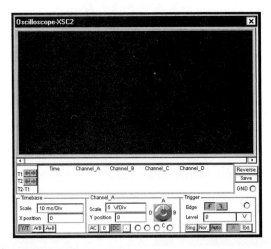

and an Agilent 54622D oscilloscope:

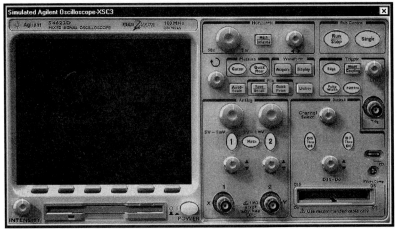

The two- and four-channel oscilloscopes function nearly the same except for the number of channel inputs. The Agilent oscilloscope operates as it does in the lab, so a student would press the same buttons in the same sequence whether in the lab or using Multisim 7. Thus, the use of the Agilent oscilloscope is a little bit different than that of the two- and four-channel oscilloscopes. Because the Agilent oscilloscope is a specialized unit and is useful only if you have the piece of equipment in your lab, and because the two- and four-channel scopes are so similar, we will only cover the operation of the two-channel oscilloscope in the following sections. The examples covered here can be easily generalized to the four-channel oscilloscope.

The settings on the Oscilloscope instrument are similar to what you would see on any analog oscilloscope you would use in the lab. Its main settings are the timebase or time per division for the horizontal axis, volts per division for the two inputs that are displayed on the vertical axis, and the trigger setting. If you know how to use an analog oscilloscope, then you already know how to use these settings. (Note that there is no "Auto-Set" or "Auto-Scale" button for the Multisim Oscilloscope instrument.) If you are unfamiliar with using oscilloscopes, then using the Oscilloscope instrument in Multisim will give you the skills you need to use an oscilloscope in the lab.

We will start with a simple circuit and show how to display a waveform. Wire the circuit below:

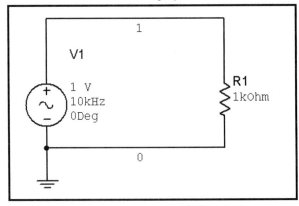

This circuit uses the AC voltage source and a virtual resistor. The value of the resistor is not important. Since the source is ideal, its voltage is not affected by the resistor. The resistor just absorbs power. The amplitude of the source is 1 volt, and its frequency is 10 kHz. To change the properties of the source, double-click the **LEFT** mouse button on the source graphic:

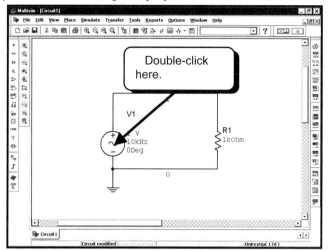 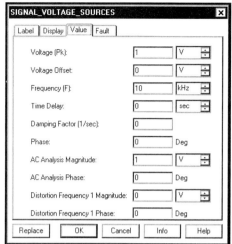

Set the parameters of your source to match those shown above. Also, make sure that your circuit is grounded similar to the circuit shown above.

Next, we must place the Oscilloscope instrument in your circuit. Click the **LEFT** mouse button on the Oscilloscope button as shown. The Oscilloscope instrument will become attached to your mouse:

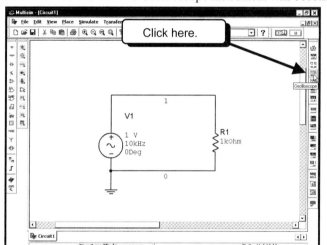 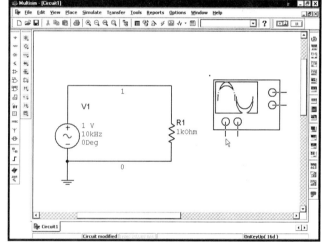

Place the Oscilloscope instrument in a convenient location:

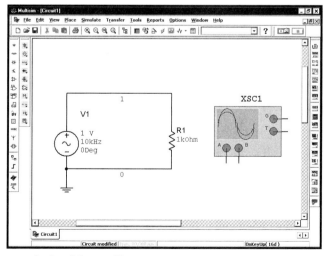

We will look closely at the input terminals of the oscilloscope:

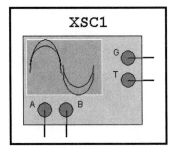

A, **B**, **G**, and **T** are terminals on the oscilloscope to which we can make connections. The ground terminal (**G**) allows you to connect the oscilloscope to ground. Most lab oscilloscopes are grounded through the three-terminal plug that you plug into the wall outlet. All oscilloscope probes have a measuring pin and a ground clip. Because the oscilloscope is grounded through the three-terminal plug, and a standard oscilloscope measures all voltage relative to ground, you do not have to use the ground clip on the oscilloscope probe to measure a voltage. However, experience will show you that using the ground clip on the oscilloscope probe greatly reduces the amount of noise you measure with the scope. Most oscilloscopes also have a ground lug somewhere on the front of the scope. This ground lug is usually a knurled knob that you can unscrew and attach a wire to, or a terminal to which you can connect a banana plug. This terminal provides you with another way to ground the oscilloscope to your circuit. Usually you will not use this terminal since the oscilloscope is already grounded through the three-terminal plug. Thus, it does not matter if you use the ground terminal on the front of the oscilloscope. The **G** terminal on the Multisim oscilloscope is analogous to the ground terminal on the front of your oscilloscope. If you do not hook up the **G** terminal, the Oscilloscope instrument will still be grounded, and will still measure voltages relative to ground. Thus, hooking up the **G** terminal to ground is optional, and does not affect the simulation of the circuit. Like oscilloscopes in the lab, the Multisim oscilloscope is automatically grounded.

The **A** and **B** terminals are the measurement inputs of the scope. These two terminals are connected to your circuit, and the voltages they measure are displayed on the oscilloscope screen. Lab oscilloscope inputs use 10:1 probes that have an input pin and a ground clip. The Multisim oscilloscope does not have the equivalent of a 10:1 probe, and there are no ground clips. On real oscilloscopes, the 10:1 probes with ground clips are necessary to reduce noise and allow the measurement of high-frequency waveforms. 10:1 probes and ground clips are not necessary in a simulation because all we are really doing with the Oscilloscope instrument is plotting the numerical data from a simulation. Thus, the Oscilloscope instrument has only a single connection for the **A** input and a single connection for the **B** input. Wiring either of these inputs to a node in the circuit causes the Oscilloscope instrument to display the voltage of that node relative to ground. The procedure is basically the same as you would do in the lab except that there is no 10:1 probe, and you do not need to connect the ground clip.

The **T** input is an external trigger input for the oscilloscope, and is analogous to the external trigger input of a real oscilloscope. Normally, oscilloscopes generate their trigger input from the waveform measured on either **Channel A** or **Channel B**. When measuring large waveforms, using the **A** or **B** input as the trigger input works well; however, if the measured waveforms on channel **A** or **B** are very small, the oscilloscope may have trouble generating a trigger from the small waveforms and the displayed waveforms will dance across the oscilloscope screen. One method to avoid this problem is to use the external trigger input (**T**) of the oscilloscope. The signal on the trigger could be a very large voltage waveform allowing the oscilloscope to easily generate a trigger, while the inputs on the **A** and **B** inputs are very small. The signal on the trigger input has to be synchronized to the waveforms being measured on the **A** and **B** inputs, but it can be much larger and allow the oscilloscope to easily generate a trigger. The external trigger input (**T**) on the Multisim Oscilloscope instrument has the same function as the external trigger input of a real scope. We will not use it in any of the simulations in this chapter because all of the waveforms we will measure are very large, and the oscilloscope can easily generate a trigger from either the **A** or **B** inputs.

We will first use **Channel A** to measure the voltage of the source. Wire the oscilloscope to your circuit as shown:

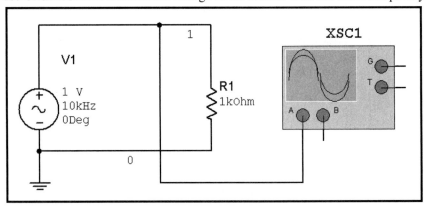

Remember that the oscilloscope is automatically grounded. To run the simulation, click the *LEFT* mouse button on the **Run / stop simulation** button as shown:

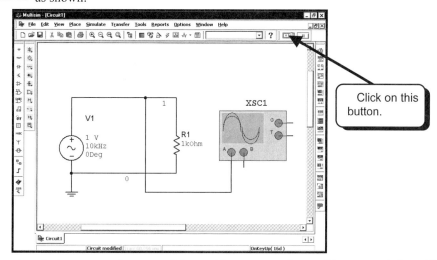

The simulation has started running, but we cannot see anything until we open the oscilloscope window. To open the oscilloscope window, double-click on the Oscilloscope instrument:

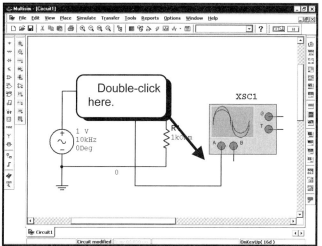
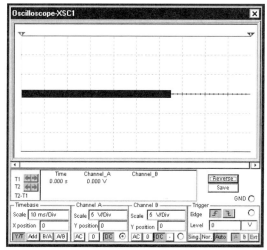

Your window may look similar to the one shown or not, depending on your oscilloscope's settings. After we change all of the settings, we should all see the same display.

6.A.1. Timebase

The first thing we will do is change the timebase. The present setting of the **Timebase** on my oscilloscope is **10 ms/Div**:

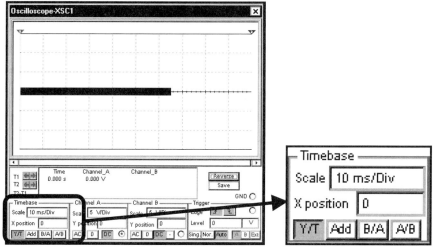

Note that the **Y/T** button is selected. This means that the horizontal (x) axis is the time axis, and all waveforms are displayed versus time. The y-axis is what is measured on **Channel A** and/or **Channel B**. The horizontal axis is time, and the **Scale** is **10 ms** per major division.

We are measuring a signal that has a 10 kHz frequency. The period of a signal is the reciprocal of its frequency, so the period of the measured signal is 0.1 ms, or 100 µs. The period of a signal is how long it takes the signal to repeat itself. Because our **Timebase** is set to 10 ms/div, and our measured signal repeats itself every 0.1 ms, there are 100 sine waves in every division on the horizontal axis. This is why the displayed waveform appears to be a solid thick black line.

Suppose we would like to display one sine wave of the measured waveform in one major division of the horizontal axis. (We would like one period to fit into one square.) One period is 0.1 ms or 100 µs, so we need to change the **Timebase** to 100 µs/div. To change the setting, click the *LEFT* mouse button in the **Scale** field. A cursor will be placed in the field, and up and down arrows will be displayed in the field:

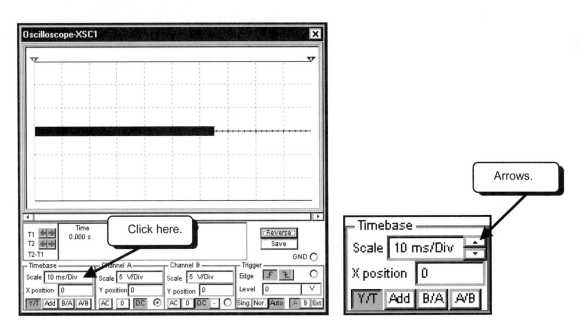

You can use the mouse to click on the arrows and adjust the settings, or you can use the arrows (⬇⬆) on the keyboard to adjust the setting. Adjust the setting to 100 µs/div (not 100 **ms**/div). I will press the down arrow ⬇ on the keyboard to change the setting:

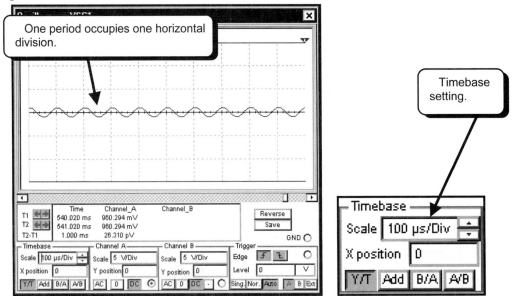

You can change the **Timebase** as you see fit to display the waveform to your tastes. I will display two sine waves over the entire screen:

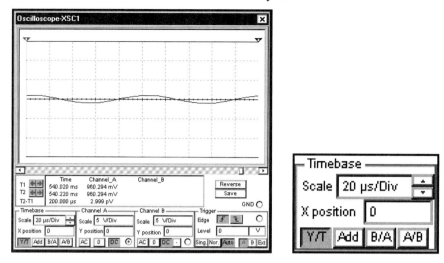

The **Add**, **B/A**, and **A/B** buttons will be discussed later.

6.A.2. Channel A and Channel B Volts per Division Settings

The volts per division setting works the same for both channels A and B. Because we are only measuring a single waveform on **Channel A**, we will confine our discussion to the **Channel A** controls. When you hook up the **Channel B** input, you can use the **Channel B** volts per division settings the same way as shown here.

The waveform we are measuring is a 1-volt-amplitude sine wave. This means that the signal's largest positive value is +1 volt, and its largest negative value is –1 volt. When we look at the **Channel A** volts per division setting, we see that it is set at **5 V/Div**:

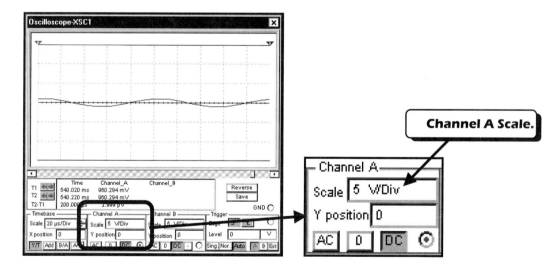

The voltage measured on **Channel A** is displayed on the vertical or y-axis. With a **5 V/Div Scale**, each major division in the vertical direction represents 5 volts. Because our signal only has an amplitude of 1, the signal occupies only 1/5 of a division in the y-direction.

To make the measured waveform appear larger in the oscilloscope window, we need to change the **Channel A** volts per division setting. To have the waveform occupy one division in the positive direction and one division in the negative direction, we would set the **Channel A** volts per division setting to 1 V/div. To change the setting, click the *LEFT* mouse button in the **Scale** field. A cursor will be placed in the field, and up and down arrows will be displayed in the field:

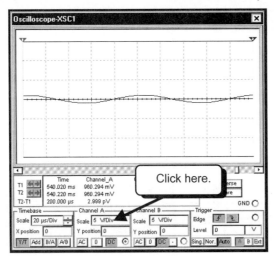
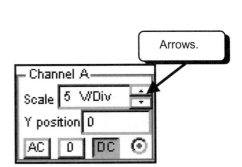

You can use the mouse to click on the arrows and adjust the setting, or you can use the arrows (⬆⬇) on the keyboard. Adjust the setting to 1 V/div. I will press the down arrow ⬇ on the keyboard to change the setting:

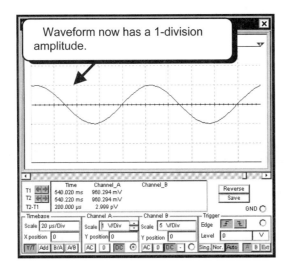
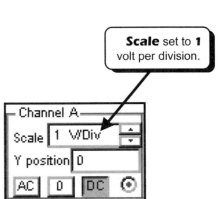

You can adjust the **Scale** setting to display the waveform to suit your tastes. The **Scale** setting below is 500 mV/Div.

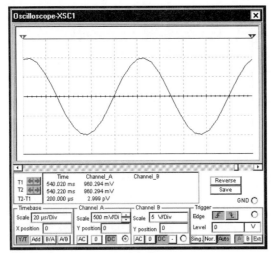
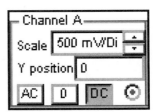

The **Y position** field is equivalent to the offset knob on an oscilloscope and it allows you to move the waveform up or down on the screen. For example, a **Y position** of 1 will shift the waveform up by 1 division:

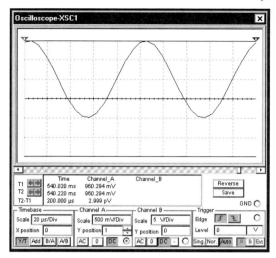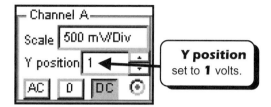

A **Y position** of –1 will shift the waveform down by 1 division:

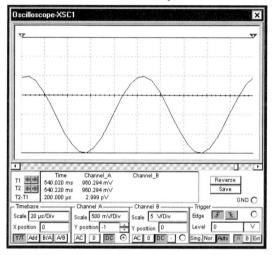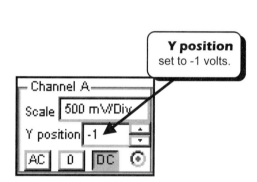

The offset knob on a real scope, and the **Y position** setting on the Multisim oscilloscope allow you to move an oscilloscope trace around the screen and to display the waveforms the way you think best. For a single waveform, the **Y position** is usually set to zero. When you use **Channel A** and **Channel B**, you may want to separate the two traces so that they do not overlap, and you can use the **Y position** settings to move the traces around the oscilloscope window and display them to your liking.

The **AC**, **0**, and **DC** buttons are the coupling settings for the oscilloscope channel:

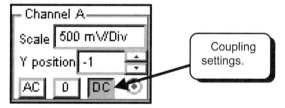

They are the same as the AC, ground, and DC coupling settings on a real oscilloscope. Setting the channel coupling to **0** (or ground on a real oscilloscope) sets the input of that channel to zero, even though that channel is connected to a node and measuring a voltage. With coupling set to **0**, the oscilloscope channel would then measure and display a constant value of zero volts. The **0** setting enables you to identify where the zero voltage reference is on your oscilloscope display, and you can then use the **Y position** to move the zero reference if you want.

The **DC** coupling setting displays everything contained in a waveform. If you are measuring a constant or DC voltage, the oscilloscope will display it. If you are measuring a pure AC waveform, such as the one we are measuring in this example, the oscilloscope will display it. If you are measuring a voltage that has an AC waveform with a DC offset, the oscilloscope will display both the waveform and the offset. Basically, the DC setting shows you everything: both the constant part and the time-varying part.

The **AC** coupling setting filters off the constant portions of a signal and displays only the AC portion of a waveform. If you are measuring a constant or DC voltage, the oscilloscope will display it as zero. If you are measuring a voltage that has an AC waveform with a DC offset, the oscilloscope will remove the DC part and display only the AC part. Basically, the **AC** setting shows you only the time-varying part of the waveform.

To illustrate the differences between the coupling settings, we will add a DC source to our circuit:

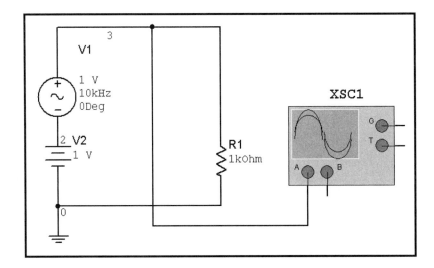

Because the two sources are in series, the sources add. Thus, the waveform being measured by the oscilloscope is 1 V DC plus a 10 kHz sine wave with an amplitude of 1 volt. We will display this waveform on the scope. First, we will set the oscilloscope coupling to **0**. This will show us where the 0 reference is for the oscilloscope display:

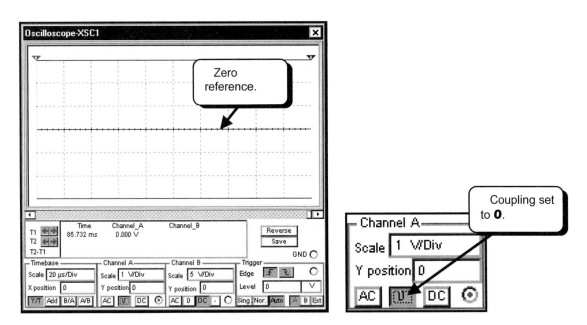

It is difficult to see, but the zero reference is now right in the center of the oscilloscope screen. You can change the **Y position** field to see the constant line shift up and down. When the **Y position** field is set to **0**, as in the screen capture above, the constant line will be right in the center of the oscilloscope display. Also note that I have set the **Scale** to **1 V/Div** in my **Channel A** setup.

If we change the coupling to **DC**, a sine wave will be displayed. It will be centered on 1 volt rather than zero because of the 1 V DC voltage source we added to the circuit. Note that the volts per division setting is set at **1 V/Div** for the display below. To have the oscilloscope display correctly, you may need to set the trigger **Level** to **1** volt and the trigger mode to **Nor**mal. We see that when the oscilloscope is set to **DC** coupling, the oscilloscope displays both the DC offset and the AC waveform:

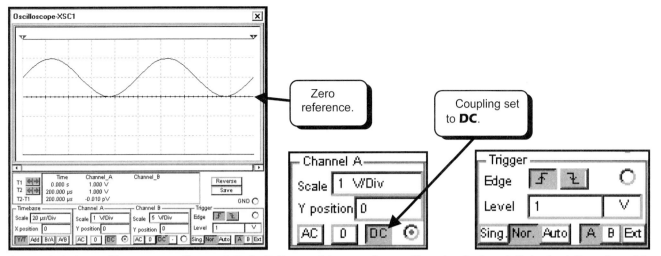

If we switch the oscilloscope coupling to **AC**, the DC part of the waveform will not be displayed. Only the AC portion of the waveform will be displayed, and it will be centered on the zero reference of the oscilloscope. Here we see that AC coupling removes the DC offset and shows us only the time-varying (AC) part of the measured waveform:

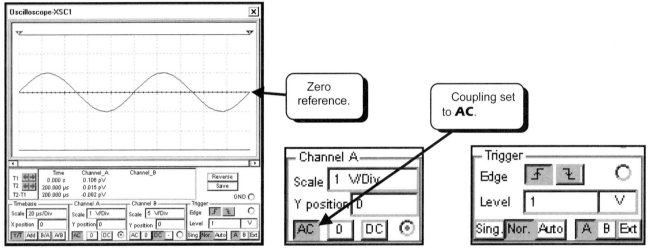

We see that DC coupling shows us both the AC and DC parts of a waveform, while the AC coupling only shows us the AC part and removes the constant or DC part.

As a last example of coupling, we will measure the voltage of the DC source using AC and DC coupling. Rewire the circuit as shown:

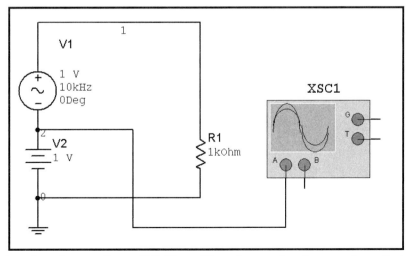

Next, simulate the circuit and set the coupling to DC. You will need to set the trigger mode to **Auto** to see anything on the oscilloscope screen:

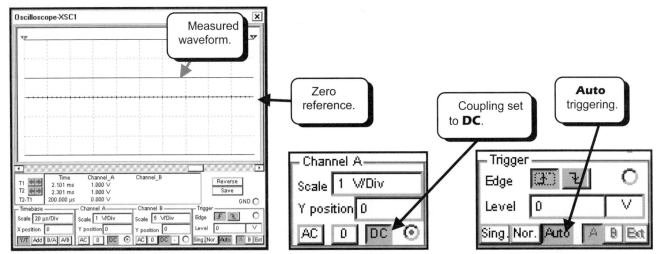

We see that the oscilloscope measures a constant voltage of 1 volt. If we switch the coupling to AC, the oscilloscope ignores the DC part of the waveform, and displays what is left. In this case, the waveform we are measuring is pure DC, so nothing is left, and the oscilloscope displays the measurement as zero:

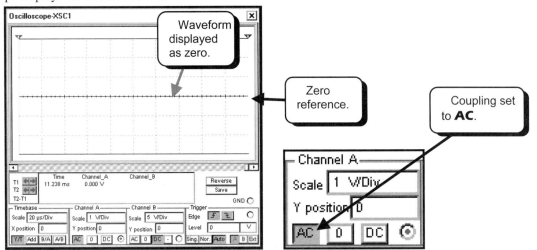

The coupling setting you use depends on what you are attempting to measure. If the offset (DC portion) of a waveform is important, you should use DC coupling. If all you are interested in is the time-varying portion of a waveform, and you do not care what the DC part is, you can use AC coupling. DC coupling shows you everything, while AC coupling shows you only the time-varying part.

6.A.3. Trigger Settings

Triggering is one of the most difficult skills to master with a real oscilloscope. With an oscilloscope that uses a cathode ray tube (CRT), the trace is drawn on the screen by an electron beam that impacts the CRT screen. As electrons impact the screen, the phosphors on the screen glow and are seen by the user. You can think of the electron beam as a pen that writes on the screen. The pen starts at the left side of the screen and moves to the right. The **Timebase** setting tells the oscilloscope how fast to move the trace from the left to the right. The **Channel A** input tells the pen how to move up and down in the vertical direction. The trigger tells the pen when to start moving.

When drawing a line on the oscilloscope screen, the following steps happen. The pen stays on the left side of the screen and does not move until it receives a trigger. When a trigger occurs, the pen starts moving to the right and draws a visible line on the screen. When the pen reaches the right side, it stops drawing a trace and positions the pen on the left side of the screen. The pen remains stationary until it receives another trigger signal.

Thus, the trigger tells the oscilloscope when to start drawing the trace and it synchronizes the trace sweep with the measured waveform. If the oscilloscope cannot generate a trigger, the pen will never move, and a trace will not be drawn on the screen. In this case, a trace will not be displayed and the oscilloscope screen will be blank. If the oscilloscope randomly generates trigger signals that are not synchronized with the measured waveform, the measured waveform will be displayed, but it will dance on the screen and never appear to stand still. When an oscilloscope is properly triggered, a waveform will be displayed and it will appear still on the screen.

For all of the trigger modes discussed here, the oscilloscope can use the waveform on **Channel A**, **Channel B**, or the External Trigger input. You select the source with the buttons shown below:

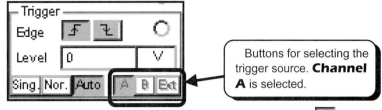

Buttons for selecting the trigger source. **Channel A** is selected.

In the screen capture, the waveform on **Channel A** is selected. The **Edge** is set to positive going and the trigger **Level** is set to **0** volts. The oscilloscope generates a trigger by comparing the waveform of **Channel A** to the trigger level. When the waveform crosses the level going positive, a trigger is generated. In our example screen capture above, when the waveform on **Channel A** crosses 0 volts going positive, the pen will move from right to left and draw a trace on the screen. The resulting waveform is displayed for the circuit below. Note in the circuit that we have changed the peak amplitude of our source to 3 V for illustration purposes. Note also that for these examples, **Nor**mal triggering is selected:

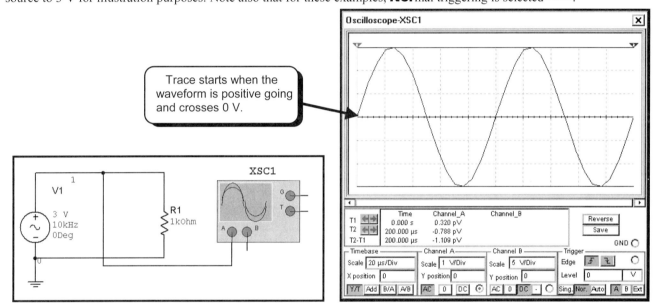

We see that the trace starts when the waveform crosses zero in the positive going direction.

Next, leave all of the settings the same but change the trigger **Level** to **1 V**. This will cause the oscilloscope to trigger when the waveform is positive going and crosses 1 volt:

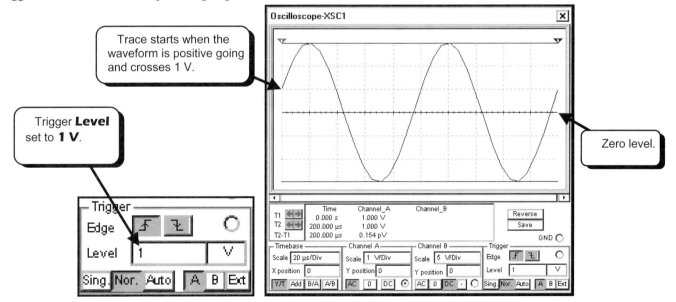

We see that the trace starts when the waveform crosses 1 volt in the positive going direction. Next, leave all of the settings the same but change the trigger **Level** to **2 V** and set the **Edge** to negative . This will cause the oscilloscope to trigger when the waveform is negative going and crosses 2 volts:

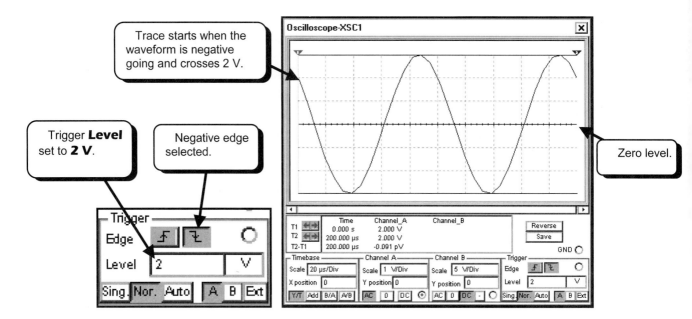

Note that the trace starts when the waveform crosses positive 2 volts. If we wanted the trace to start when the waveform is negative, we would need to set the level to a negative value, –2 volts for example:

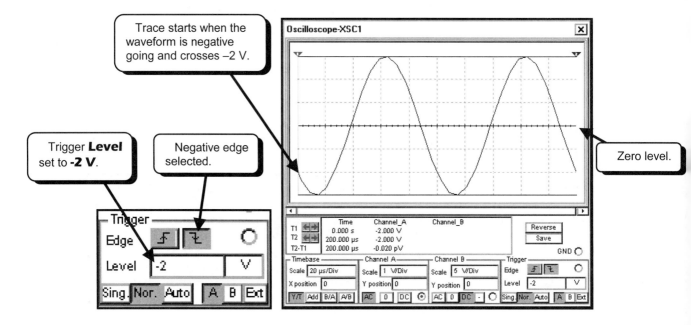

As a last example, we will set the trigger level to 4 volts. Because the waveform never crosses 4 volts, a trigger will never be generated and a trace will not be displayed on the screen:

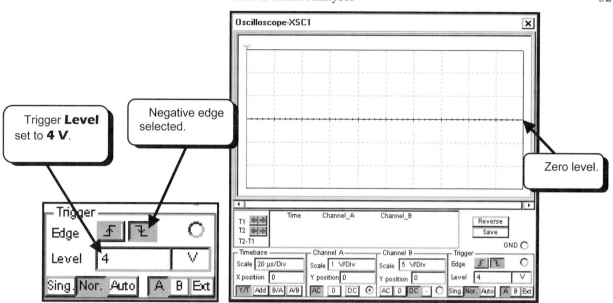

No trace is shown. Since we now understand oscilloscope triggering, we know that we need to decrease the level so that the waveform on **Channel A** crosses the **Level** setting.

The Multisim oscilloscope has three trigger modes, Normal, Single, and Auto:

Normal Triggering (Nor.) – All of the examples above used normal triggering. In this mode, the pen waits at the left side of the oscilloscope window for a trigger. A trigger is generated when the specified trigger source (**Channel A**, **Channel B**, or the External Trigger input) crosses the **Level**. At the trigger, the pen moves to the right and draws a trace on the screen. When the pen reaches the right side, it stops drawing the trace and then positions itself back on the left side of the window. It remains stationary until the next trigger is generated. With this mode, a measured waveform will appear stationary on the oscilloscope window.

Single Trigger (Sing.) – Single trigger mode works the same as Normal trigger mode except that only a single trace is drawn across the screen. When you press the Sing. button, the oscilloscope becomes armed and waits for a trigger. When a trigger occurs, a single trace is drawn on the screen, and then the trigger becomes disarmed. No more traces are drawn until the oscilloscope is once again armed by the user and a trigger is generated. This mode is usually used in situations where waveforms are not repetitive, such as a pulse that occurs only once.

If you have a conventional analog oscilloscope that has a single trigger mode, you would see a single trace drawn on the screen, and then the trace would slowly fade away as the phosphors on the screen slowly stopped glowing. A digital oscilloscope in single trigger mode knows that you want to measure a single event. After it records the single trace, it saves the waveform and continues to display it. The waveform is not erased until you arm the oscilloscope and record another single trace, or you switch the mode of the oscilloscope. Before the advent of digital scopes, there were analog oscilloscopes called "storage oscilloscopes." These oscilloscopes would allow you to measure a waveform in single trigger mode, but would continue to display the waveform long after the trigger occurred. In a conventional analog oscilloscope, the trace would slowly fade. With a storage oscilloscope, the single trace waveform was displayed on the screen as long as you needed it. With conventional analog scopes, digital scopes, and analog storage scopes, the single trigger mode works the same. The oscilloscopes just differ on how the trace is displayed and stored.

Auto Triggering (Auto) – With auto triggering, a trigger is not generated by comparing one of the inputs to the trigger. Instead, the trigger is generated internally. On many scopes, auto triggering uses the 60 Hz signal from the power line to generate a trigger. No matter how the trigger is generated, it is not synchronized to the waveform you are measuring. This will cause a measured waveform to move on the oscilloscope. The waveform you are measuring has nothing to do with the trigger. Thus, the point on your waveform where the trace starts is random and changes. As an example, if we use our previous circuit and switch the trigger mode to **Auto** Auto, a waveform will be displayed, but the starting point of the waveform will not be constant. Three screen captures are shown below to illustrate this point:

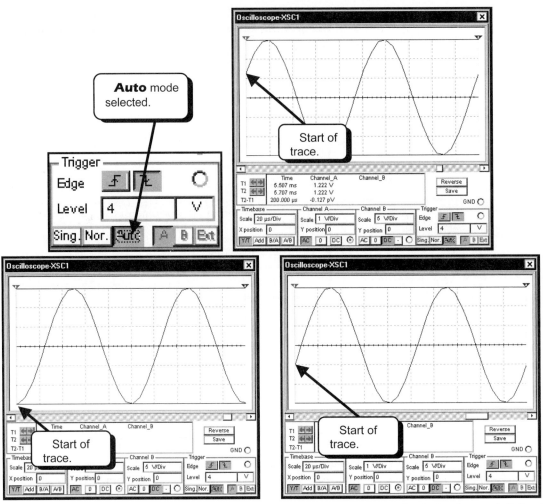

Auto trigger is usually used in two situations. The first is when you are measuring DC voltages. A DC voltage never changes. Therefore, the measured signal can never cross the level and a trigger is never generated if you are using single trigger mode or normal trigger mode. The only way to display a DC waveform is to switch to **Auto** triggering and have the trigger generated internally.

The second situation in which you would use **Auto** triggering is if you are having trouble getting the oscilloscope to trigger. If you cannot get the oscilloscope to generate anything in single or normal mode, switch the oscilloscope to Auto trigger mode. This will allow you to display something on the oscilloscope. Once you see some waveforms, you can measure the waveforms and see how to set the level to the appropriate value to get the oscilloscope to trigger.

We will not stress triggering in future examples. However, when using an oscilloscope, you must set the oscilloscope trigger properly or you will not see any waveforms on the oscilloscope screen.

6.A.4. Using the Cursors

The oscilloscope provides two cursors to allow us to make measurements of waveforms displayed on the oscilloscope window. We will continue with the previous circuit:

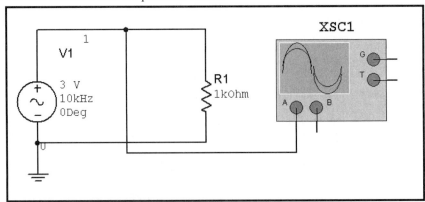

Simulate the circuit and set the **Timebase** to **50 μs/Div**. You may see one of the screens shown below:

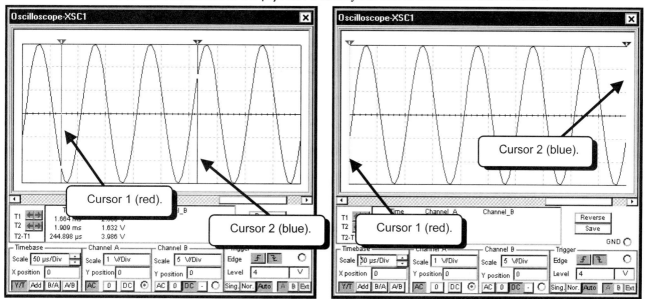

My screen shows two cursors: cursor 1 is shown in red and cursor 2 is shown in blue. Your cursors may not be as easily seen as those above. It may be that both cursors are located at the leftmost portion of the oscilloscope window as shown:

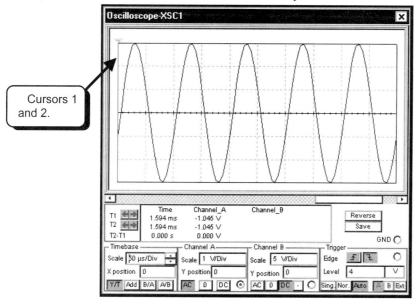

Because the cursors are in the same location, they are displayed in green. We can easily separate the cursors if they are positioned as shown above. Click **and hold** the *LEFT* mouse button on the green triangle as shown below:

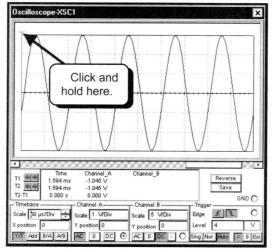

While continuing to hold down the mouse button, drag the mouse. Cursor 1 will move with the mouse. When you release the mouse button, cursor 1 will be repositioned:

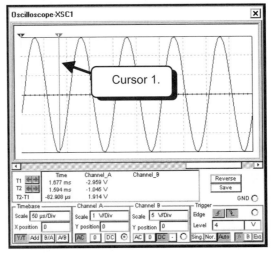

The cursors are now separated and easily identified.

We can now position each cursor separately using the method shown above. To move cursor 1, drag the red triangle. To move cursor 2, drag the blue triangle. As a first measurement, we will measure the period of this waveform. Position cursor 1 (red) as close to a zero crossing as possible by using your eye. Next, place cursor 2 close to a zero crossing, but just do not be that careful. Your cursors should be as shown below:

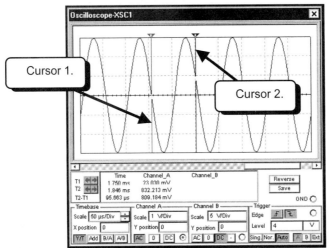

Cursor 1 is pretty close because our eye is pretty good. However, Multisim 7 provides a few facilities so that we can hit the zero crossing exactly. Right-click the mouse on the blue triangle for cursor 2 as shown below. A menu of choices should appear:

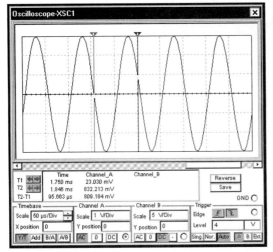 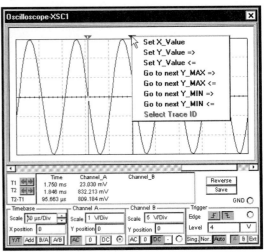

Select **Set Y_Value** => to specify a value for Multisim 7 to find and Multisim places the cursor at the next value it finds going to the right. Select this menu selection and then specify 0 for the value to find:

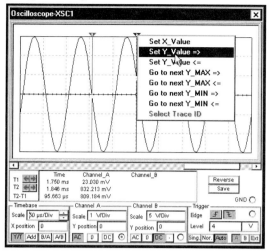
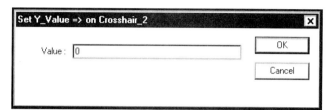

When we click the **OK** button, Multisim 7 will place cursor 2 at the zero crossing:

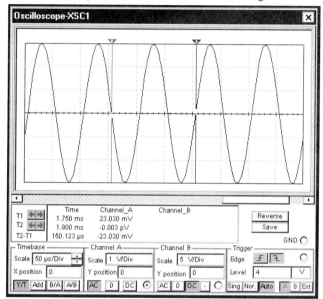

It did not place cursor 2 at the zero crossing I had intended, but I will use the same technique to have Multisim 7 move cursor 1 (red) to the next zero crossing:

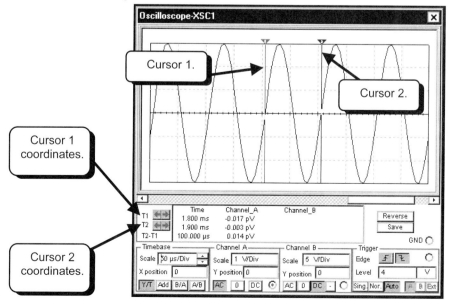

The cursor information is displayed in the text window below the oscilloscope screen. The coordinates of each trace are displayed for each cursor:

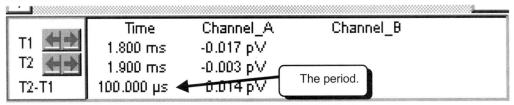

T1 is the coordinate information at cursor 1. **T2** is the coordinate information at cursor 2. The **Time** column lists the time coordinates of each cursor. The **Channel_A** column lists the voltage coordinates for each cursor for the waveform displayed on channel A. The **Channel_B** column lists the voltage coordinates for each cursor for the waveform displayed on channel B. Because there is no waveform on channel B, this column is empty. **T2-T1** is the difference between the two cursors. We see that the time difference between the two cursors is 100 µs. This is the period of the waveform and corresponds to a frequency of 10 kHz.

We will next measure the peak-to-peak swing of the waveform. We can either drag the cursors to the approximate position or right-click on a cursor and use one of the menu selections. I will right-click on cursor 1 and select **Go to next Y_MAX =>** to automatically place cursor 1 (red) at the peak of the sine wave:

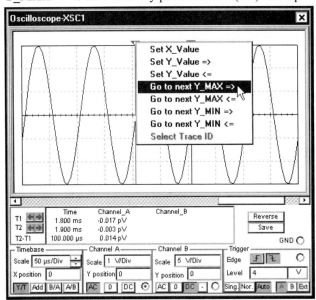
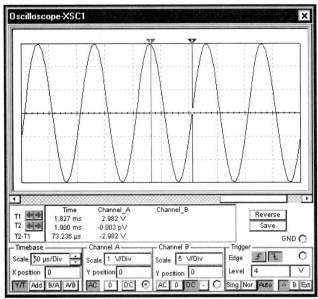

Next, I will right-click on cursor 2 and select **Go to next Y_MIN =>** to automatically place cursor 2 (blue) at the trough of the sine wave:

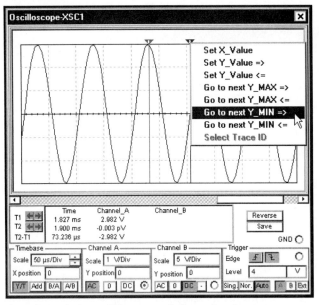
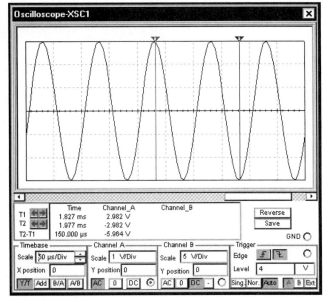

An enlarged view of the cursor information is:

	Time	Channel_A	Channel_B
T1	1.827 ms	2.982 V	
T2	1.977 ms	-2.982 V	
T2-T1	150.000 µs	-5.964 V	← Peak-to-peak voltage of the waveform.

We see that the difference between the voltages of cursor 2 and cursor 1 is **5.964** volts. This is the peak-to-peak voltage of the waveform. The peak-to-peak value is twice the amplitude, so this measurement agrees with the 3-volt amplitude specified in the schematic.

6.B. Phase Measurements of a Capacitive Circuit

For a simple example, we would like to measure the magnitude and phase of the capacitor voltage in the circuit below:

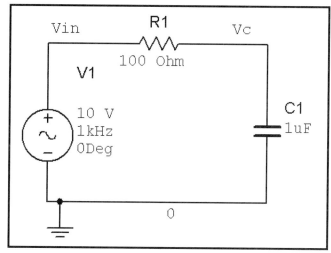

Before we measure the circuit with the oscilloscope, we will make some calculations. The impedance of the capacitor at 1000 Hz is

$$Z_C = \frac{1}{j\omega C} = \frac{1}{j2\pi(1000\,\text{Hz})(10^{-6}\,\text{F})} = -j159.155\ \Omega = 159.155\angle -90°.$$

We will use the input voltage as our reference, so the phase of the input is taken as zero degrees, $V_{IN} = 10\angle 0°$ volts. Using a voltage divider, we can obtain the capacitor voltage:

$$V_C = V_{IN}\left(\frac{Z_C}{Z_C + R}\right) = (10\angle 0°\,\text{volts})\left(\frac{-j159.155\,\Omega}{-j159.155\,\Omega + 100\,\Omega}\right) = (7.17 - j4.505)\,\text{volts} = 8.467\angle -32.142°\,\text{volts}$$

From these examples, we expect the magnitude of the capacitor voltage to be about 8.5 volts and the phase of the capacitor voltage to lag the input voltage by about 32 degrees.

We will first measure the magnitude of the capacitor voltage. Add the Oscilloscope instrument as shown:

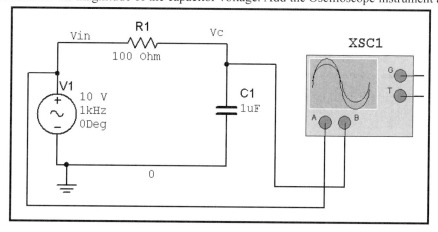

We are measuring the input voltage with **Channel A** and the capacitor voltage with **Channel B**. Click on the **Run / stop simulation** button to start the simulation and then double-click on the Oscilloscope instrument to display the oscilloscope window:

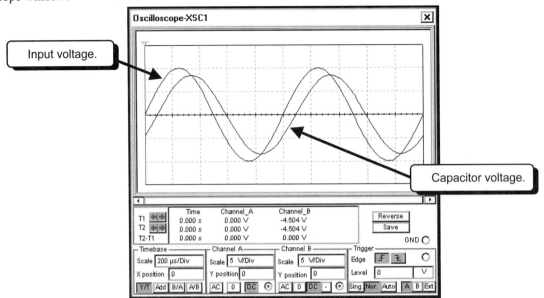

It may not be obvious which trace is which on the oscilloscope. However, from our calculations we know that the capacitor voltage is smaller in magnitude than the input voltage, so we can identify the traces by their magnitude.

Next, we will use the cursors to measure the magnitude of the capacitor voltage. See section 6.A.4 for instructions on using the oscilloscope cursors. Place the cursors as shown:

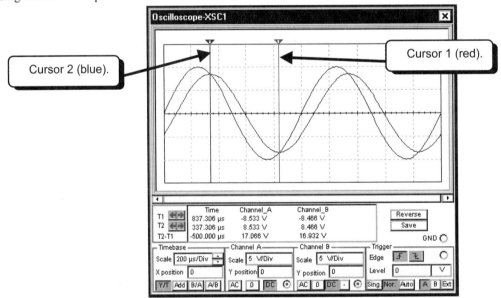

The cursors are located at the peaks of the capacitor voltage waveforms. The measurements displayed by the cursors are:

	Time	Channel_A	Channel_B
T1	837.306 μs	-8.533 V	-8.466 V
T2	337.306 μs	8.533 V	8.466 V
T2-T1	-500.000 μs	17.066 V	16.932 V

Peak-to-peak capacitor voltage.

We are measuring the capacitor voltage with **Channel B**, so the **T2-T1** measurements correspond to the capacitor peak-to-peak voltage. We see that the capacitor peak-to-peak voltage is **16.932** volts. The peak-to-peak value is twice the amplitude, so a peak-to-peak measurement of **16.932** volts corresponds to an amplitude of 8.466 volts. The calculated value of the capacitor voltage magnitude was 8.467 volts. Thus, our measurement agrees with our calculation.

Next, we will measure the phase of the capacitor voltage relative to the phase of the input voltage. We will first look at our present oscilloscope display and identify which trace is leading:

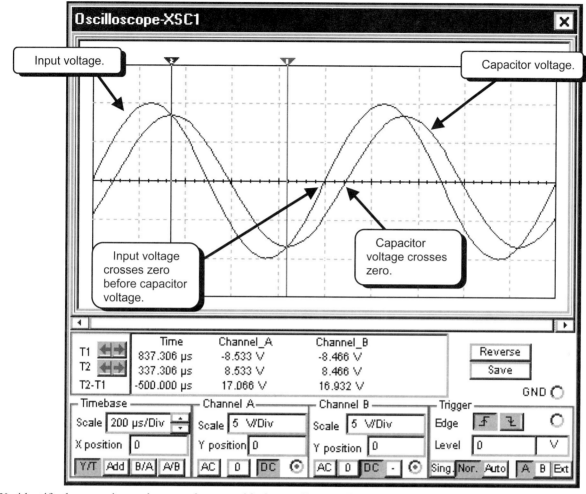

We identify the capacitor voltage as the one with the smaller amplitude. From the screen above, we conclude that the input voltage is leading the capacitor voltage, or that the capacitor voltage is lagging the input voltage. When we calculate the phase angle of the capacitor voltage, we will write it as a negative number because the capacitor voltage is lagging the input voltage.

To measure the phase angle, we need to measure the time difference between the two traces when they cross zero. To get an accurate measurement, we need to blow up the time axis. Change the **Timebase** to **20 μs/Div** and position the cursors at the zero crossing points:

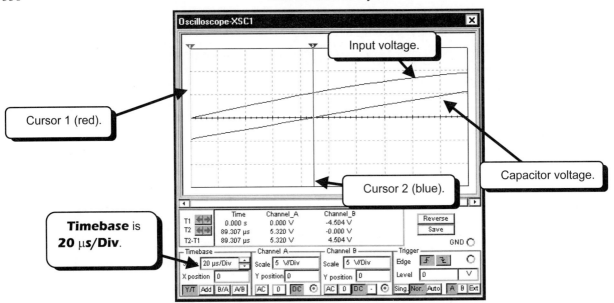

The difference in time is used to calculate the phase angle. The cursor readings are shown below:

We see that the time difference between the cursors is Δt = **89.307** μs. We must now convert this time difference into a phase angle. One period or one cycle of a sine wave corresponds to 360°. Because we know the period of our signal, we can use the relationship:

$$\frac{\Delta t}{T} = \frac{\theta}{360°}$$

where T is the period in seconds and θ is the phase angle in degrees. Our source has a frequency of 1 kHz, so the period of our waveforms is 1 ms. We can easily solve this equation for θ:

$$\theta = 360° \frac{\Delta t}{T} = 360° \frac{89.307 \, \mu s}{1 \, ms} = 32.15°$$

Earlier we found that the capacitor voltage lags the input voltage, so the phase angle above is negative. Thus, our measured phase angle is –32.15°. Our calculated value from page 335 was –32.14°, so once again the measured result agrees well with the calculated result.

EXERCISE 6-1: Simulate the RC circuit below and find the magnitude and phase of the capacitor voltage.

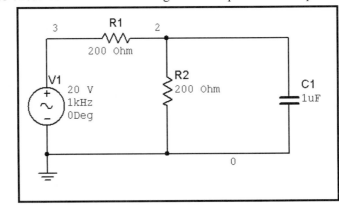

ANSWER: The calculated value of the capacitor voltage is: $8.467 \angle -32.142°$ volts.

EXERCISE 6-2: Simulate the RC circuit below and find the magnitude and phase of the capacitor voltage.

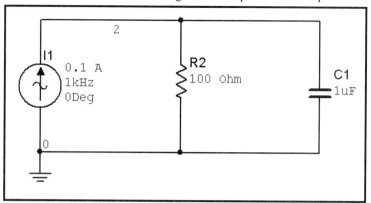

ANSWER: The calculated value of the capacitor voltage is: $8.467 \angle -32.142°$ volts.

6.C. Phase Measurements of an Inductive Circuit

In this section we will look at the magnitude and phase response of an inductive circuit. From a simulation point of view, observing the magnitude and phase of an inductive circuit is no different than what we did in the previous circuit. To make this simulation a little different, we will display the inductor current rather than the inductor voltage. To display the current we need a sensor to sense the current and create a voltage signal that corresponds to the measured current. Multisim recommends using a current-controlled voltage source with a gain of 1. That is, this device senses a current, and the output voltage is 1 times the measured current. An example circuit that uses this method is:

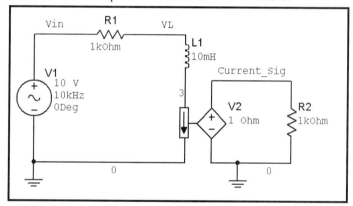

Source **V2** is a current-controlled voltage source. The output of V2 is 1 Ω times the sensed current, which is the inductor current in this case. The gain of the sensor is 1 Ω. The units must be ohms because the output of the sensor is the gain times the sensed current, which has units of amps. For the output to have units of volts and the sensed quantity to have units of amps, the gain must have units of ohms. **R2** was added to prevent the error of having only one element connected to a node. The value of R2 does not affect anything and is arbitrary.

The problem with this method is that in the lab you probably do not have a current sensor like the one shown, and thus this virtual simulation would not closely match how you would do this in the lab. Several current sensors are available, but most are expensive. We will use a very inexpensive sensor called a current-sensing resistor. An example circuit is:

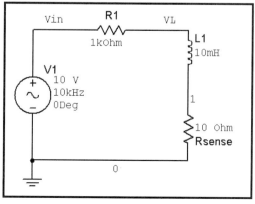

Resistor **Rsense** was added to the circuit and is a current-sensing resistor. The value of a current-sensing resistor is small, so adding it to the circuit does not greatly affect the circuit's operation. However, the current through **Rsense** is the same as the current you want to measure. To measure the current, you measure the voltage across the sensing resistor and then divide by the resistance.

The value of the current-sensing resistor depends on two factors. (1) It must be small compared to other impedances in the circuit so that it does not affect the current you are trying to measure. (2) It must not be so small that the measured voltage is tiny and hard to measure with your instruments. Thus, the value of a current-sensing resistor is a compromise between the effect the resistor has on the circuit and the size of the voltage produced by the resistor. In our example circuit, **Rsense** is 10 Ω and R1 is 1 kΩ. Thus **Rsense** introduces an error of 1% in our measurements. The current we measure will be on the order of 10 mA, so the voltage across **Rsense** will be on the order of 100 mV. This voltage is easily measured with an oscilloscope.

We will now calculate the magnitude and phase of the inductor current. We will ignore the current-sensing resistor, because we are interested in the original circuit. **Rsense** was added in the lab as a method to measure the current. We will calculate the current for the circuit below:

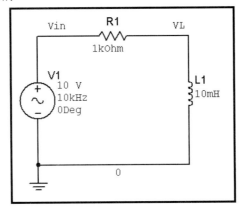

The impedance of the inductor at 10 kHz is

$$Z_L = j\omega L = j2\pi(10000\,\text{Hz})(0.01\,\text{H}) = j628.319\,\Omega = 159.155\angle 90°$$

We will use the input voltage as our reference, so the phase of the input is taken as zero degrees, $V_{IN} = 10\angle 0°$ volts. Because this is a series circuit:

$$I_L = \frac{V_{IN}}{Z_L + R} = \frac{10\angle 0°\,\text{volts}}{1000\,\Omega + j628.319\,\Omega} = (7.17 - j4.505)\,\text{mA} = 8.467\angle -32.142°\,\text{mA}$$

From these examples, we expect the magnitude of the current to be about 8.5 mA and the phase of the current to lag the input voltage by about 32 degrees.

We will first measure the magnitude of the current. Add the Oscilloscope instrument and current-sensing resistor as shown:

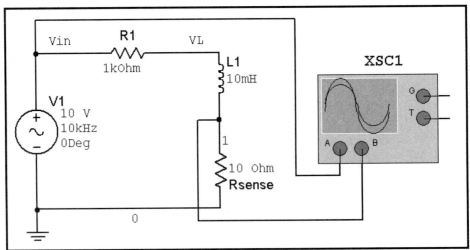

We are measuring the input voltage with **Channel A** and the current-sensing resistor voltage with **Channel B**. Click on the **Run / stop simulation** button to start the simulation and then double-click on the Oscilloscope instrument to display the oscilloscope window:

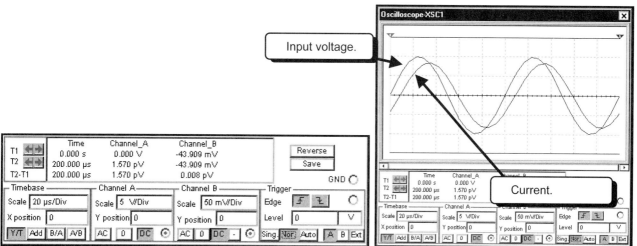

The two traces are easy to identify when you change the volts per division setting for **Channel B**. The voltage across the current-sensing resistor is very small. As you change the **Channel B** volts per division setting, you can see which trace changes.

Next, we will use the cursors to measure the magnitude of the current. See section 6.A.4 for instructions on using the oscilloscope cursors. Place the cursors as shown:

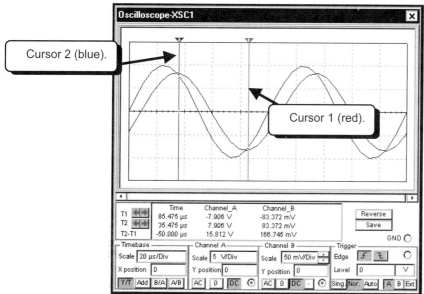

The cursors are located at the peaks of the current waveform. The measurements displayed by the cursors are:

	Time	Channel_A	Channel_B
T1	85.475 µs	-7.906 V	-83.372 mV
T2	35.475 µs	7.906 V	83.372 mV
T2-T1	-50.000 µs	15.812 V	166.745 mV

Peak-to-peak value of the current signal.

We are measuring the current-sensing resistor's voltage with **Channel B**, so the VB measurements correspond to the current. We see that the resistor's peak-to-peak voltage is **166.745 mV**. The peak-to-peak value is twice the amplitude, so a peak-to-peak measurement of **166.745 mV** corresponds to an amplitude of 83.37 mV. To find the current, we need to divide the voltage across the current-sensing resistor by its resistance:

$$Current = \frac{V_{RSense}}{R_{Sense}} = \frac{83.37\,\text{mV}}{10\,\Omega} = 8.337\,\text{mA}$$

The calculated value of the magnitude of the current from page 340 was 8.467 mA. The measured value of 8.337 mA is within 1.5% of the calculated value. A slight error is expected because we neglected the current-sensing resistor in our calculations and because the current signal is a fairly small signal. Remember that adding the current-sensing resistor to our circuit introduces a 1% error into the measurement.

Next, we will measure the phase of the current relative to the phase of the input voltage. We will first look at our present oscilloscope display and identify which trace is leading:

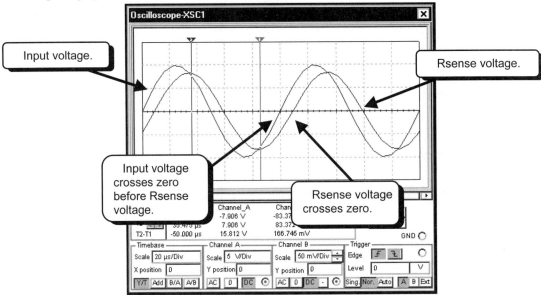

From the screen above, we conclude that the input voltage is leading the current, or that the current is lagging the input voltage. When we calculate the phase angle of the current, we will write it as a negative number because the current is lagging the input voltage.

To calculate the phase angle, we need to measure the time difference between the two traces when they cross zero. To get an accurate measurement, we need to blow up the time axis. Change the **Timebase** to **2 μs/Div** and position the cursors at the zero crossing points:

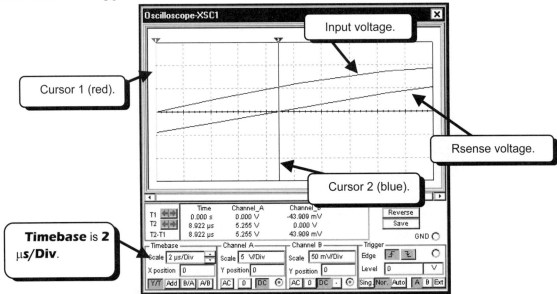

The difference in time is used to calculate the phase angle. The cursor readings are shown below:

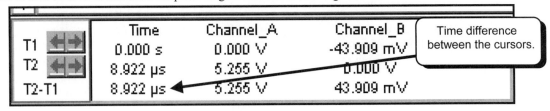

We see that the time difference between the cursors is Δt = **8.922 μs**. We must now convert this time difference into a phase angle. One period or one cycle of a sine wave corresponds to 360°. Because we know the period of our signal, we can use the relationship:

$$\frac{\Delta t}{T} = \frac{\theta}{360°}$$

where T is the period in seconds and θ is the phase angle in degrees. Our source has a frequency of 10 kHz, so the period of our waveforms is 0.1 ms. We can easily solve this equation for θ:

$$\theta = 360° \frac{\Delta t}{T} = 360° \frac{8.922\,\mu s}{0.1\,ms} = 32.12°$$

Earlier we found that the current lags the input voltage, so the phase angle above is negative. Thus, our measured phase angle is −32.12°. Our calculated value from page 340 was −32.14°, so once again the measured result agrees well with the calculated result.

EXERCISE 6-3: Simulate the LR circuit of this section, but use a current-controlled voltage source with a gain of 10 Ω rather than a current-sensing resistor. Show that the results are nearly the same as those we saw in this section and that measured values agree with the calculated values.

SOLUTION: Simulate the circuit below:

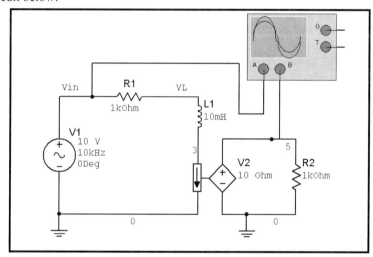

EXERCISE 6-4: Simulate the RC circuit below and find the magnitude and phase of the capacitor current.

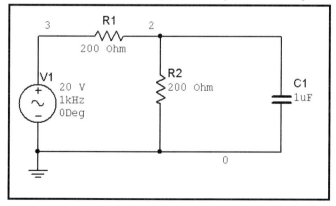

ANSWER: The calculated value of the capacitor current is: $0.0532 \angle 57.86°$ amps.

6.D. Series LCR Resonant Circuit

As a last example of magnitude and phase measurements, we will look at a series RLC circuit at the resonant frequency:

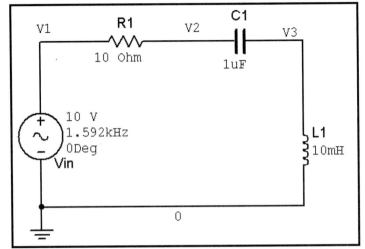

Note that we have labeled the nodes as **V1**, **V2**, and **V3**. This was done so that we can easily talk about the nodes. You do not need to label the nodes if you do not want to. For this example, we are interested in the input voltage, the resistor voltage, the capacitor voltage, and the inductor voltage. All oscilloscopes measure voltages relative to ground. Thus, we can easily measure the voltages V1, V2, and V3 with an oscilloscope. In this circuit, the input voltage and the inductor voltage are easy to measure because the input voltage is equal to V1 and the inductor voltage is equal to V3. Measuring the resistor voltage and the capacitor voltage is a little more difficult. The resistor voltage is equal to V1 − V2, and the capacitor voltage is equal to V2 − V3. Most oscilloscopes provide a subtraction function so that you can display one **Channel** minus the other, **Channel A** minus **Channel B** for example. These oscilloscopes could display the resistor or capacitor voltage. The problem is that with some scopes, when you display **Channel A** minus **Channel B**, you cannot display anything else. Thus you could display the resistor voltage, but you may not be able to compare it to the input voltage because the oscilloscope can display only the subtraction by itself. This limitation depends greatly on your equipment. With an advanced four-channel oscilloscope, you could display all voltages of interest at the same time.

The Multisim oscilloscope is limited in that it displays only a single trace when it displays **Channel A** minus **Channel B**. To display the resistor and capacitor voltages relative to other voltages in the circuit, we use a trick to reference these voltages to ground. This will be shown later.

Before we measure the circuit with the oscilloscope, we will make some calculations. We are interested in measuring the voltages at the resonant frequency, so the first thing we need is the resonant frequency ω_o:

$$\omega_o = \frac{1}{2\pi\sqrt{LC}} = \frac{1}{2\pi\sqrt{(0.01\,\text{H})(1\times 10^{-6}\,\text{F})}} = 1592\,\text{Hz}$$

We can now calculate the impedance of the capacitor and inductor at resonance:

$$Z_C = \frac{1}{j\omega C} = \frac{1}{j2\pi(1592\,\text{Hz})(1\times 10^{-6}\,\text{F})} = -j100\,\Omega = 100\angle -90°$$

$$Z_L = j\omega L = j2\pi(1592\,\text{Hz})(0.01\,\text{H}) = j100\,\Omega = 100\angle 90°$$

Note that at resonance, the impedance of the inductor and the impedance of the capacitor add to zero.

We will use the input voltage as our reference, so the phase of the input is taken as zero degrees, $V_{IN} = 10\angle 0°$ volts. Using a voltage divider, we can obtain the voltage of all the elements in the circuit:

$$V_R = V_{IN}\left(\frac{R}{Z_C + Z_L + R}\right) = (10\angle 0° \text{ volts})\left(\frac{10\Omega}{-j100\Omega + j100\Omega + 10\Omega}\right) = 10 \text{ volts} = 10\angle 0° \text{ volts}$$

$$V_C = V_{IN}\left(\frac{Z_C}{Z_C + Z_L + R}\right) = (10\angle 0° \text{ volts})\left(\frac{-j100\Omega}{-j100\Omega + j100\Omega + 10\Omega}\right) = -j100 \text{ volts} = 100\angle -90° \text{ volts}$$

$$V_L = V_{IN}\left(\frac{Z_L}{Z_C + Z_L + R}\right) = (10\angle 0° \text{ volts})\left(\frac{j100\Omega}{-j100\Omega + j100\Omega + 10\Omega}\right) = j100 \text{ volts} = 100\angle 90° \text{ volts}$$

We see that the resistor voltage is equal to the input voltage (same phase and magnitude). The capacitor voltage lags the input voltage by 90°, and the inductor leads the input voltage by 90°. The inductor and capacitor voltages are equal in magnitude and 180 degrees out of phase. Thus, the inductor voltage and the capacitor voltage add to zero.

We will now modify the circuit to easily measure the resistor and capacitor voltages. We will add voltage-controlled voltage sources with a gain of 1 to the circuit:

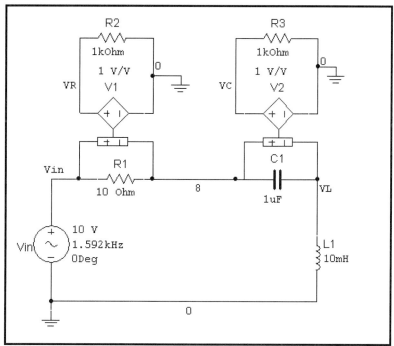

An enlargement of a voltage-controlled voltage source is:

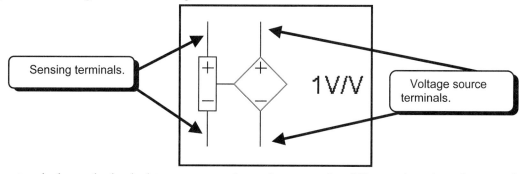

The sensing terminals can be hooked to any two nodes and measure the difference in voltage between those nodes. The voltage of the source is equal to the gain times the sensed voltage. Because we set the gain of our sources to 1, the output voltage of the source is equal to the measured voltage. In the circuit above, the sources measure the resistor voltage and the capacitor voltage. The negative terminals of the sources are grounded. This references the sources to ground, allowing the voltage of the sources to be measured with the oscilloscope. We have labeled the important nodes in the circuit as Vin, VR, VC, and VL because they are the voltages in which we are interested, and they can all be easily measured by an oscilloscope.

You might say that the method we are showing here is purely academic because we cannot use a voltage-controlled voltage source in the lab. This is true, but we can create an OPAMP circuit that implements the same function. We are basically using the voltage-controlled voltage source as a difference amplifier (an amplifier that subtracts one voltage from another). One method of implementing a difference amplifier is shown below:

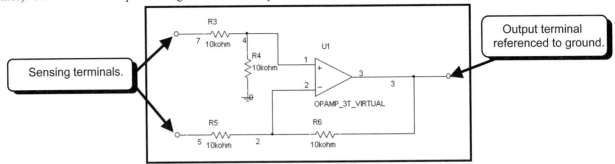

If you want to, you can use this OPAMP circuit rather than the voltage-controlled voltage source. You will need to make two copies, one for the resistor voltage and one for the capacitor voltage. Using the circuit offers no advantage as far as the simulation is concerned, and it will take a long time to draw. The OPAMP circuit shown can handle only small voltages in the range of about –15 to +15 volts. You can use this circuit in the lab, but you will need additional circuitry to reduce the voltages seen by the circuit.

The voltage-controlled voltage source is located in the **CONTROLLED_VOLTAGE_SOURCES Family** in the **Sources Group**:

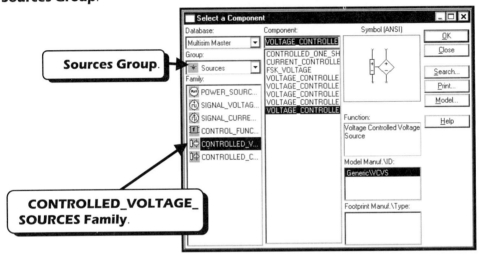

We will first measure the input voltage and the resistor voltage. Add the Oscilloscope instrument as shown:

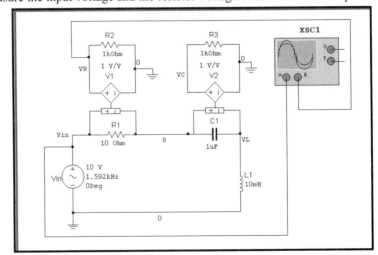

We are measuring the input voltage with **Channel A** and the resistor voltage with **Channel B**. Click on the **Run / stop simulation** button to start the simulation and then double-click on the Oscilloscope instrument to display the oscilloscope window:

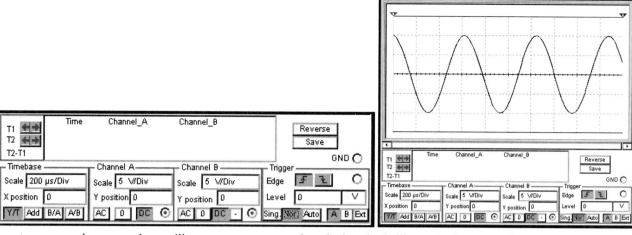

Two traces are shown on the oscilloscope screen even though they look like a single trace. The two traces are nearly identical, so they appear as a single trace. If you change the volts per division or the **Y position** of one of the channels, you will see that two traces are displayed. I will change the **Scale** for **Channel B** to 10 V/Div:

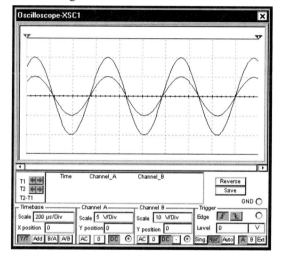

We can now see that two traces are displayed. In our first oscilloscope capture, the settings for both channels were the same and the traces were identical, so they looked like a single trace on the oscilloscope screen. Our simulation shows that **VIN** and **VR** are equal within the measurement accuracy of the oscilloscope. This agrees with our calculations that **VIN** was equal to **VR**.

Next, we will measure the magnitude of the capacitor voltage and compare it to the input voltage. Connect **Channel B** of the oscilloscope to node **VC** and display the waveforms:

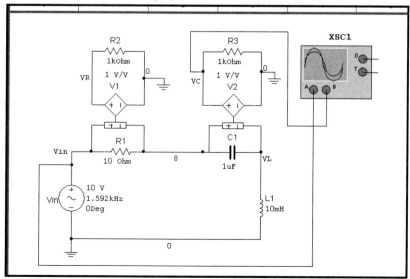

We are measuring the input voltage with **Channel A** and the capacitor voltage with **Channel B**. Click on the **Run/stop simulation** button to start the simulation and then double-click on the Oscilloscope instrument to display the oscilloscope window. Note in the screen capture below that I have not yet changed the **Scale** setting for **Channel B**:

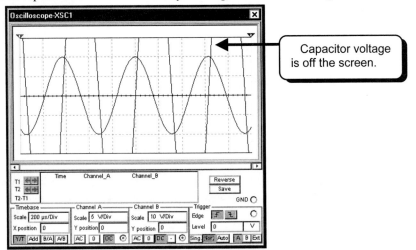

Capacitor voltage is off the screen.

The capacitor voltage is much larger than the resistor voltage of the previous measurement. The voltage is so large that it cannot be displayed with the **Scale** set to 10 V/Div. Change the **Scale** for **Channel B** to **50 V/Div**.

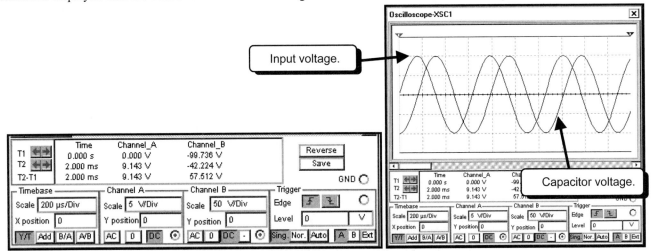

Input voltage.

Capacitor voltage.

A quick look at the oscilloscope screen shows that the two waveforms have about the same amplitude in terms of the number of divisions on the oscilloscope graticule, but the **Scale** setting for **Channel B** is ten times as large as the **Scale** setting for **Channel A**. We conclude that the capacitor voltage is about 10 times as large as the input voltage.

Next, we will use the cursors to measure the magnitude of the capacitor voltage. See section 6.A.4 for instructions on using the oscilloscope cursors. Place the cursors as shown:

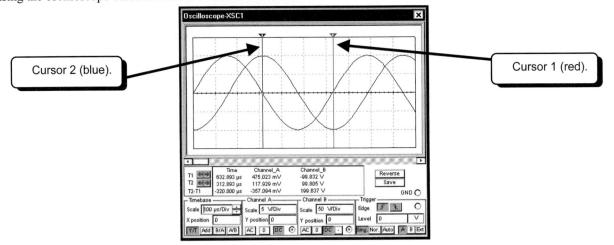

Cursor 2 (blue).

Cursor 1 (red).

The cursors are located at the peaks of the capacitor voltage waveforms. The measurements displayed by the cursors are:

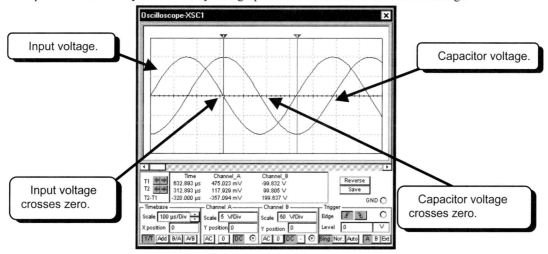

We are measuring the capacitor voltage with **Channel B**, so the **T2-T1** measurements for **Channel B** correspond to the capacitor peak-to-peak voltage. We see that the capacitor peak-to-peak voltage is **199.6** volts. The peak-to-peak value is twice the amplitude, so a peak-to-peak measurement of 199.6 volts corresponds to an amplitude of 99.8 volts. The calculated value of the capacitor voltage magnitude was 100 volts. Thus, our measurement agrees with our calculation.

Next, we will measure the phase of the capacitor voltage relative to the phase of the input voltage. From our calculations, we expect the capacitor voltage to be 90° out of phase with the input. Instead of using the technique presented in sections 6.B and 6.C, we remember that if two sine waves are 90° out of phase, one sine wave crosses zero when the other sine wave is at its peak. We can easily check this by lining up the cursors at one of the zero crossings:

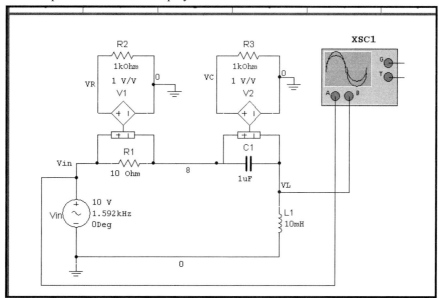

It does appear that the zero crossings of one waveform line up with the peaks of the other waveform, so we conclude that the waveforms are approximately 90° out of phase.

Next, we will measure the magnitude of the inductor voltage and compare it to the input voltage. Connect **Channel B** of the oscilloscope to node **VL** and display the waveforms:

We are measuring the input voltage with **Channel A** and the inductor voltage with **Channel B**. Click on the **Run/stop simulation** button to start the simulation and then double-click on the Oscilloscope instrument to display the oscilloscope window:

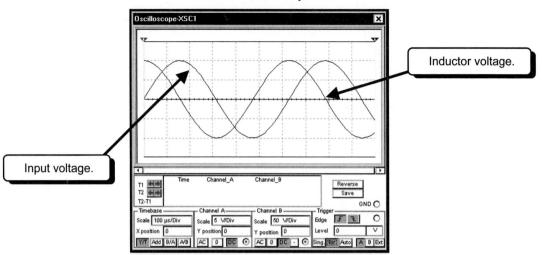

The oscilloscope settings are the same as they were for the last capacitor measurement. The inductor voltage looks nearly the same as the capacitor voltage. If you identify the traces, you will see that the inductor voltage leads the input voltage, while in the previous example, the capacitor voltage lags the input voltage.

Next, we will use the cursors to measure the magnitude of the inductor voltage. See section 6.A.4 for instructions on using the oscilloscope cursors. Place the cursors as shown:

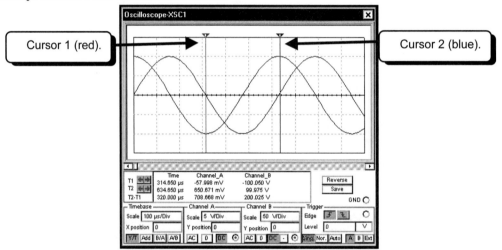

The cursors are located at the peaks of the inductor voltage waveforms. The measurements displayed by the cursors are:

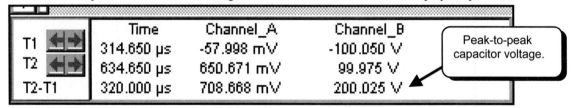

We are measuring the inductor voltage with **Channel B**, so **T2-T1** for **Channel B** corresponds to the inductor peak-to-peak voltage. We see that the inductor peak-to-peak voltage is **200.025** volts. The peak-to-peak value is twice the amplitude, so a peak-to-peak measurement of **200.025** volts corresponds to an amplitude of about 100 volts. The calculated value of the inductor voltage magnitude was 100 volts. Thus, our measurement agrees with our calculation.

Next, we will measure the phase of the inductor voltage relative to the phase of the input voltage. From our calculations, we expect the inductor voltage to be 90° out of phase with the input. Using the same technique we used for the capacitor, we verify that the two waveforms are 90° out of phase. In the previous screen captures, we note that the zero crossings of one waveform line up with the peaks of the other waveform, so we conclude that the waveforms are approximately 90° out of phase.

Next, we will measure the inductor voltage and the capacitor voltage. Change the circuit as shown so that you measure **VC** with **Channel A**:

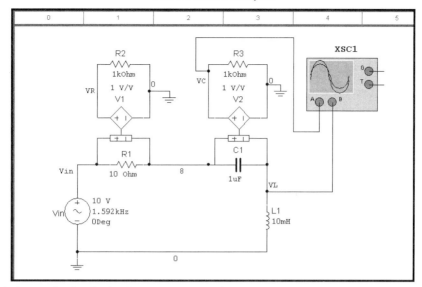

Run the simulation and change the **Scale** for **Channel A** to **50 V/Div**:

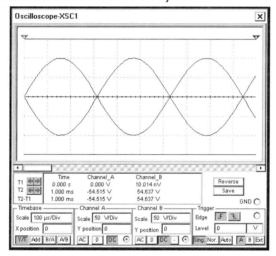

We see that the amplitudes of the waveforms are approximately equal, and that the two waveforms are 180° out of phase, which is the expected result.

For our last measurement, we will measure the input voltage with **Channel A** and the voltage across the capacitor and inductor with **Channel B**. I have connected **Channel B** to the node labeled **Vmid** in my circuit. You do not need to label the node in your circuit. You only have to connect **Channel B** to the same node in your circuit.

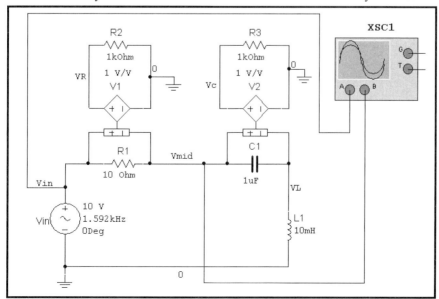

When you run the simulation, you will note that the voltage measured by **Channel B** is very small. I have changed the **Scale** on **Channel B** to **100 mV/Div**. The **Scale** setting for **Channel A** is **5 V/Div**:

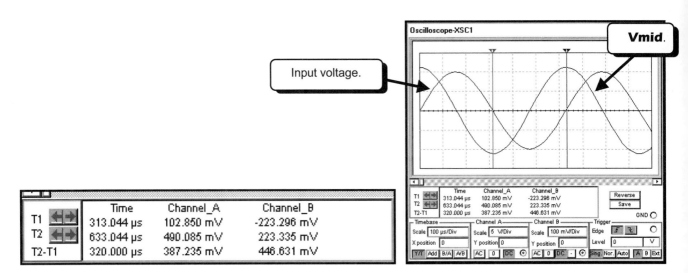

The calculated voltage across the inductor and capacitor is supposed to be zero. In our simulation, we measure a peak-to-peak value of 446.6 mV, corresponding to an amplitude of 223 mV. The voltage is not exactly zero because our signal source is close to, but not exactly at the resonant frequency. The measured voltage is small enough that we conclude that the circuit is working properly, and that our calculations that the voltage goes to zero are also correct.

EXERCISE 6-5: Simulate the parallel RLC circuit below and find the magnitude and phase of the current through each element, and the voltage across the resistor.

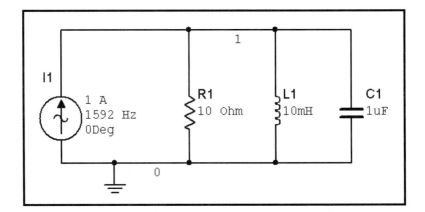

ANSWER: Calculated values: The resistor current is $1\angle 0°$ amps, the capacitor current is $0.1\angle 90°$ amps, the inductor current is $0.1\angle -90°$ amps, and the resistor voltage is $10\angle 0°$ volts.

6.E. Regulated DC Power Supply

We would like to simulate the operation of a voltage-regulated DC power supply. The complete circuit is shown below:

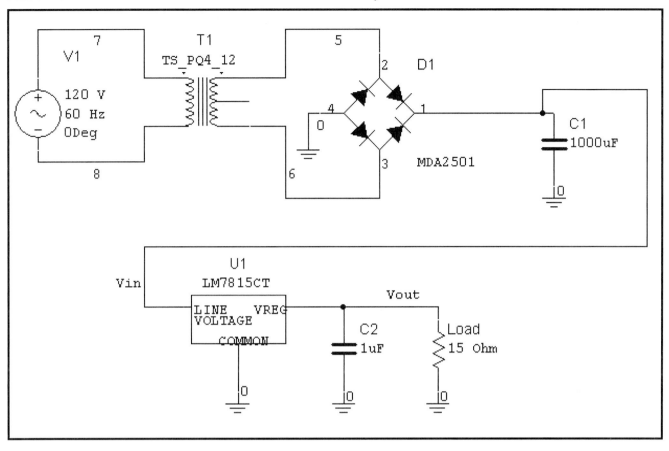

This is a 15 V DC power supply designed for a maximum output current of 1 A. The MDA2501 is a 25-amp, 100-volt bridge rectifier. This part has a much larger current specification than needed for this application, but it is a standard part provided with the Multisim program. This bridge is a standard part made by several manufacturers including International Rectifier (GBPC2501A), Diodes Incorporated (GBPC2501), and General Semiconductor (GBPC2501). The part number MDA2501 is a Motorola number, and is now obsolete.

Because this circuit contains three new parts, we will show how to find them. To place the transformer, click the **LEFT** mouse button on the **Basic** parts button, to select the **Basic Group** of parts:

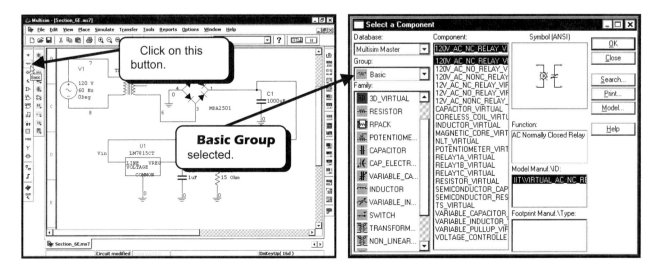

Next, we must select the **TRANSFORMER Family** to see the list of available of transformers:

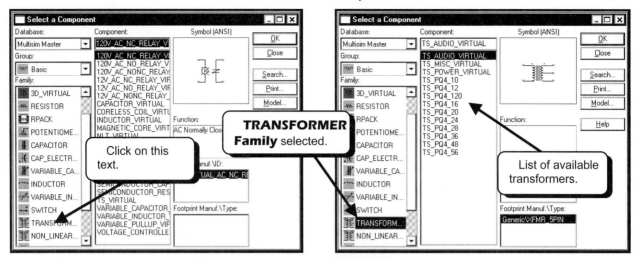

We would like to use a 24 V RMS transformer, so select the **TS_PQ4_12**:

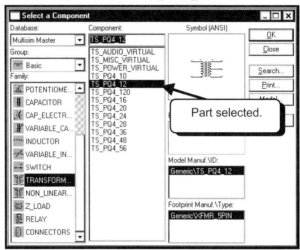

Click on the **OK** button and place the part in your circuit. The part number terminology convention used by Electronics Workbench for transformers is that for a 12-volt part, the RMS voltage between the center-tap and one of the outside terminals is the voltage specified in the part number, in this case 12 Vrms. Thus, for this part the voltage between the two outside terminals is 24 Vrms, and if we use the two outside terminals, we can use it like a 24 V transformer. This is fairly nonstandard terminology. Usually the voltage between the two outside terminals is specified, and the voltage of the center-tap is understood to be half the terminal voltage.

To place the bridge component, click the **LEFT** mouse button on the **Diode** button:

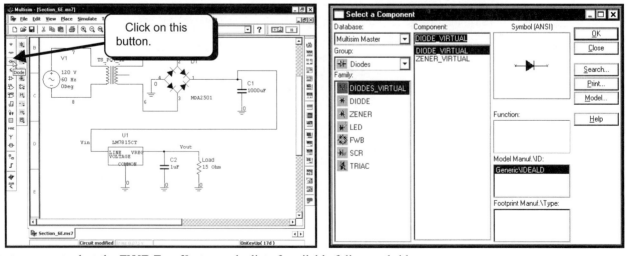

Next, we must select the **FWB Family** to see the list of available full-wave bridges:

Multisim 7   Time Domain Analyses   355

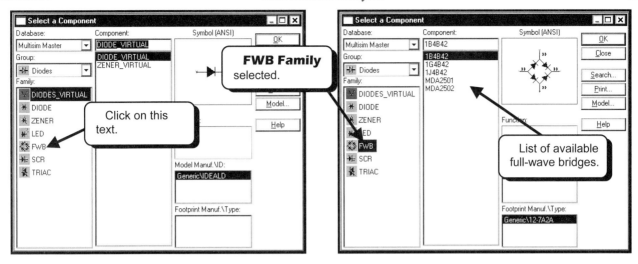

Select the **MDA2501**:

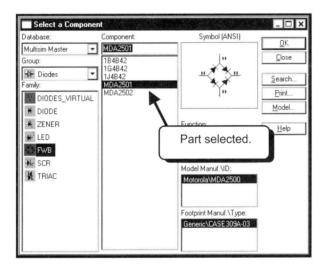

Click on the **OK** button to place the part in your circuit.

    The last new part we will show how to place is the 7815 linear voltage regulator. This part is in the Miscellaneous group. Click the *LEFT* mouse button on the **Misc** button:

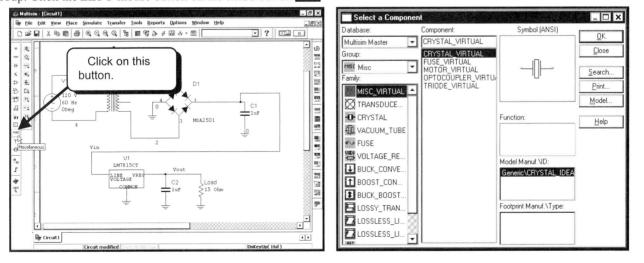

Next, we must select the **VOLTAGE_RE**gulator **Family** to see the list of available full-wave bridges:

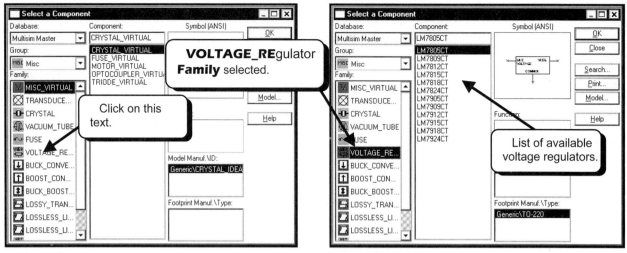

Select the **LM7815CT**:

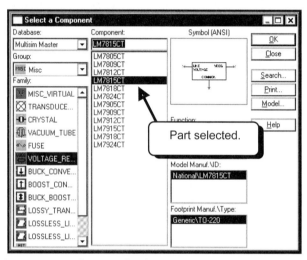

Click on the **OK** button to place the component in your circuit.

We can simulate this circuit with either the virtual lab simulation using an Oscilloscope instrument or the SPICE Transient Analysis. The virtual simulation gives us a feel for what we should see in the lab, while the Transient Analysis is a more powerful simulation that can give us information about every voltage and current in the circuit. We will show both here.

6.E.1. Regulated DC Power Supply – Virtual Lab Simulation

We will first run the virtual simulation. To add the Oscilloscope instrument to your circuit, click the **LEFT** mouse button on the Oscilloscope instrument button :

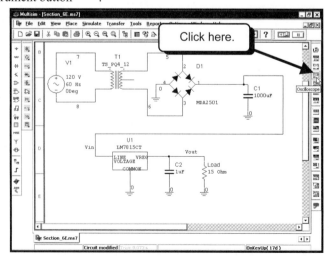

Place the oscilloscope in your circuit and wire it as shown:

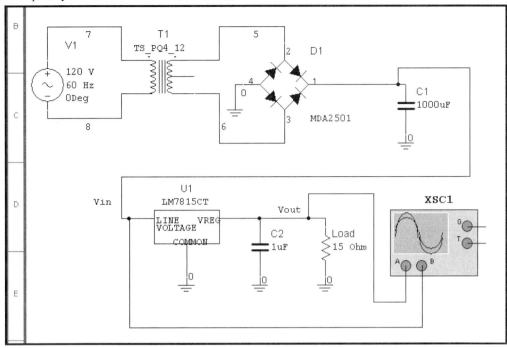

We are now ready to run the simulation, so click on the **Run / stop simulation** button to start the simulation.

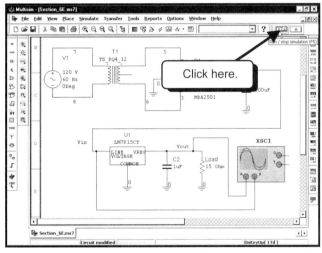

The simulation will start running, but we cannot see the results until we open the oscilloscope window. Double-click the *LEFT* mouse button on the oscilloscope to open its window:

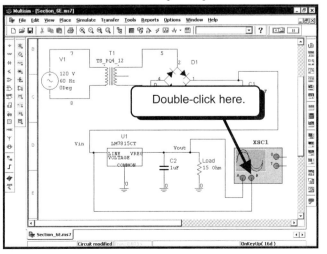 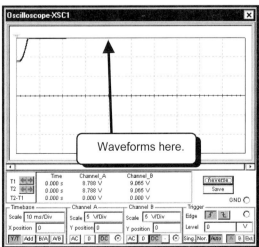

Waveforms are shown, but they are off the screen. If your oscilloscope is set like mine, the **Channel A** and **Channel B** volts per division are set to 5 volts per division (**5 V/Div**). The controls are enlarged below:

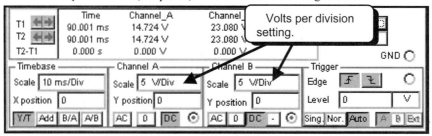

Change the setting to **10 V/Div** for both **Channel A** and **Channel B**:

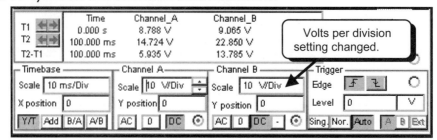

As you change the settings, the display of the waveforms on the oscilloscope screen will also change:

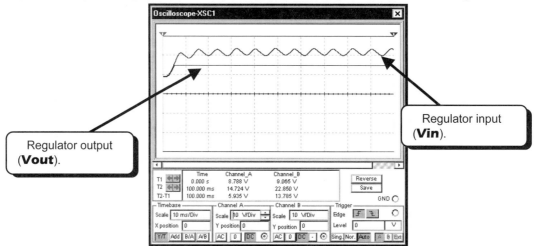

The output appears to be relatively constant at 15 V or so, and the input has a large amount of 120 Hz ripple. This is the function of a voltage regulator. As long as the input is large enough, the output of the regulator will be constant even though there is a large amount of ripple on the input.

If we want to see the ripple on the output of the regulator, we need to specify a small number of volts per division (say 10 mV/Div) for the output trace. With this setting, a 15 V DC signal would be way off the oscilloscope screen. To see the AC portion of the signal and filter off the DC portion we need to set the oscilloscope coupling to AC. Click the *LEFT* mouse button on the AC button for **Channel A** as shown below:

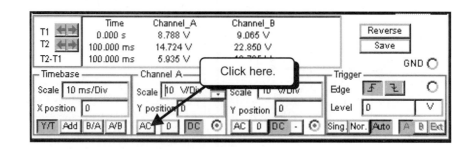

The trace will shift down to the center of the screen:

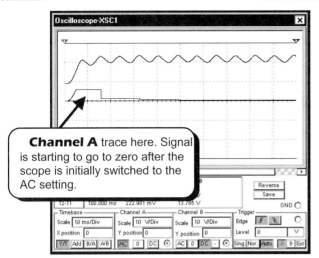
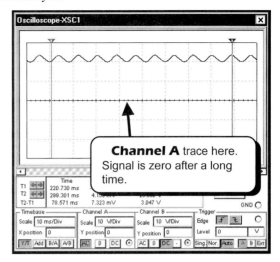

To see the ripple, we need to change the volts per division setting for **Channel A**. Change the setting to blow up the ripple so that it can easily be viewed on the oscilloscope screen. I chose 2 mV per division and also shifted the trace down slightly using the **Y position** setting:

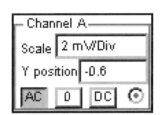
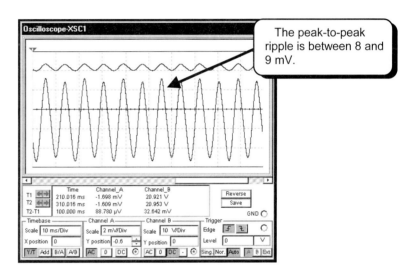

We see that there is ripple at the output, but it is very small when compared to the 15 V DC output.

To get a better idea of how the ripple on the input relates to the ripple on the output, we can change the **Timebase** setting using the same method we used to change the volts per division. I will change the setting to two **2 ms/Div**. Also change the Scale and Y position for Channel A as shown:

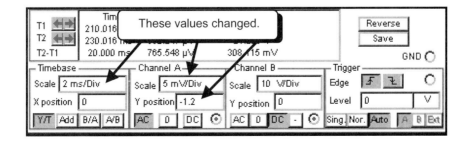

You should see the oscilloscope display below:

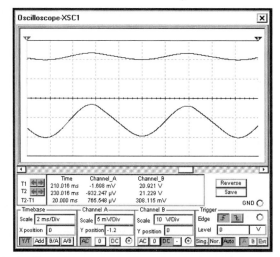

To see what the ripple on the input looks like, change the settings for **Channel B** to **AC** coupling and **2 V/Div**:

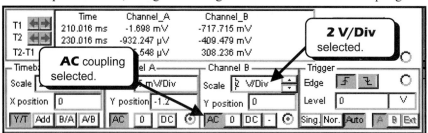

These settings show us that the ripple on the input and output has nearly the same shape:

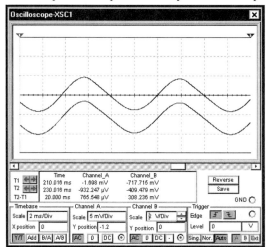

If we change the **Y position** of **Channel A** back to zero, we will see that the two traces overlap:

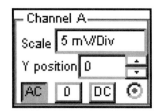

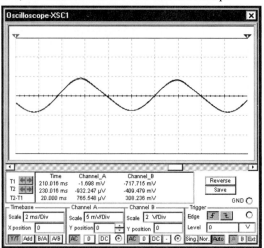

Because the waveforms have the same amplitude, and the volts per division setting for **Channel B** is 400 times larger than the volts per division setting for **Channel A**, we conclude that the ripple at the output is 400 times smaller than the ripple at the input for this regulator.

The specification for the relationship between the ripple on the output as a function of the ripple on the input is called the ripple rejection ratio, and is defined as:

$$\text{Ripple Rejection Ratio}_{dB} = 20\log_{10}\left(\frac{\Delta Vin}{\Delta Vo}\right)$$

For our simulation, we found that the ratio $\frac{\Delta Vin}{\Delta Vo}$ was equal to 400. This corresponds to a ripple rejection ratio of 52 dB. The National Semiconductor data sheet for an LM7815C specifies the ripple rejection ratio to be a minimum of 54 dB and a typical value of 70 dB. So we conclude that the Multisim model for the LM7815CT is very close to the minimum specifications for the National Semiconductor version of the LM7815C.

The last thing we might want to check on our circuit is the actual DC voltage at the output. We could use the oscilloscope cursors to find this value. Instead, we will use the Multimeter to measure the DC voltage at the output. Turn off the simulation by clicking on the **Run / stop simulation** button. Close the oscilloscope window and place the Multimeter instrument as shown. See section 3.A.2 for instructions on using the Multimeter instrument.

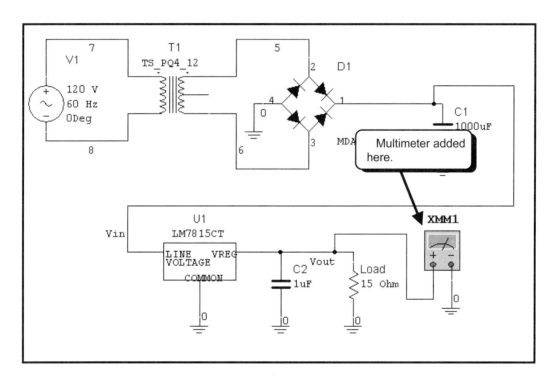

Start the simulation by clicking on the **Run / stop simulation** button. Double-click the *LEFT* mouse button on the Multimeter to open its window. Select the DC voltage setting. After a few moments the meter reading will become steady and display the DC voltage at the output. My meter reads a voltage of **14.721** volts:

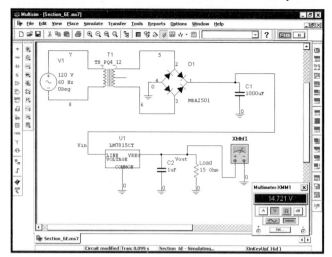
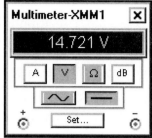

The National Semiconductor datasheet for the LM7815C specifies the output voltage to be between 14.4 and 15.6 volts, so the output voltage is within range for this device.

EXERCISE 6-6: Simulate a power supply using a half-wave rectifier and a regulator. Use the same circuit as on page 353, except use a half-wave rectifier instead of a full-wave rectifier. Use the virtual diode because Multisim does not have a discrete diode that can handle the required current in the library. The remainder of the circuit should be the same.

SOLUTION:

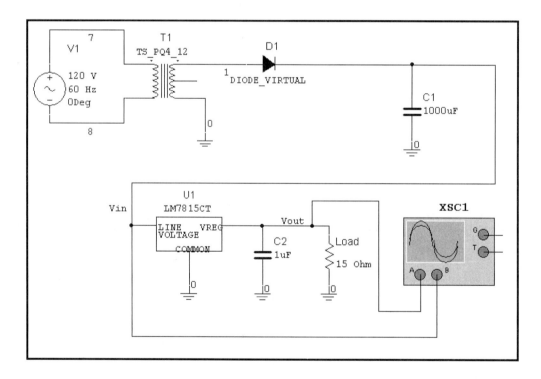

Run the simulation and then display the results on the oscilloscope window. You will need to run the simulation for a long time because it takes the capacitor a long time to reach steady state. When you first start the simulation, you will see waveforms similar to the ones shown below for the input and output of the regulator:

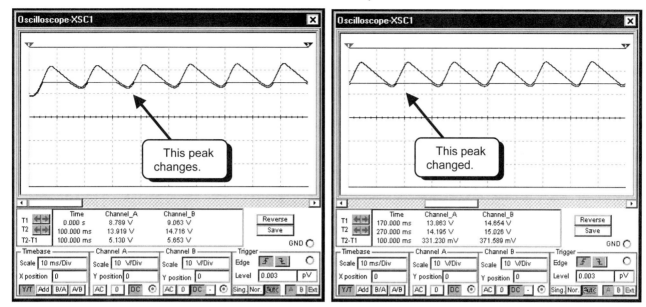

As the simulation runs, the minimum peak of the capacitor voltage increases:

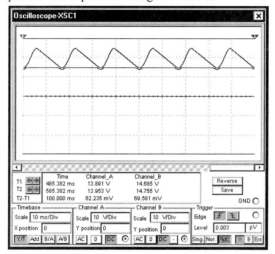

We see that the ripple on the output of the regulator becomes less because the input to the regulator is larger.

As we let this circuit run for a long time, the output of the regulator never becomes pure DC. That is, the drop-out we see on the output of the regulator never disappears because the voltage at the input of the regulator is too small:

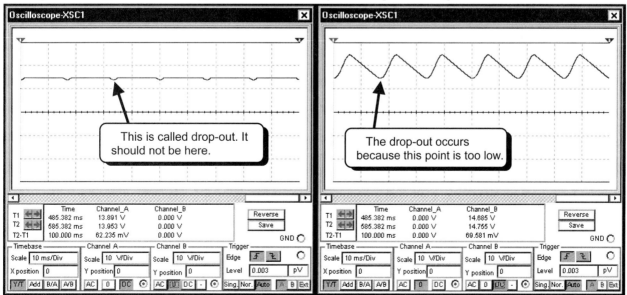

This is because the capacitor we are using for a half-wave rectifier is too small. We can fix the problem by either choosing a higher voltage transformer or a larger capacitor. Some standard size capacitor values are 1200 µF, 1500 µF, 1800 µF, and 2200 µF. I will change the capacitor to a 2200 µF:

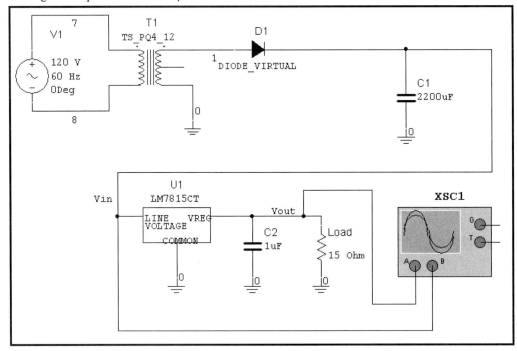

Simulations show that the output is now 15 with a very small amount of ripple on the output. You will need to let the simulation run for a long time for the circuit to reach steady state:

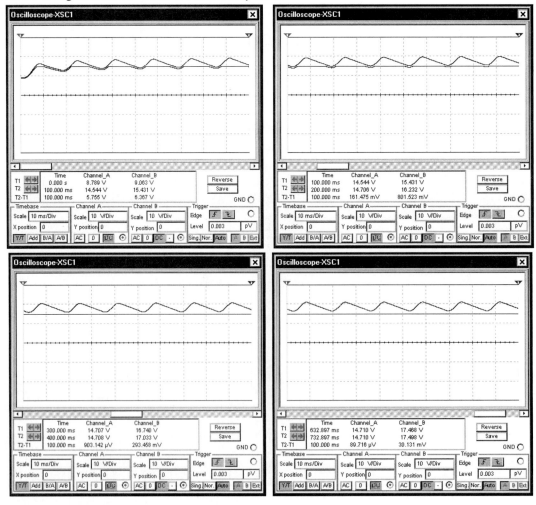

It appears that the circuit functions properly since the output of the regulator is constant at 15 V. However, it took several cycles for the circuit to reach a constant output voltage. If we switch the oscilloscope coupling to AC, we can see the ripple on the input and output of the regulator:

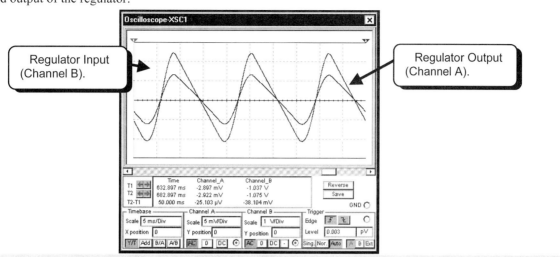

6.E.2. Regulated DC Power Supply – SPICE Transient Simulation

The Multisim virtual simulation shows us what to expect if we built the DC power supply in the lab and measured voltages using the oscilloscope and the Multimeter. These simulations give us practice in how to connect the oscilloscope and Multimeter instruments, and how to adjust these instruments in the lab so that we can measure the desired waveforms and voltages. We must connect and set up the virtual instruments the same way we would the real instruments in the lab. Thus, the virtual simulation serves several purposes: (1) It gives us practice in connecting a circuit using the same techniques we would in the lab, (2) it gives us practice connecting and adjusting instruments using the same techniques we would use in the lab, and (3) we get useful simulation results that can tell us how the circuit works. The simulation results can tell us if the circuit works properly or if our circuit operates within our design specifications.

The Transient Analysis simulation is used only to simulate the circuit, and the results are plotted on a graph rather than displayed on an instrument screen. The purpose of the Transient Analysis is to analyze the circuit and tell us if our design is working properly. Most SPICE simulators have only a Transient Analysis and do not have a virtual lab simulation using an oscilloscope. The results will be the same for both simulations. We are showing both the virtual simulation and the Transient Analysis to bridge the gap between the laboratory testing, the virtual lab simulations of Multisim, and the traditional SPICE Transient Analysis simulations.

We will start with the circuit below, which is the circuit developed in Section 6.E:

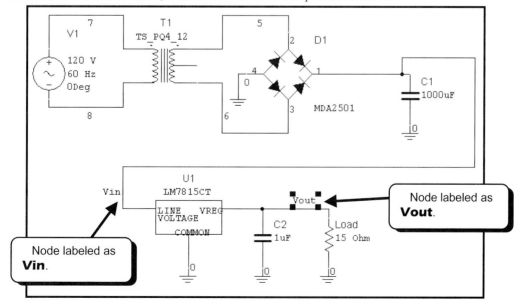

Notice that the nodes at the input and output of the regulator have been labeled as **Vin** and **Vout**. To label a node, double-click the **LEFT** mouse button on a wire and then enter a name for the wire. Your labels may not be displayed in the same locations as shown in the screen capture above, but you do need to label the two nodes.

Next, we need to set up the Transient Analysis. Select **Simulate**, **Analyses**, and then **Transient Analysis** from the menus:

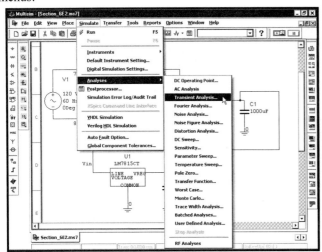 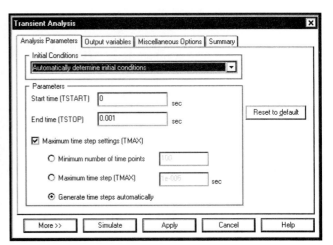

The **End time** is the total time of the simulation. The simulation runs from the **Start time** to the **End time**. We would like this simulation to run for 10 cycles of the 60 Hz waveform. The period of a 60 Hz waveform is 1/60 second. Ten periods is approximately 160 ms or 0.16 s.

The **Maximum time step settings** instruct Multisim on how to choose the time between simulation points. If you choose the option to **Generate time steps automatically**, Multisim will choose the time between simulation points to be as large as possible while keeping the simulation error below a specified maximum. This choice generates the least number of simulation points and allows the simulation to finish as quickly as possible. This is a good choice for long simulations that take several minutes or several hours to complete. The problem in using the **Generate time steps automatically** option is that simulation points are connected to one another using straight lines. With too few simulation points, curves tend to look jagged and you may question the accuracy of the simulation results since the actual waveforms are not jagged.

To avoid the problem of jagged curves resulting from too few simulation points, we can force the simulation to use more points using two options. The **Minimum number of time points** option allows us to specify a minimum number of points for the entire simulation. For example, if we specify this option to be 10,000, there will be at least 10,000 points during the simulation. Using this option instructs Multisim to use a maximum time between simulation points of

$$\text{maximum time between simulation points} = \frac{\text{TSTOP} - \text{TSTART}}{\text{minimum number of time points}}$$

For example, if we set **TSTART** to 0, **TSTOP** to 1 second, and specified a minimum of 1000 points, the largest time between simulation points would be 1 ms.

A second way to do the same thing is to specify a value for the **Maximum time step (TMAX)** option. This option directly specifies the maximum time between simulation points and is a standard parameter in most SPICE Transient Analyses. Thus, if we wanted the maximum time between simulation points to be 1 ms, we would specify this option to be 0.001.

For our first simulation of this DC power supply, we will not specify either of these options and will let Multisim choose the time step automatically. This will allow us to see how well Multisim chooses the time step and also see the effect of not specifying a maximum time between simulation points. Fill out the dialog box as shown below. Because each cycle takes about 16 ms, the simulation will run for about 10 cycles of the 60 Hz input voltage waveform.

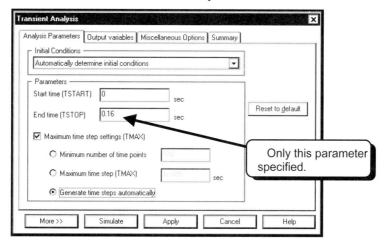

Next, we must specify which traces we would like to look at. Select the **Output variables** tab:

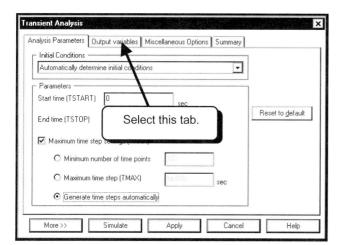
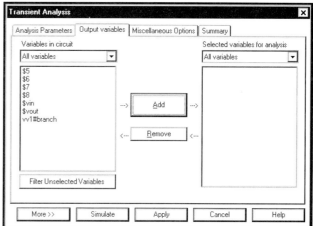

Select to plot **$vin** and **$vout** during the simulation:

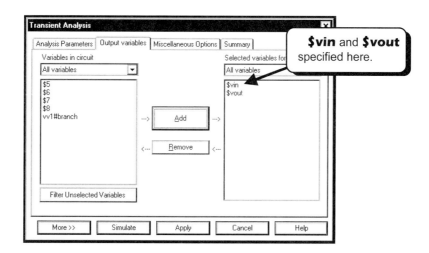

Click on the **Simulate** button to run the simulation. **$vin** and **$vout** will be displayed automatically by the Grapher:

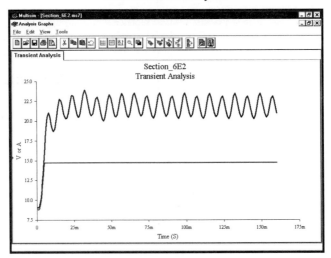

The output voltage appears to be holding very constant at about 15 V as expected, but the input waveform does not appear to be quite right. The 120 Hz ripple is expected, but it is not constant, and the envelope of the ripple seems to slowly oscillate. If we run the same simulation but change the end time to 0.4 seconds, we see that the ripple on the envelope does not appear to die out:

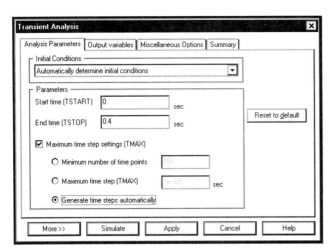
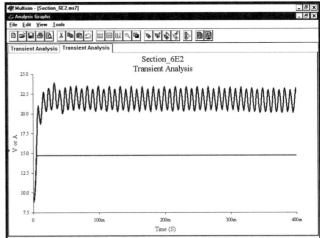

We see that the capacitor ripple never repeats itself, and this is not correct. From our understanding of the circuit, the observations in the lab, and the virtual simulation of the previous section, we know that the ripple at the input to the regulator should be periodic with a frequency of 120 Hz.

This simulation is incorrect because Multisim chooses too large a time between simulation points. We will fix this problem by specifying a maximum time step. Close the Grapher window by clicking on the ☒ as shown below:

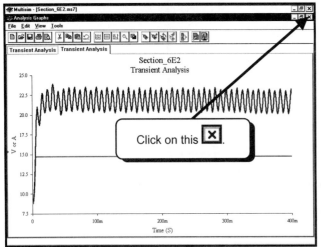
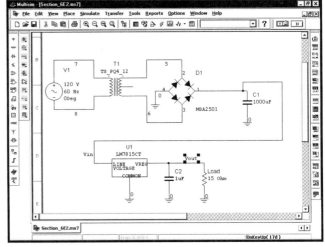

Select **Simulate**, **Analyses**, and then **Transient Analysis** from the menus to obtain the Transient Analysis setup dialog box:

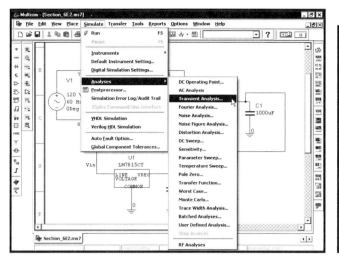
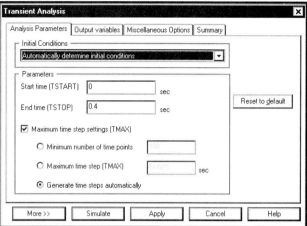

We would like to specify a maximum time between simulation points, but how do we choose a value? A general method is to choose a large value for the maximum time and then run the simulation. Observe the results, and then rerun the simulation for a smaller value of the maximum time between points (a factor of 10 smaller is a good choice). At first you will notice a difference between the simulation results. As you make the maximum time step smaller, the results will eventually converge and you will notice no difference between simulation results even if you continue to reduce the maximum time between simulation points. When you reach this point, you have found a range of good values for the maximum time step, and you should use the largest value of the maximum time step such that the simulation results are the same. You can use a smaller value if you wish, but this could increase the amount of round-off error, take more simulation time, and enlarge the files that contain the calculated data.

I will not use this method here. Because I have a lot of experience simulating DC power supplies, I have a good idea of the value I would like to use (experience with running simulations also helps you choose the correct parameters for a simulation). I would like to have at least 100 simulation points in every 120 Hz cycle. The period of a 120 Hz waveform is 1/120 second, or 8.33 ms. To have 100 points occur in 8.33 ms, the maximum time between simulation points should be 8.33 ms divided by 100, or 0.083 ms, or 83e-6. Modify the dialog box as shown:

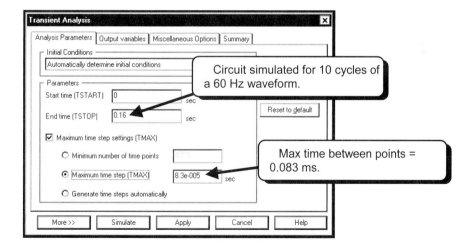

The dialog box is set up to simulate the circuit for approximately 10 cycles of a 60 Hz waveform, and the maximum time between simulation points is 0.083 ms, or about 100 points in a 120 Hz waveform.

The output variables should be specified from our previous simulation. However, we will check to make sure. Click on the **Output variables** tab and check to see if **$vin** and **$vout** are still specified to be plotted:

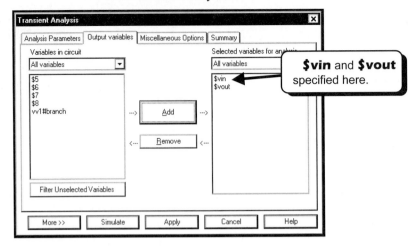

Click on the **Simulate** button to run the simulation. **$vin** and **$vout** will be plotted automatically:

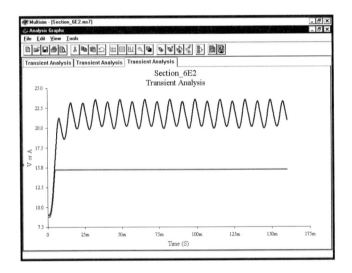

The simulation looks better, but there does appear to be a slight variation in the envelope. A side-by-side comparison of the two simulations is shown below. The simulation on the left was run with Multisim 7 choosing the step size. The simulation on the right was with a step size specified at 8.3e-5 seconds:

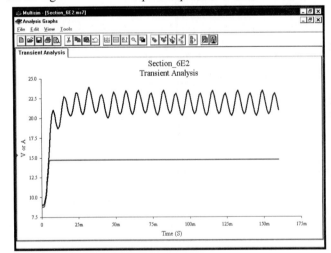
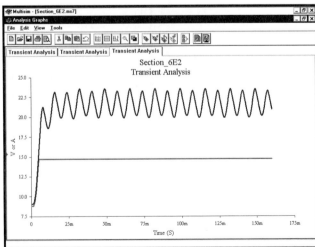

You can see a small difference in the top traces.

Next, I will rerun the same simulation but change the end time to 0.4 seconds.

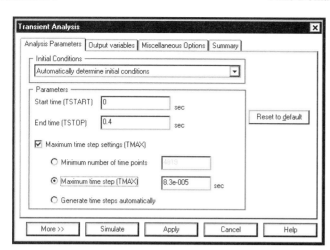
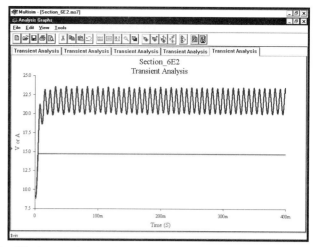

The envelope of **vin** looks fairly constant, but to be careful, I will change the maximum time step to 10 μs and rerun the simulation. The results are shown below:

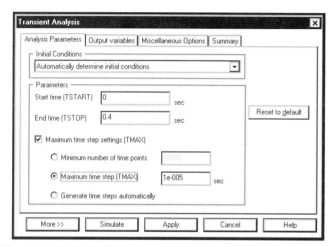
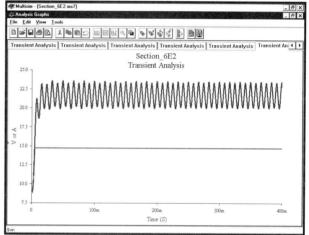

These results look about the same as when we had a maximum time step of 0.083 ms, so we conclude that the time step of 0.083 ms is sufficient. Closer inspection of the results may yield a difference, but for the plots shown, the results for a maximum time step of 10 μs and 0.083 ms appear to be about the same. Thus, we will rerun the simulation for a maximum time step of 0.083 ms and then examine the waveforms in a bit more detail. You can use the data generated with a time step of 10 μs, but your computer may take longer to plot the graphs because more data is being plotted.

We will now examine the ripple more closely. We would like to display the ripple on the input and the output and see how they relate to one another. To do this, we will create a new plot with the Postprocessor and plot the input and the output on separate graphs. If you were following the previous example, close the Grapher window by typing **CTRL-G**. We will start with Multisim displaying the circuit. We will assume that you have already run the simulation and that the data is available to the Postprocessor:

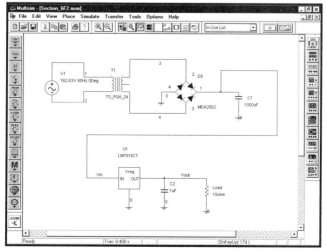

Select **Simulate** and then **Postprocessor** from the menus:

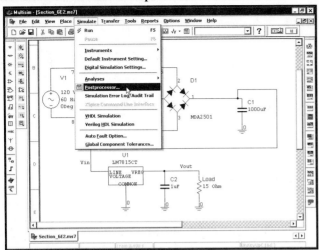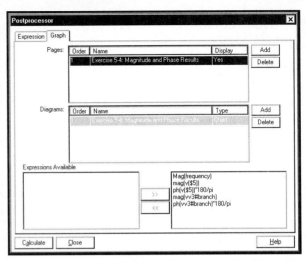

Use the Postprocessor to plot the input and output voltage waveforms of the voltage regulator. If you are unfamiliar with the Postprocessor, see Part 2.

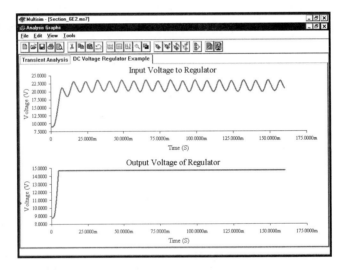

We would like to see the ripple at the output of the voltage regulator, so we need to change the y-axis **Scale** of the bottom graph. To do this, click the **RIGHT** mouse button on the lower plot:

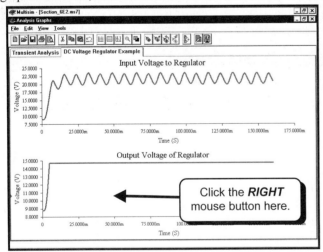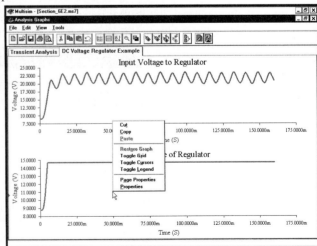

Select **Properties**:

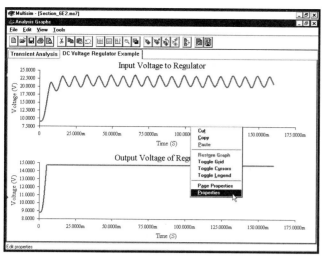
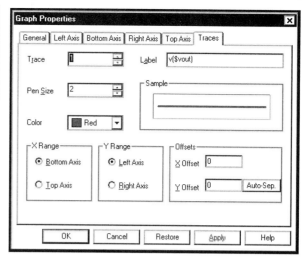

This dialog box allows us to change all of the properties of the lower graph. For this example, all we need to do is change the y-axis scale. To do this, select the **Left Axis** tab:

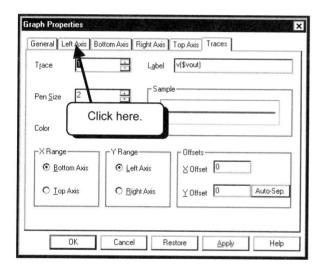
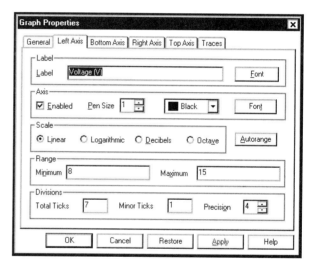

From our virtual lab analysis of this circuit, we know that the average value of the output was 14.721 (see page 361) and peak-to-peak ripple on the order of 10 mV (see page 359). From these numbers, we get a good idea of the y-axis range we need to see the ripple. Without this information, it may take a few guesses at the y-axis range before we find a suitable value. Specify the y-axis range to have a minimum of 14.6 and a maximum of 14.8 volts as shown below:

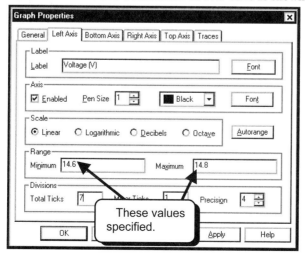

Click on the **OK** button to view the graph:

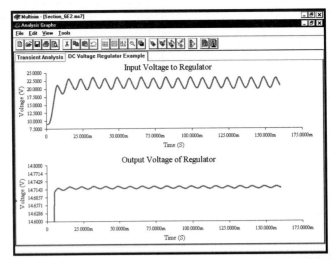

You can zoom in further if you wish. You can also use this technique to scale the time axis. To change the time axis, click the **RIGHT** mouse button on the graph and then select the **Bottom Axis** tab instead of the **Left Axis** tab. In the plot below, I have scaled the time axis of the top and the bottom graphs. The y-axis scales for the two plots are different, but the time axes for both plots have the same scale.

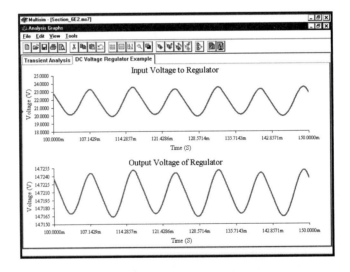

To get numerical values from the graph, you can use the cursors. To display the cursors, click the **RIGHT** mouse button on the bottom graph and then select **Toggle Cursors**. Use the mouse to place the cursors at the peaks of the waveforms.

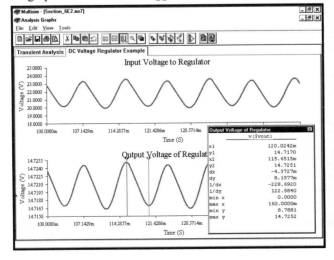

From the **dy** value of the cursors, we see that the ripple at the regulator output is **8.1577 mV**. To remove the cursors from the bottom graph, click the **RIGHT** mouse button on the bottom graph and then select **Toggle Cursors** again. Next, click the **RIGHT** mouse button on the top graph and then select **Toggle Cursors** to display the cursors on the top graph:

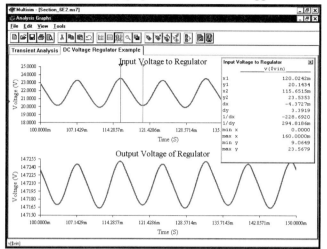

The **dy** value of the cursors shows us that the ripple at the regulator input is **3.3919** V.

Using these values to calculate the regulator ripple rejection ratio, we get

$$\text{Ripple Rejection Ratio}_{dB} = 20\log_{10}\left(\frac{\Delta Vin}{\Delta Vo}\right)$$

$$= 20\log_{10}\left(\frac{3.3919\,\text{V}}{8.1577\,\text{mV}}\right)$$

$$= 52.19\,\text{dB}$$

This value compares favorably to the value of 52 dB obtained using the virtual simulation on page 361.

EXERCISE 6-7: Run a power supply using a half-wave rectifier and a regulator. Use the same circuit as on page 353, except with a half-wave rectifier instead of a full-wave rectifier. Use the virtual diode because Multisim does not have a discrete diode that can handle the required current in the library. The remainder of the circuit should be the same.

SOLUTION:

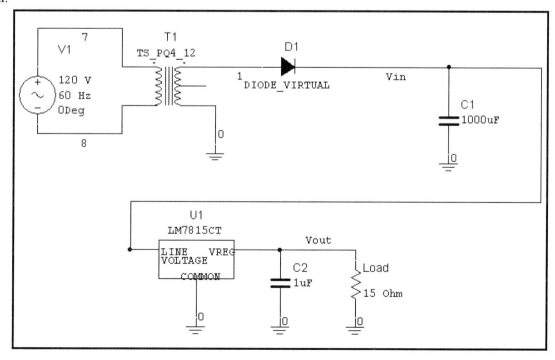

Set up a Transient Analysis. Set the maximum time step to 10 μs to get accurate results. Also, set the end time to simulate several cycles to allow the capacitor to charge to a large enough voltage so that the regulator operates correctly:

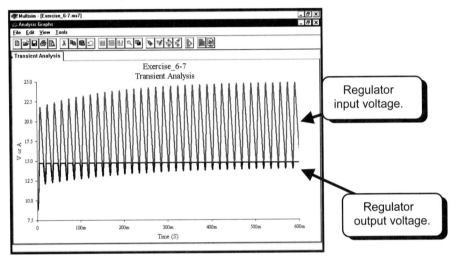

The simulation results are:

It is a bit hard to distinguish the input and output traces on this screen capture, so I will use the Postprocessor to plot the traces on separate graphs:

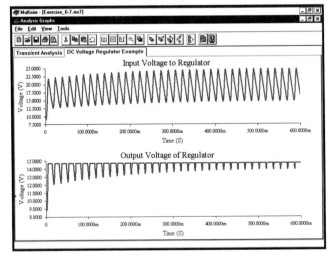

The output of the regulator never becomes pure DC. We see what is called "drop-out" in the output of the regulator because the voltage at the input of the regulator is too small. This is because the capacitor we are using for a half-wave rectifier is too small. We can fix the problem by either choosing a higher voltage transformer or a larger capacitor. Some standard size capacitor values are 1200 μF, 1500 μF, 1800 μF, and 2200 μF. I will change the capacitor to a 2200 μF:

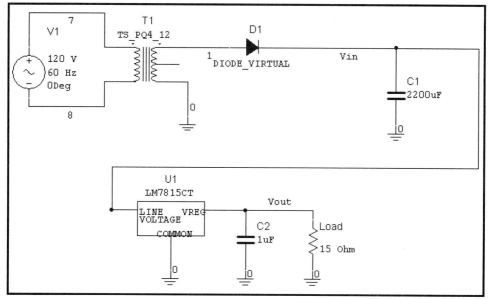

The Transient Analysis shows that the output of the regulator is now constant at 15 volts, but it seems that the regulator output is dropping out just a small amount when the input is at its minimum input:

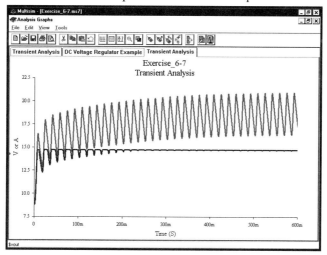

We will zoom in on both traces a bit more to see what is going on at the output of the regulator:

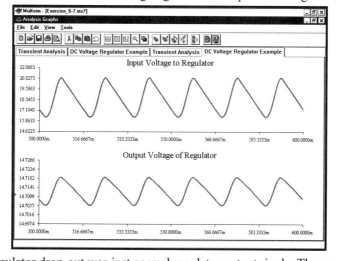

What originally looked like regulator drop-out was just normal regulator output ripple. The regulator now appears to work normally: a large amount of ripple on the input is reduced to only a small amount of ripple on the output.

6.F. Zener Clipping Circuit – SPICE Transient Analysis

In this section we will demonstrate the use of a Zener diode, as well as the piece-wise linear (PWL) voltage source. The PWL source can be used to create an arbitrary voltage waveform that is connected by straight lines between voltage points. Wire the circuit shown:

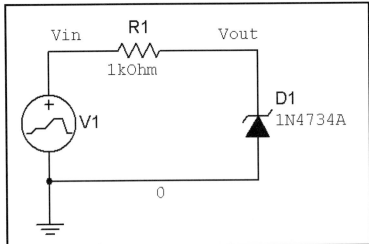

This circuit contains two new parts, the PWL voltage source and the Zener, so we will show how to place them in our circuit. First we will place the PWL voltage source. Click the **LEFT** mouse button on the **Show Signal Source Components Bar** button :

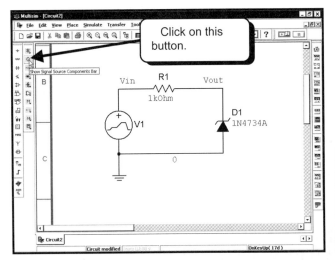

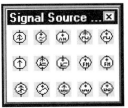

Click the **LEFT** mouse button on the **PIECEWISE_LINEAR_VOLTAGE_SOURCE** button . The source will become attached to the mouse:

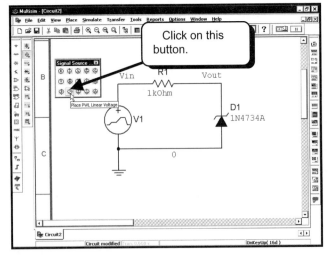
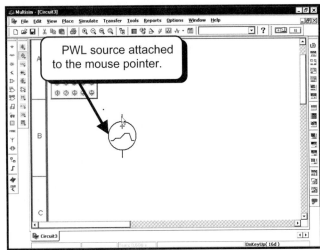

Multisim 7   Time Domain Analyses   379

You can now place the PWL source in your circuit.

To place the Zener diode, click the *LEFT* mouse button on the **Diode** button:

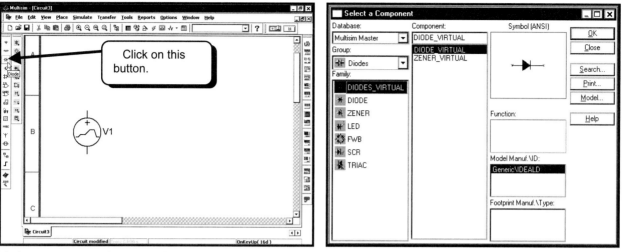

Click the *LEFT* mouse button on the text **ZENER** to select the **Family**:

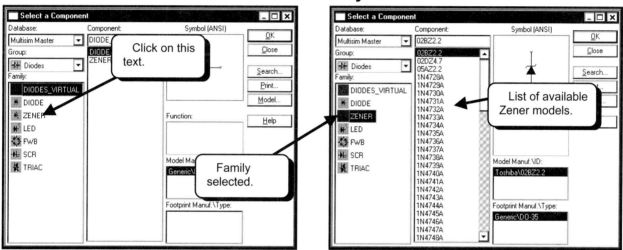

For this example we are using a **1N4734A** Zener diode, so select the diode and then click on the **OK** button. The diode will become attached to the mouse:

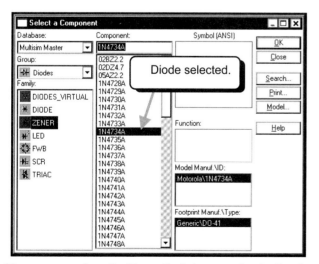
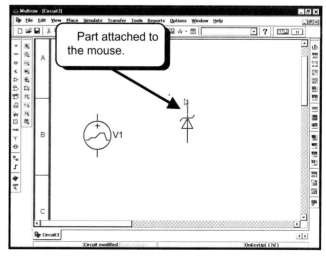

You can now place the Zener diode in your circuit. Continue drawing the circuit until you have the circuit shown below:

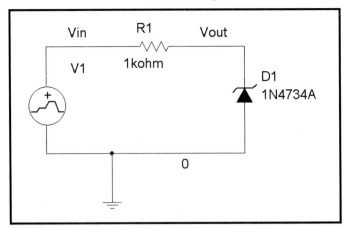

We now need to define the voltage waveform of the PWL voltage source. With a PWL voltage source, the time and voltage of points on the waveform are defined and then the points are connected by straight lines. This type of source is useful for creating waveforms made up of straight lines such as a saw tooth, ramp, or triangle wave. It is also useful if you have measured data of a waveform saved as time and voltage coordinates. The PWL source can be programmed with this data and the source will then reproduce the waveform. For this example, we will use the PWL source to create one cycle of a triangle wave.

To define the PWL waveform, double-click the **LEFT** mouse button on the PWL graphic:

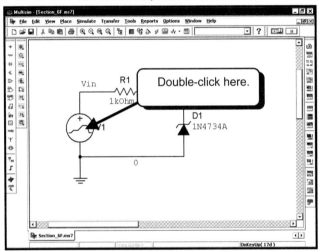
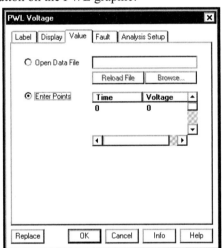

We can define the source by either reading in a text file of data points or entering the data points manually using this dialog box. If you want to define the PWL source using a data file, the file should be a text file where each line contains a single point. The values on a single line must be separated by a space or a tab. The first value specified is the time, the second value is the voltage. An example is shown below:

```
0        0
0.001    15
0.003   -15
0.004    0
0.005    15
0.007   -15
0.008    0
0.009    15
0.011   -15
0.012    0
```

In the example above, the values are separated by a tab character. This data was created using Microsoft Excel and saved as a text file delimited by tabs. This is one of the standard formats available with Excel. You could also create this data file using the Windows Notepad and typing the data in manually. The data defines the following voltage waveform: At time = 0, the voltage is zero; at time = 0.001 s, the voltage is 15 volts; at time = 0.003 s, the voltage is –15 volts; and at time = 0.004, the voltage is zero volts. The waveform then repeats itself.

It appears that the PWL source does not have an easy way to generate a repeating waveform. To create a long repeating waveform using the PWL source, you would need to use Excel to create a large table of points and save the file as a tab-delimited text file. In Excel, you can easily create the points for a single period, and then use spreadsheet functions to make the data repeat periodically. We will not discuss this here.

For our example, we will simulate the circuit for a single period of a triangle wave. Thus, we will use the PWL source to generate a single triangle wave. We will enter the data points manually using the dialog box for the PWL source. The first time point (time = 0, voltage = 0) is already defined for us. To enter the second point, click the **LEFT** mouse button as shown below:

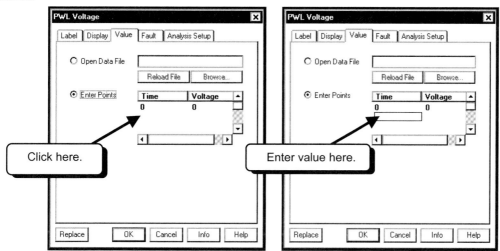

An empty cell appears. Enter the time coordinate for the second point. Do not press the **ENTER** key:

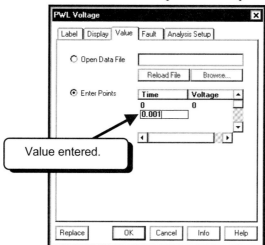

We now need to enter the voltage coordinate. Click the **LEFT** mouse button as shown below:

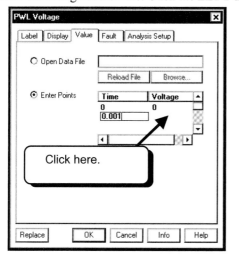
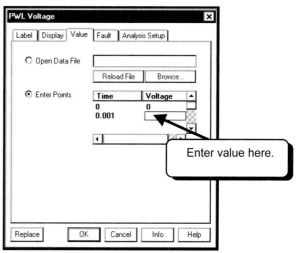

Type a value for the voltage coordinate of the point:

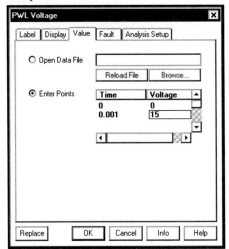

Repeat the process to add two more points: At time = 0.003 s, the voltage is –15 volts, and at time = 0.004 s, the voltage is 0 volts. The completed dialog box is shown below:

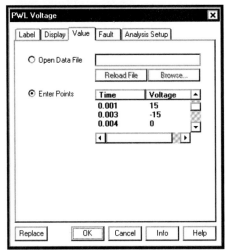

The data we entered defines the triangle wave below:

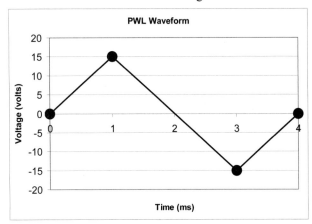

Table 6-1			
Time	**Time**	**Time**	**Time**
0	0.001	0.003	0.004
Voltage	**Voltage**	**Voltage**	**Voltage**
0	15	-15	0

Click on the **OK** button to accept the settings for the PWL source.

We would like to run a Transient Analysis for 4 ms. To set up the analysis select **Simulate**, **Analyses**, and then **Transient Analysis** from the Multisim menus:

Time Domain Analyses

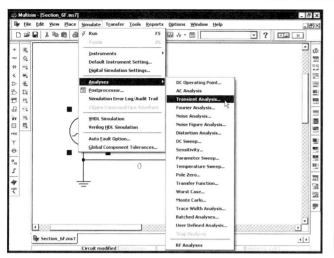
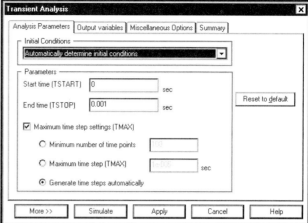

We would like to run the Transient Analysis for 4 ms and to have at least 1000 simulation points during the simulation. Fill in the dialog box as shown:

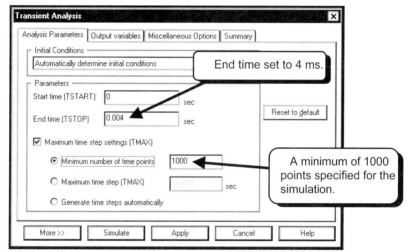

Next, we must specify the waveforms we would like to see. Select the **Output variables** tab and specify that **$vin** and **$vout** be plotted during the simulation.

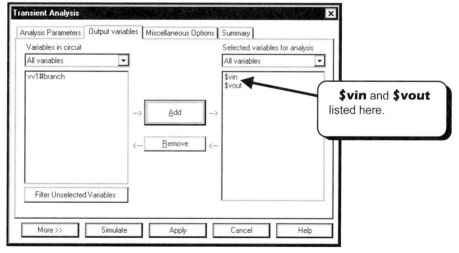

Click on the **Simulate** button to run the simulation:

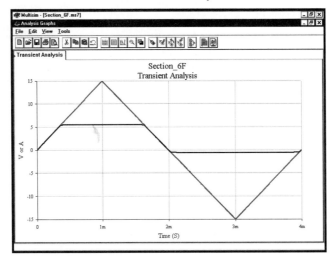

I have turned on the grid by clicking the **RIGHT** mouse button on the graph and then selecting **Toggle Grid**. If we turn on the cursors (right mouse button and then **Toggle Cursors**) we can measure the maximum and minimum values of the output waveform:

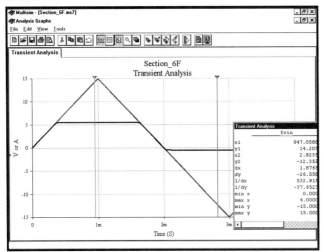

Cursor 1 (red) shows that the output is limited to a positive maximum value of **5.5594** volts and cursor 2 (blue) shows that the output is limited to a maximum negative voltage of **-563 mV**.

EXERCISE 6-8: Plot the input and output voltage waveforms using the Transient Analysis for the circuit below. Let the input be a ±15 volt triangle wave.

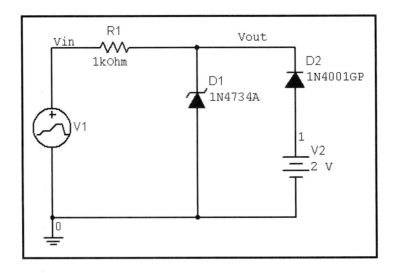

SOLUTION:

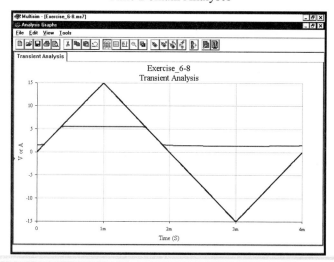

EXERCISE 6-9: Plot the input and output voltage waveforms using the Transient Analysis for the "Dead-Zone" clipping circuit below. Let the input be a ±15 volt triangle wave.

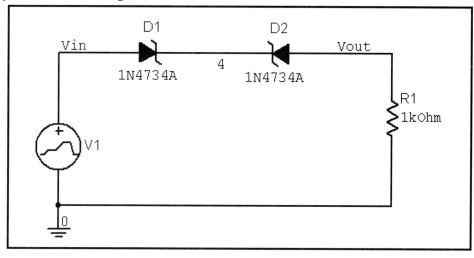

SOLUTION:

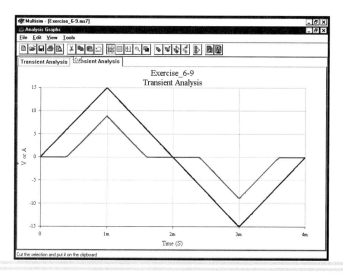

6.G. Zener Clipping Circuit – Virtual Lab Simulation

We will run the same simulation as in section 6.F, but we will use a square-wave voltage source to generate a triangle wave and use the oscilloscope to view the waveforms. Wire the circuit shown:

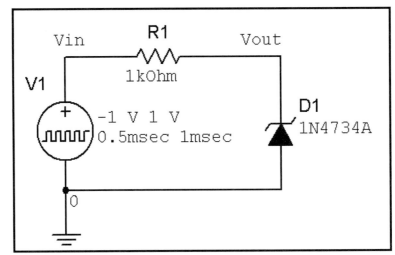

This circuit contains the pulse voltage source, which we have not yet used; so we will show how to place it in our circuit and set up its parameters. Click the **LEFT** mouse button on the **Show Signal Source Components Bar** button:

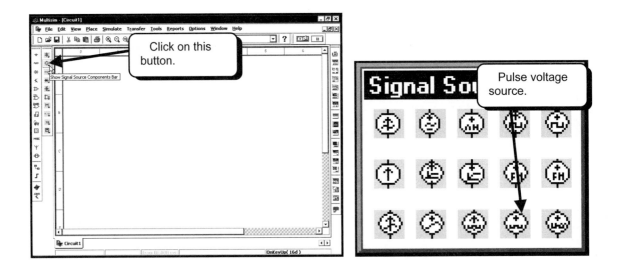

Click the **LEFT** mouse button on the **PULSE_VOLTAGE_SOURCE** button. The source will become attached to the mouse:

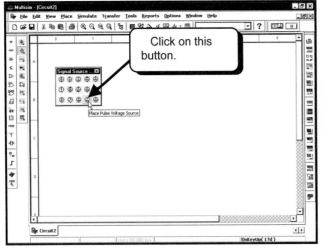
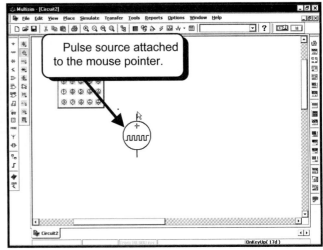

You can now place the source in your circuit. Continue drawing until you have the circuit shown below:

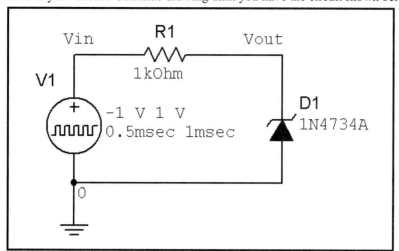

We now need to set the properties of the pulse voltage source. Double-click the **LEFT** mouse button on the pulse source graphic:

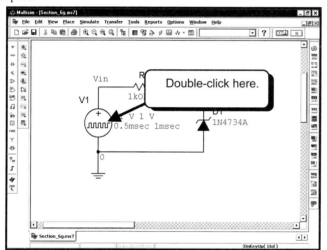
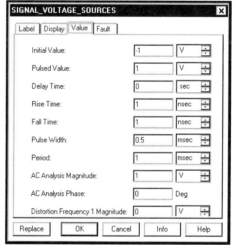

The **Initial Value** is the voltage level before the pulse starts, and it is also the voltage level after the pulse is finished. The **Pulsed Value** is the value of the voltage during the pulse. The **Pulse Width** is the amount of time the voltage stays at the **Pulsed Value**. The **Period** is the period of the waveform and is the amount of time over which the waveform repeats itself. The **Rise Time** is the amount of time it takes the voltage to go from the **Initial Value** to the **Pulsed Value**. The **Fall Time** is the amount of time it takes the voltage to go from the **Pulsed Value** to the **Initial Value**.

At the beginning of a simulation, the voltage starts at the **Initial Value**. The voltage then immediately starts rising towards the **Pulsed Value**. After the **Rise Time**, the voltage holds constant at the **Pulsed Value** for an amount of time equal to the **Pulse Width**. We see that the pulse starts immediately at the beginning of the simulation. If you want to delay the pulse at the beginning of a simulation, use the **Delay Time**. The **Delay Time** is the amount of time between the beginning of the simulation and the beginning of the first pulse. Its default value is zero so that the first pulse starts at the beginning of the simulation.

We will be running a Transient Analysis. The parameters **AC Analysis Magnitude, AC Analysis Phase**, and **Distortion Frequency 1 Magnitude** are used in the AC and Distortion analyses and are ignored by the Transient Analysis. Thus, we do not need to specify them here.

We would like to define a triangle wave that goes from –15 volts to 15 volts and has a **Period** of 4 ms. This triangle wave spends 2 ms going from –15 V to +15 V, thus we will specify the **Rise Time** as 2 ms. This triangle wave spends 2 ms going from +15 V to –15 V, so we will specify the **Fall Time** at 2 ms also. We will set the **Pulse Width** to zero because once the voltage reaches 15 V, it immediately reverses direction and heads back toward –15 V. If we specify the **Pulse Width** as something other than 0, when the voltage reaches 15 V, it will remain at 15 V for a length of time equal to the **Pulse Width**. This creates a flat line at +15 V that we do not want. The dialog box below defines a 15-volt amplitude triangle wave with a period of 4 ms:

Click on the **OK** button to accept the changes.

We will run a virtual simulation and view the results with the oscilloscope. Add the oscilloscope to your circuit as shown:

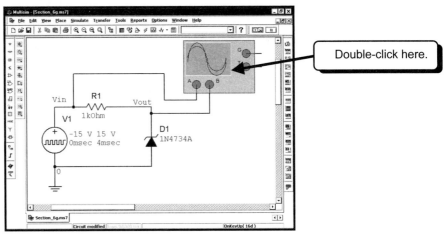

Run the simulation and then double-click on the oscilloscope to view the results:

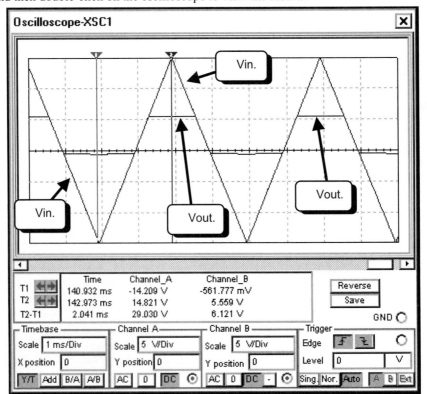

The cursors show that the maximum positive value of Vout is **5.559 V** and the maximum negative value is **-561.7 mV**. These values agree with the values obtained using the Transient Analysis on page 384.

6.G.1. Plotting Transfer Curves

When you have a clipping circuit, as in the last example, a transfer curve may be desired. A transfer curve is a plot of the output versus the input of a circuit. In the lab, this can be easily done using the X-Y capabilities of an oscilloscope. In normal mode, an oscilloscope displays **Channel A** versus time and **Channel B** versus time. When you switch the oscilloscope to X-Y mode, the oscilloscope will display **Channel A** versus **Channel B** or, depending on your scope, **Channel B** versus **Channel A**. The Multisim oscilloscope has the ability to display either **Channel A** versus **Channel B** or **Channel B** versus **Channel A**.

With a transfer curve, you usually plot the output voltage versus the input voltage. In our circuit, we are measuring **Vin** with **Channel A** and **Vout** with **Channel B**. We would like to plot **Vout** on the y-axis and **Vin** on the x-axis. This plot can be viewed on the oscilloscope by selecting the **B/A** button:

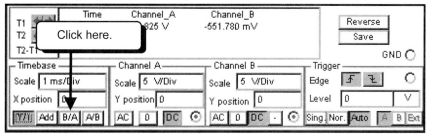

After selecting this button, you will see the transfer curve:

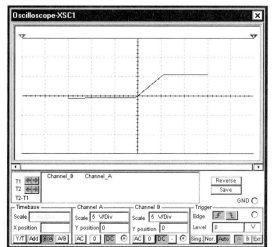

You can change the volts per division settings to enlarge the transfer curve. The **Channel A** settings control the x-axis and the **Channel B** settings control the y-axis. I will also use the cursors to measure the breakpoints of the transfer curve:

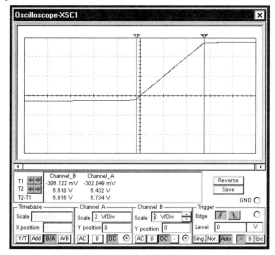

To return to the normal display of waveforms versus time, select the **Y/T** button. Note that selecting the **A/B** button will plot **Channel A** versus **Channel B**. For our connections, this would plot **Channel A** or **Vin** on the y-axis and **Channel B** or **Vout** on the x-axis. This would be a plot of **Vin** versus **Vout**, not quite the transfer curve we were looking for.

EXERCISE 6-10: Find the output voltage waveform and the transfer curve for the circuit below using the oscilloscope. Let the input be a ±15 volt triangle wave at 100 Hz. Use the pulsed voltage source to create the triangle wave.

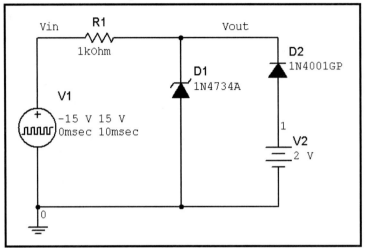

SOLUTION: The attributes of the pulsed voltage source are:

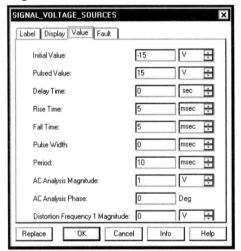

Add the oscilloscope as shown:

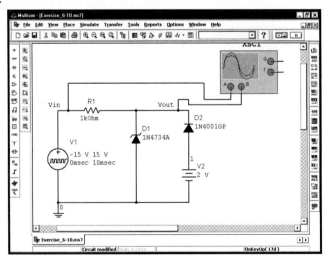

The waveforms and transfer curve are:

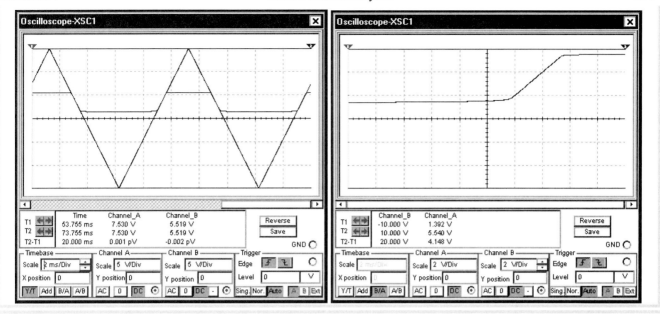

EXERCISE 6-11: For the "dead-zone" clipping circuit shown below, find the output voltage waveform and the transfer curve. Let the input be a ±15 volt triangle wave at 500 Hz. Use the pulsed voltage source to create the triangle wave.

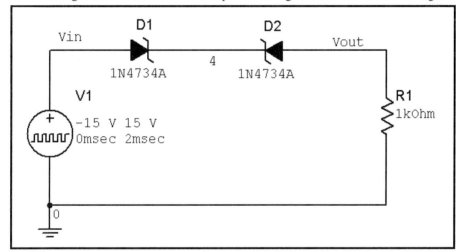

SOLUTION: The attributes of the pulsed voltage source are:

Add the oscilloscope as shown:

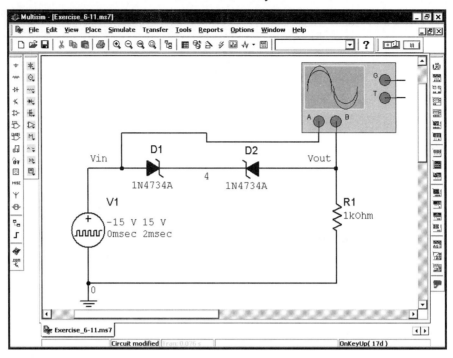

The waveforms and transfer curve are:

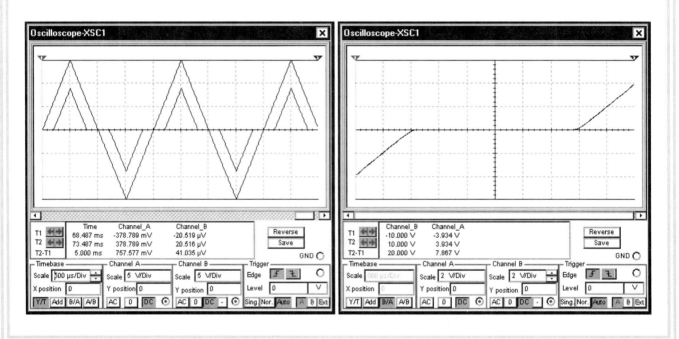

6.H. Amplifier Voltage Swing

In this section we will examine the voltage swing of the transistor amplifier used in section 5.C. In section 5.C we found that the gain of the amplifier was 45.7 dB, which corresponds to a gain of 193 V/V. We also found that the mid-band for this amplifier was from 65 Hz to 6.2 MHz. We would like to see what the maximum voltage swing of this amplifier looks like at mid-band. We will use the same circuit as in section 5.C.

Wire the following circuit:

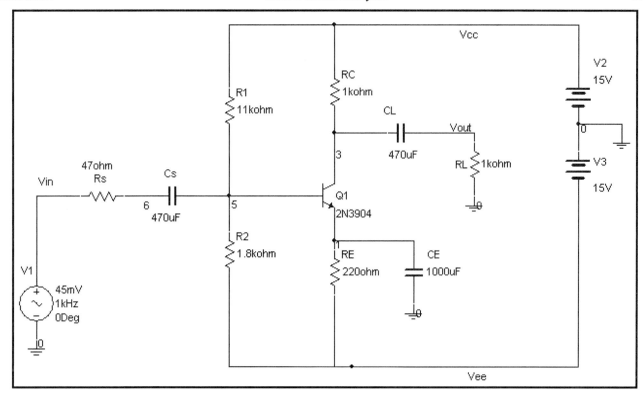

The source is a sinusoidal voltage of amplitude 45 mV and frequency 1 kHz, $V_1 = 0.045\sin(2\pi*1000t)$. The amplitude was chosen to get the maximum swing out of this amplifier. You may have to have to run the simulation several times with different amplitude sine waves for **Vin** to see what amplitude you need for **Vin** to observe the maximum swing on the output. Too small an amplitude will cause a small output swing. Too large an output will cause the amplifier to go into saturation or cutoff. The amplitude should be chosen to cause the output of the amplifier to just saturate or cutoff.

We must now set up the parameters for the Transient Analysis. Select **Simulate**, **Analyses**, and then **Transient Analysis** from the Multisim menus. Set up the Transient Analysis simulation dialog box as shown below:

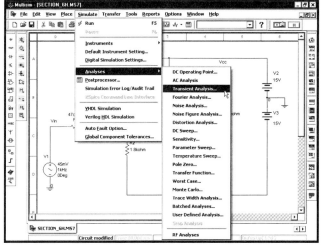
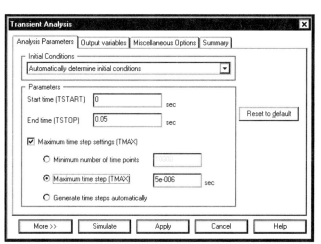

Note that we are running the simulation for 50 ms (**End time = 0.05**). Because the source is at a frequency of 1 kHz, we are simulating the amplifier for 50 cycles. This is done to allow the amplifier to reach steady-state. If we look at the amplifier output at the beginning of the simulation, we will see the output voltage waveform slowly drift because the capacitors are charging in response to the DC average of the output voltage waveform. Since we are interested in the steady-state response of the amplifier, the simulation is run for 50 ms and we will look at the results near the end of the simulation. We have also specified the **Maximum time step** to be 5 µs to produce accurate simulation results. If you omit this value, Multisim may take large time steps and the waveform will look very jagged. A 1 kHz sine wave has a period of 1 ms or 1000 µs. With a **Maximum time step** of 5 µs, there will be at least 200 simulation points for each period of the sine wave. Specifying this small time step will result in smooth waveforms, but will also cause the simulation to take a long time to run.

The last thing we must do is select the **Output variables** tab of this dialog box and select **$vin** and **$vout** to be plotted during the simulation:

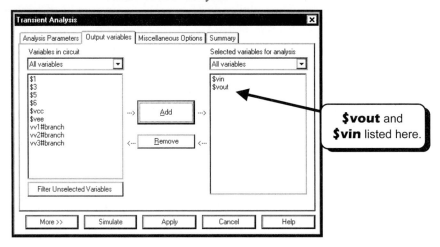

Click on the **Simulate** button to run the simulation:

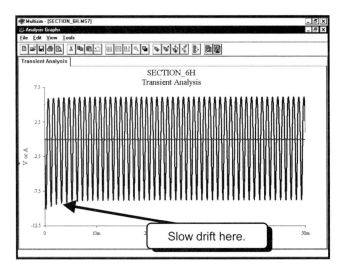

The slow drift in the output waveform is easily seen in the screen capture above. The negative peak appears to slowly drift up during the first 10 ms, and then appears constant for the remainder of the simulation. To see the waveforms more closely, we will need to zoom in on them. To zoom in, **click and hold** the *LEFT* mouse button at the upper left corner of the area in which you want to zoom:

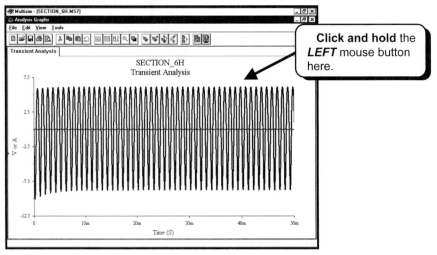

While continuing to hold down the mouse button, move the mouse pointer down and to the right. A zoom box will appear:

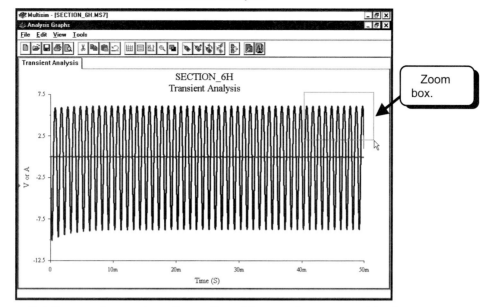

Move the mouse so that the zoom box encloses the portion of the waveform you want to view:

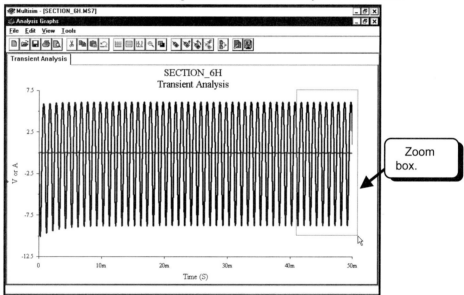

When you release the mouse button, Multisim will zoom in around the box:

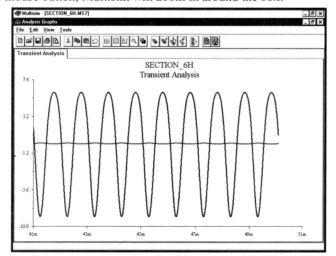

The output is so much larger than the input that you can hardly see any variation on the input. We also notice that the output does look sinusoidal, but also seems to be rounded at the top and pointed at the bottom. The output is not really a pure sine wave and contains some distortion. Distortion means that the output of the amplifier contains frequencies not present in the input signal. For a distortion-free amplifier, if we input a frequency of 1 kHz, the only frequency contained in the output will be 1 kHz. The output of our amplifier does not look like the input, thus it contains additional frequencies not present in the input. We will analyze the distortion in the output waveform in the next section using the Fourier Analysis.

If you want to see an enlarged view of the input voltage, you will need to plot the input and output voltages on separate graphs. See pages 372-374 to create two graphs and zoom in on those graphs. The input and output waveforms for the last 5 cycles of the simulation are shown below:

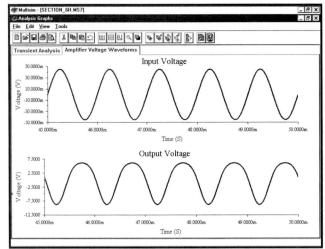

6.H.1. Fourier Analysis with Multisim

From the last simulation, we know that the amplifier adds distortion because the input to the amplifier was a pure sine wave at 1 kHz, and the output was a periodic wave, but it was not a pure sinusoid. From Fourier series analysis of periodic signals, we know that any periodic waveform is made up of a sum of sine waves at different frequencies and different phase angles. If our amplifier did not add distortion, the output would contain the same frequencies as the input. If we input a single frequency of 1 kHz, the output should contain only a single frequency at 1 kHz. Any distortion added by the amplifier would appear as additional frequencies in the output signal. The Fourier Analysis will tell us the frequency components of a signal, and thus can tell us the distortion added by the amplifier.

We will continue with the circuit from the previous section. To set up a Fourier Analysis, select **Simulate**, **Analyses**, and then **Fourier Analysis** from the menus:

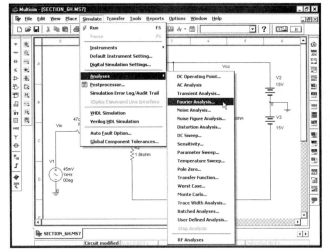 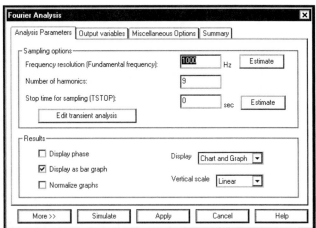

Our input signal is a 1 kHz sine wave. This will be the lowest frequency in the output, and is called the fundamental frequency. Thus the default value of **1000 Hz** for the **Fundamental frequency** is correct. If our input signal was a different value, say 1.5 kHz, we would need to specify the **Fundamental frequency** as 1.5 kHz. The output waveform will contain multiples of the fundamental called harmonics. You can specify how many harmonics you would like to view. We will use the default value of 9; however, you can specify more or less. Your parameters should be set as shown above.

In order to calculate the Fourier components, Multisim must run a Transient Analysis. Click on the **Edit transient analysis** button to set up the Transient Analysis:

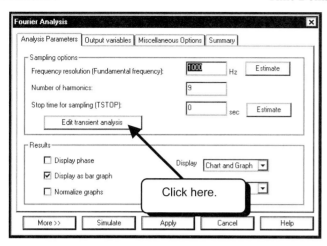
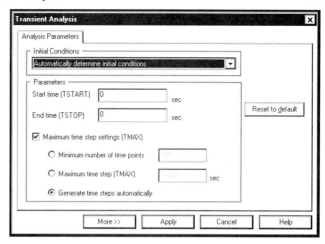

We will set up the parameters for the Transient Analysis the same as in the previous section. The **End time** is set to 50 ms and the **Maximum time step** is set to 10 μs:

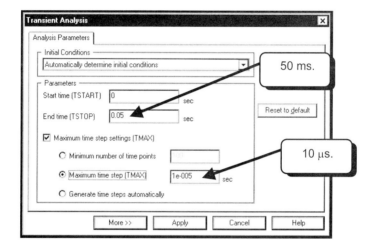

Click on the **Apply** button to return to the Fourier Analysis setup. Select the **Output variables** tab and then specify to plot **$vout** during the simulation:

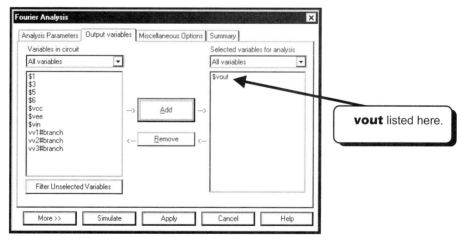

We are now ready to run the simulation, so click on the **Simulate** button:

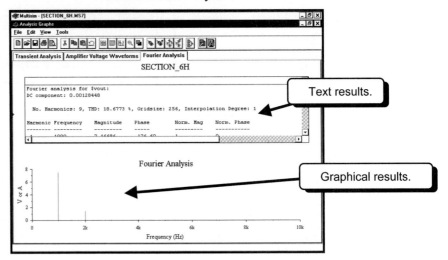

The results are displayed as text and as a graph. The graph shows us that the output is composed mainly of two frequencies, the fundamental at 1 kHz and the second harmonic at 2 kHz:

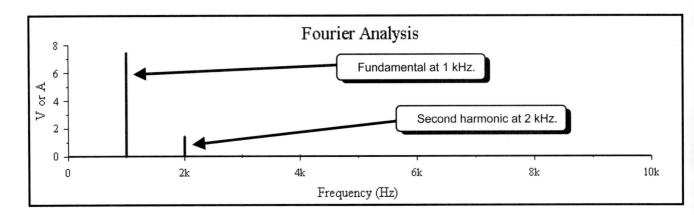

Other frequencies are contained in the output, but they are too small to see on the graph.

The text information gives us more detailed information. First, the total harmonic distortion (THD) is calculated as 18.6773%.

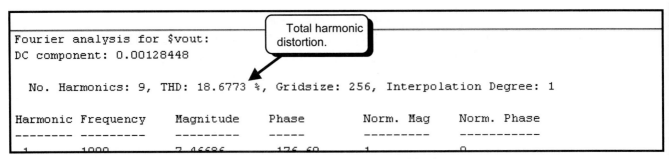

If we scroll down the text window, we get detailed information about each of the frequency components:

```
   No. Harmonics: 9, THD: 18.6773 %, Gridsize: 256, Interpolation Degree: 1

Harmonic Frequency    Magnitude     Phase       Norm. Mag    Norm. Phase
-------- ---------    ---------    -----        ---------    -----------
   1       1000        7.46686     -176.69        1             0
   2       2000        1.39336      97.9434       0.186605      274.635
   3       3000        0.0318442   166.108        0.00426474    342.799
   4       4000        0.0497600   104.31         0.00666544    281.003
```

3	3000	0.0318442	166.108	0.00426474	342.799
4	4000	0.0497699	104.31	0.00666544	281.002
5	5000	0.00261411	44.1245	0.000350094	220.816
6	6000	0.00187943	108.075	0.000251703	284.767
7	7000	0.000218378	44.3526	2.92463e-005	221.044
8	8000	7.6837e-005	112.643	1.02904e-005	289.334
9	9000	1.95674e-005	86.4119	2.62057e-006	263.104

The results of the Fourier Analysis show that the magnitude of the sine wave at 1 kHz is **7.46686** V, and the magnitude of the sine wave at 2 kHz is **1.39336**. The text data also gives us the magnitude of the frequency components too small to see on the graph, and the phase of each frequency component. From the data above, an equation for the output voltage is:

$$V_o = 7.467 \sin(2\pi \bullet 1000t - 176.69°) + 1.393 \sin(2\pi \bullet 2000t + 97.94°) + 0.0318 \sin(2\pi \bullet 3000t + 166.1°) + \ldots$$

EXERCISE 6-12: Find the magnitude of the harmonics of a ±1 volt, 1 kHz square wave using the Fourier analysis.

SOLUTION: Use the pulse voltage source. Set the rise and fall times to 1 ns so that they are much shorter than the pulse width and period of the square wave. Wire the circuit below. The attributes for the pulse voltage source are also shown:

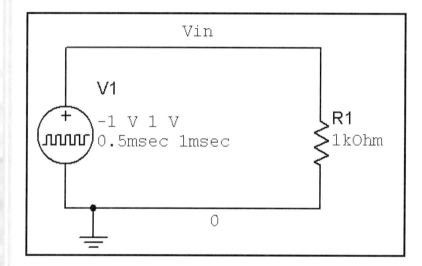
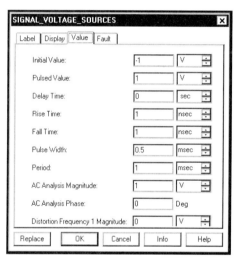

Set up the Fourier Analysis and Transient Analysis as shown:

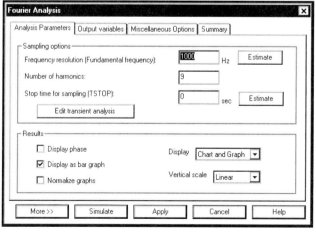
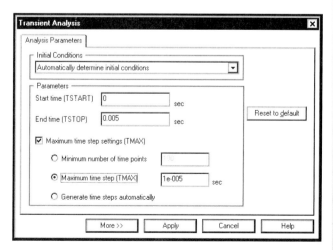

The results are:

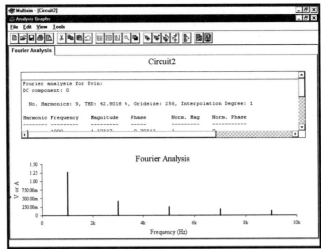

No. Harmonics: 9, THD: 42.9018 %, Gridsize: 256, Interpolation Degree: 1					
Harmonic	Frequency	Magnitude	Phase	Norm. Mag	Norm. Phase
1	1000	1.27327	-0.70312	1	0
2	2000	3.03577e-017	36.8699	2.38423e-017	37.573
3	3000	0.424509	-2.1094	0.3334	-1.4062
4	4000	4.20877e-017	164.46	3.30548e-017	165.163
5	5000	0.254808	-3.5156	0.200121	-2.8125
6	6000	6.30689e-017	125.767	4.95329e-017	126.47
7	7000	0.182115	-4.9219	0.143029	-4.2188
8	8000	7.05302e-017	-125.74	5.53929e-017	-125.04
9	9000	0.141759	-6.3281	0.111335	-5.625

EXERCISE 6-13: Find the harmonics contained in the cross-over distortion of a push-pull amplifier. Let the input to the amplifier be a 2 V amplitude, 1 kHz sine wave. The push-pull amplifier drives a 100 Ω load. Run a Transient Analysis to view the waveforms and then run a Fourier Analysis to find the distortion.

SOLUTION: Wire the push-pull amplifier below. Run a Transient Analysis for one cycle with many points per cycle:

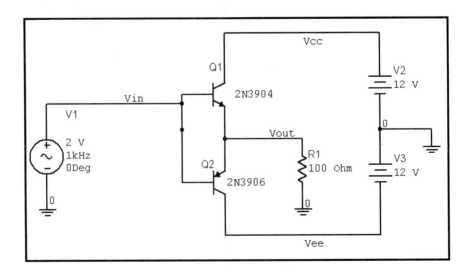

Time Domain Analyses

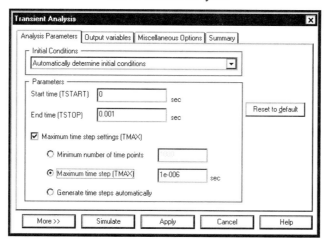

The input and output waveforms are:

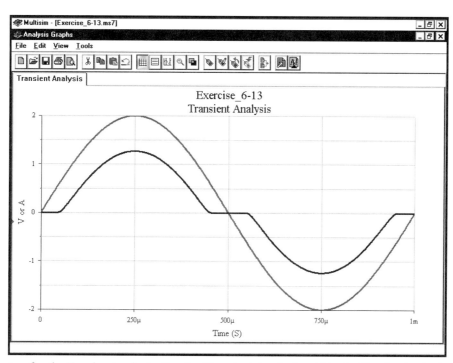

Use the following setup for the Fourier Analysis. For the Transient Analysis within the Fourier Analysis, simulate the circuit for four cycles to provide more data for the Fourier series calculations.

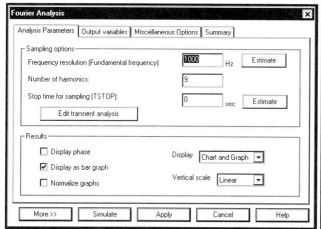 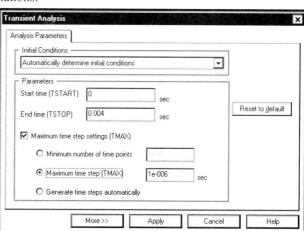

The results of the Fourier Analysis are:

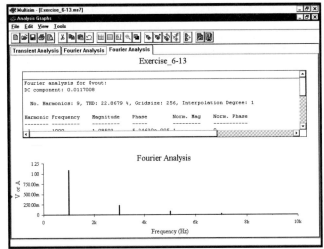

No. Harmonics: 9, THD: 22.8679 %, Gridsize: 256, Interpolation Degree: 1				
Harmonic Frequency	Magnitude	Phase	Norm. Mag	Norm. Phase
1 1000	1.08593	5.24639e-005	1	0
2 2000	0.00626112	-89.952	0.00576568	-89.952
3 3000	0.230321	179.998	0.212096	179.998
3 3000	0.230321	179.998	0.212096	179.998
4 4000	0.00453029	-89.864	0.00417181	-89.864
5 5000	0.0880167	179.992	0.081052	179.992
6 6000	0.00249281	-89.669	0.00229555	-89.669
7 7000	0.0283589	179.98	0.0261148	179.98
8 8000	0.000715386	-89.018	0.000658778	-89.018
9 9000	0.00140767	179.861	0.00129628	179.861

We see that major frequency peaks of the output are at 1 kHz, 3 kHz, 5 kHz, and 7 kHz. Since the output contains frequencies not contained in the input, the amplifier adds distortion. The total harmonic distortion (THD) was **22.8679%**.

EXERCISE 6-14: Using feedback, we can reduce the distortion of the push-pull amplifier. Repeat the previous exercise using the amplifier below. An ideal OPAMP is used because a real OPAMP cannot drive this push-pull amplifier. Note that the power supplies were changed to ± 15 V for the push-pull amplifier.

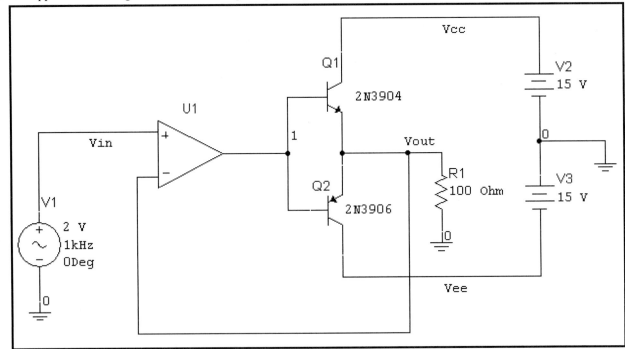

SOLUTION: Run a Transient Analysis for one cycle with many points per cycle. The input and output waveforms are:

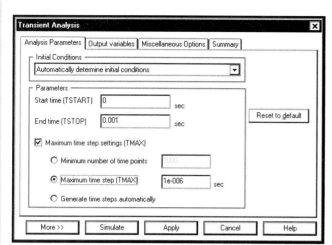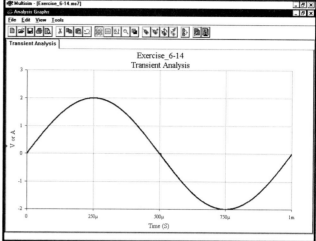

We see that the input and output waveforms look exactly the same. This is because when you enclose an amplifier within a feedback loop, the feedback reduces the amplifier's distortion.

Use the following setup for the Fourier Analysis. For the Transient Analysis within the Fourier Analysis, simulate the circuit for four cycles to provide more data for the Fourier series calculations.

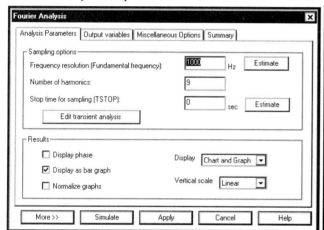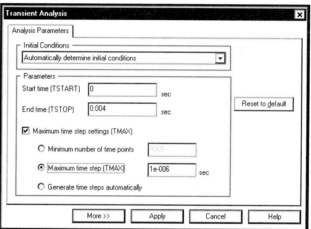

The results of the Fourier Analysis are:

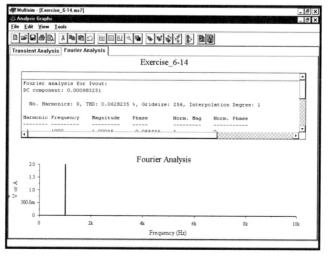

```
No. Harmonics: 9, THD: 0.0628235 %, Gridsize: 256, Interpolation Degree: 1

Harmonic  Frequency    Magnitude       Phase        Norm. Mag      Norm. Phase
--------  ---------    ---------       -----        ---------      -----------
   1        1000       1.99935         -0.055436    1              0
   2        2000       2.62526e-005    -97.236      1.31305e-005   -97.181
   3        3000       0.000557384     -90.438      0.000278782    -90.383
   3        3000       0.000557384     -90.438      0.000278782    -90.383
   4        4000       7.10635e-006    -93.237      3.55432e-006   -93.182
   5        5000       0.000562046     -82.484      0.000281114    -82.428
   6        6000       0.000148985     10.4093      7.45165e-005   10.4647
   7        7000       0.000726638     -101.83      0.000363437    -101.78
   8        8000       0.000109421     -60.98       5.47285e-005   -60.925
   9        9000       0.000623066     -105.64      0.000311634    -105.58
```

We see that most of the distortion of the push-pull amplifier has been removed. The total harmonic distortion is about 0.0628%. This amplifier has two problems. The first is that most OPAMPs cannot drive this push-pull amplifier when it is driving a large load. The second is that the transistors turn on and off. This adds a phase delay, which can cause the amplifier to oscillate. A push-pull amplifier biased such that both transistors are on works better. This example was chosen to show how feedback can reduce distortion.

6.I. Ideal Operational Amplifier Integrator

In this section we will demonstrate the use of the three-terminal virtual operational amplifier (OPAMP) and the pulsed voltage waveform. Wire the circuit shown below. The virtual OPAMPs are located in the Analog Components toolbar and can be configured as ideal OPAMPs, which are required for the circuit of this section to work properly.

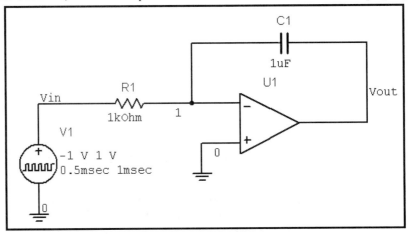

To place the OPAMP, click the **LEFT** mouse button on the **Show Analog Components Bar** button as shown below:

This toolbar allows you to select virtual 3- or 5-terminal OPAMPs, or a virtual comparator. Select the 3-terminal virtual OPAMP and place it in your circuit. Create the remainder of the circuit.

To see the properties of the virtual OPAMP, double-click the **LEFT** mouse button on the OPAMP graphic:

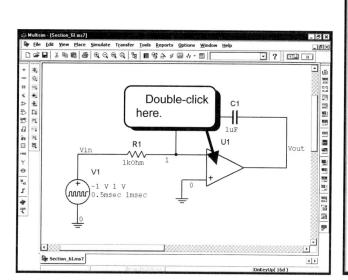
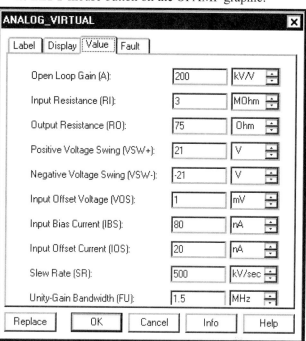

In order for this ideal integrator circuit to work, our OPAMP must not have any bias currents or offset voltages. We will set the **Input Offset Voltage (VOS)**, **Input Bias Current (IBS)**, and **Input Offset Current (IOS)** to zero. These are the nonideal parameters that make this circuit fail. By setting these parameters to zero, we make the OPAMP appear ideal, and the ideal integrator circuit will work properly.

The drawback of using an ideal OPAMP model is that the nonideal properties are not modeled. In this example, if a nonideal OPAMP model were used in the simulation, the integrator would not work because of bias currents. If this circuit were tested in the laboratory, it also would not work because of bias currents. Thus, the circuit simulation with a nonideal OPAMP matches the results in the lab, but the circuit simulation with an ideal OPAMP does not match the lab results. For this example, the ideal model is not a good choice for simulation because it does not match the results in the lab. We will use it here for demonstration purposes only. See **EXERCISE 6-16** to learn how this integrator performs using non-ideal OPAMPs. **In general, you should always use the nonideal OPAMP models if possible**. Note that the virtual OPAMP models are not ideal OPAMP models. They are OPAMP models for which we can easily change many of the operating parameters, and if we wish, make those parameters appear ideal.

Change the properties of the OPAMP as shown below. We have set the bias currents and offset voltage to zero:

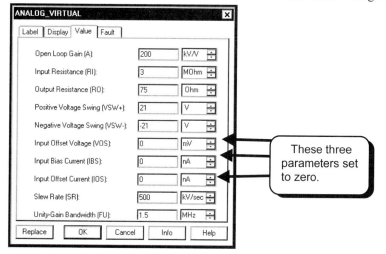

The pulse voltage source can be used to create an arbitrary pulse-shaped waveform. We will use it to create a 1 kHz square wave with a 5-volt amplitude (the wave goes from −5 V to +5 V). Double-click on the pulse voltage source graphic to modify its parameters. Set the parameters as shown below:

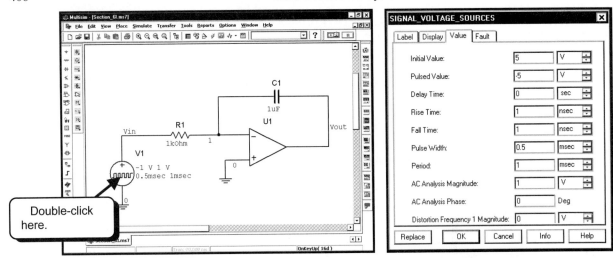

The parameters for the pulse voltage source were discussed on pages 386 to 388. The voltage waveform described by this dialog box will produce the following waveform: At the start of the simulation, the voltage will be set to the **Initial Value** of **5** volts. After the **Delay Time** (zero seconds in this case) the voltage will flip to the pulsed voltage of **−5** volts. From the **Delay Time** (t = 0) to 0.0005 s, the voltage will be equal to the **Pulsed Value** of **−5** volts. From 0.0005 s to t = 0.001 s, the voltage will be equal to the **Initial Value** of **5** V. At t = 0.001 s, the voltage will switch back to the **Pulsed Value** and repeat the cycle.

It is important to note that when Multisim calculates the bias to determine the initial condition of the capacitor, it sets the voltage of the pulsed source to its initial voltage. For our pulse waveform, Multisim will calculate the initial capacitor voltage assuming that Vin = 5 V. This will cause the integrator to have an infinite initial voltage. To prevent this, we must set the initial condition of the capacitor to 0 V. This can be done in several ways. We can set the initial condition of the capacitor to zero and then specify that Multisim use the specified initial conditions, or we can instruct Multisim to set all initial conditions to zero. When we set up the Transient Analysis, we will instruct Multisim to set all initial conditions to zero.

We would like to simulate the circuit for 10 cycles of the square wave using a Transient Analysis. Select **Simulate**, **Analyses**, and then **Transient Analysis** from the menus. Fill in the parameters as shown:

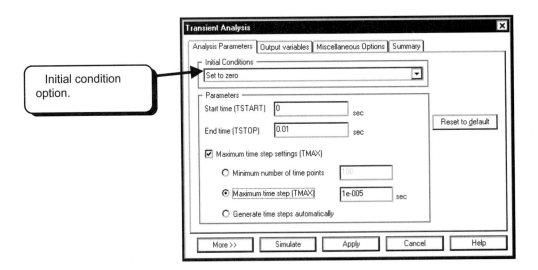

Note that we have specified that the initial conditions are set to zero. We will simulate the circuit for 10 cycles of the 1 kHz square wave or 10 ms. We have also specified a maximum step size of 10 μs to get an accurate simulation.

Select the **Output variables** tab and plot **$vout** and **$vin** during the simulation. Click on the **Simulate** button to run the Transient Analysis:

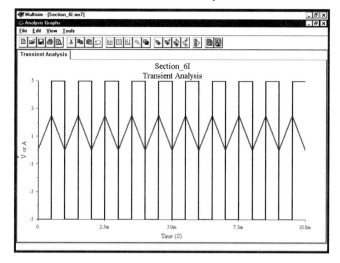

Note that the time integration of a square wave is a triangle wave.

EXERCISE 6-15: Run the integrator shown below. Set the initial condition of the capacitor to zero volts. Show the output waveform for a 1 kHz square wave input voltage. Note that this circuit will not work if the resistors are not matched exactly. This is a good circuit for simulation, but never use it in practice. You may also wish to run this circuit using nonideal OPAMPs to see the effects of bias currents and offset voltages.

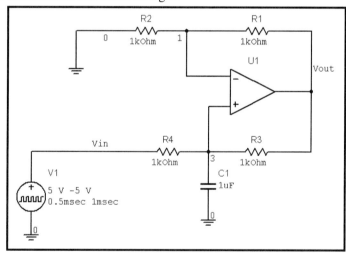

SOLUTION: The results of the Transient Analysis are:

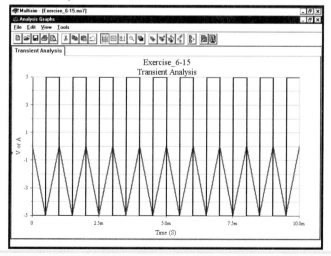

EXERCISE 6-16: Run the integrator of the previous section using a nonideal OPAMP model like the LM124AJ.

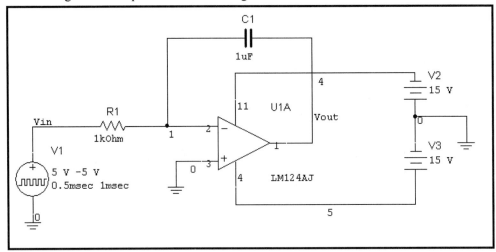

SOLUTION: Make sure you set the initial condition of the capacitor to zero. First, run the simulation for 10 ms:

The output voltage is:

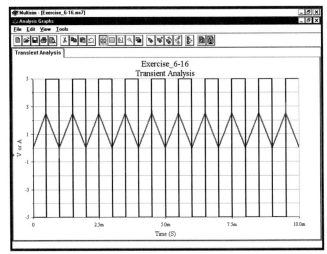

At first glance, the graph looks the same as for the simulation for the ideal OPAMP. However, if you use the cursors to look at the positive peaks, you will see that they are slowly drifting positive:

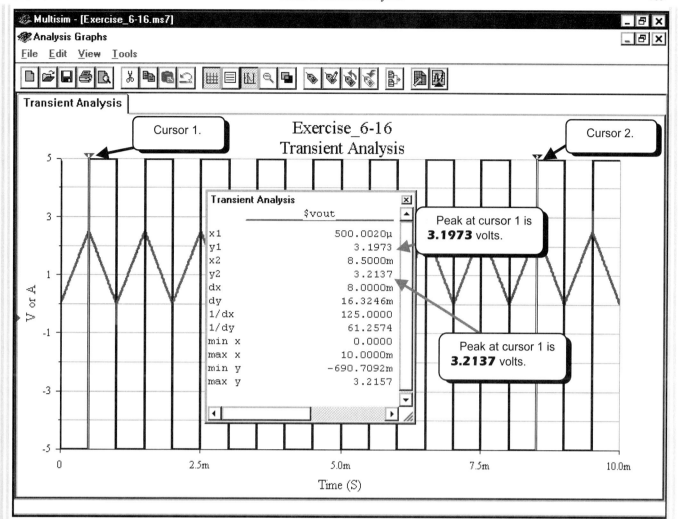

At cursor 1, the peak value of the triangle wave is **3.1973** volts. At cursor 2, the peak of the triangle wave is **3.2137** volts, 0.0164 volts higher than at the first cursor. Run the simulation again with the same settings as before, but set the end time to 100 ms and plot the output:

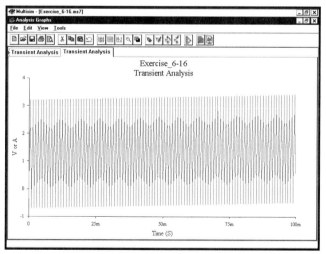

We see that the output is drifting quickly toward the positive supply. The screen captures below show a zoomed-in view of the drift at the start of the simulation (close to time = 0) and the drift at the end of the simulation (close to time = 100 ms).

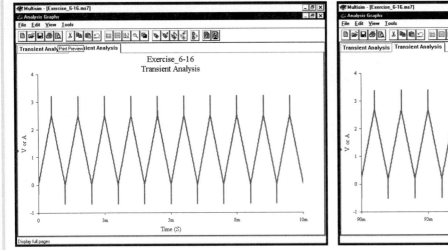
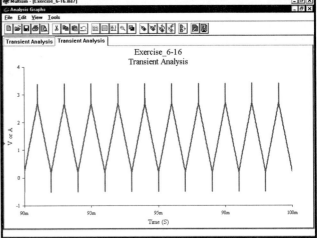

If we let the simulation run long enough, the OPAMP will eventually saturate and the output will be stuck at the positive supply rail. The output drifts up due to the bias currents of the OPAMP.

Note that the spikes at the top and the bottom of the output waveform should not be there. This may be an error in the OPAMP model:

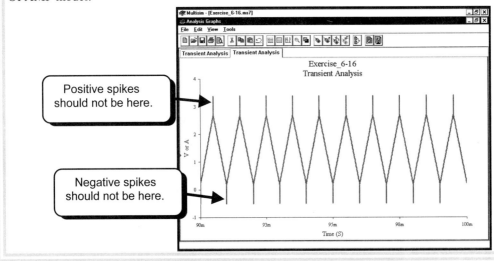

6.J. Operational Amplifier Schmitt Trigger

In this section we will use an operational amplifier to create a Schmitt Trigger. A nonideal operational amplifier must be used because the ideal OPAMP model has trouble converging when it is used as a Schmitt Trigger. Furthermore, a Schmitt Trigger requires an OPAMP with supply limits, so that the output of the circuit saturates at the supplies. Wire the circuit:

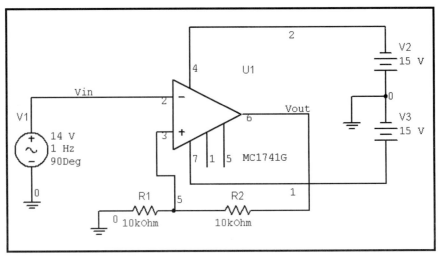

This OPAMP circuit is a Schmitt Trigger with trigger points at approximately ±7.5 V. A sinusoidal voltage source will be used to swing the input from +14 V to –14 V and from –14 V to +14 V a few times. The frequency of the source is 1 Hz to eliminate the effects of the OPAMP slew rate on the Schmitt Trigger performance. If you wish to observe slew rate effects, a higher frequency should be chosen. We would also like the sinusoidal source to start at –14 V instead of zero. This can be done by setting the phase of the source to 90°.

Because many of the attributes of the sinusoidal source have not been previously described, we will give a brief description here. The attributes of a sinusoidal source are:

The sinusoidal source is best described by an equation. We will use the following variables:

Attribute	Symbol	Units
Phase	φ	degrees
Voltage Offset	B	volts
Voltage Amplitude	A	volts
Frequency	F	Hz
Damping Factor	β	sec^{-1}
Delay Time	τ_d	seconds

Using these variables, the equation for the sinusoidal source is:

$$V_{sin} = \begin{cases} B - A\sin\left(\left(\dfrac{2\pi}{360}\right)\varphi\right) & \text{for } 0 \leq t < \tau_d \\ B + A\sin\left[2\pi F(t - \tau_d) - \left(\dfrac{2\pi}{360}\right)\varphi\right] e^{-\beta(t-\tau_d)} & \text{for } t \geq \tau_d \end{cases}$$

(This term is a constant.)

This may not be too clear. The full sinusoidal source can be exponentially damped, have a DC offset, and have a time delay as well as a phase delay. In the above equation, the phase (ϕ) is specified in degrees and is converted into radians by the constant $2\pi/360$. We note that for $\tau_d \geq t \geq 0$, V_{sin} is constant. The sinusoid does not start until $t = \tau_d$. If there are no time or phase delays, the above equation reduces to the exponentially damped sine wave:

$$V_{sin} = B + A\sin[2\pi F t]e^{-\beta t}$$

The sources available for use in the Transient Analysis are very flexible to provide the user with many possible waveforms for simulation. Unfortunately, this flexibility can also lead to confusion when using a source for the first time.

We would now like to set up a Transient Analysis that allows the sinusoidal source to complete two cycles. Because the source has a frequency of 1 Hz, two cycles will take two seconds of simulation time. Select **Simulate**, **Analyses**, and then **Transient Analysis** from the menus. Fill in the parameters as shown:

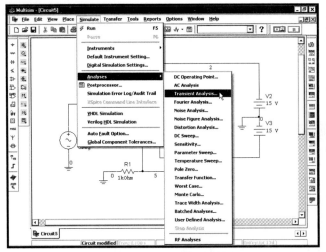 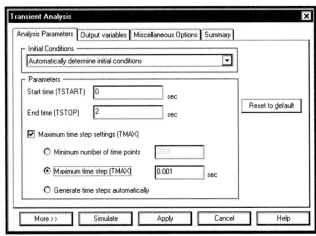

We have specified a **Maximum time step** of 1 ms to get accurate results when the Schmitt Trigger flips.

Next, select the **Output variables** tab and specify to plot **$vin** and **$vout**:

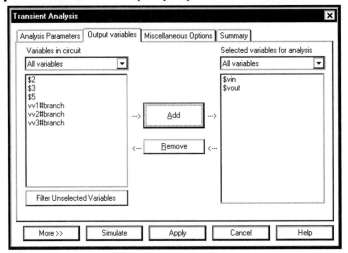

Click on the **Simulate** button to run the simulation:

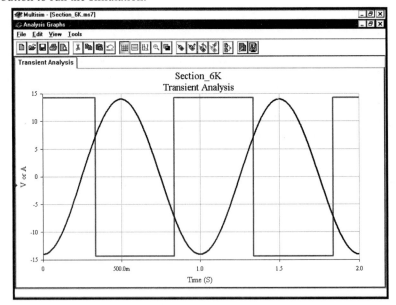

Using the cursors, we can display the trigger points of the Schmitt Trigger:

Time Domain Analyses

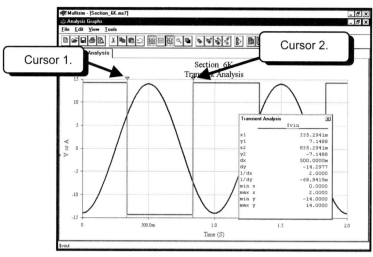

We can see that the Schmitt Trigger changes state at approximately VIN = ±7.15 V.

The next thing we would like to do is plot the hysteresis curve for this Schmitt Trigger. The hysteresis curve is a plot of V_O versus V_{IN}. To do this, we must display the waveforms on the Oscilloscope instrument and then change the mode of the scope. Add the oscilloscope to the circuit and then connect **Vin** to **Channel A** and **Vout** to **Channel B**. Change the frequency **V1** to **10 Hz**:

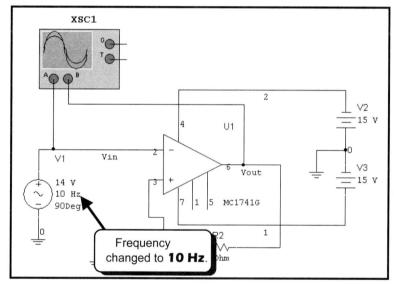

Click on the **Run / stop simulation** button to start the simulation and then double-click on the Oscilloscope instrument to view the waveforms. You will need to change the oscilloscope **Timebase** to **20 ms/Div**, and the oscilloscope window will take a long time to display the waveforms. After a while, you will see the waveforms below:

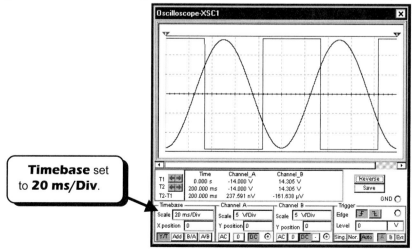

To view the hysteresis curve, all we need to do is to change the oscilloscope mode to **B/A**. Click on the **B/A** button as shown:

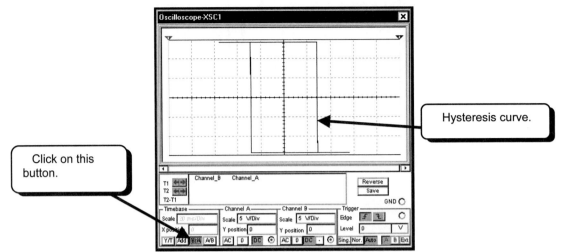

EXERCISE 6-17: Run a non-inverting Schmitt Trigger with a ±14 volt sine wave input.

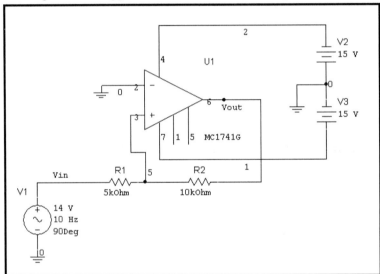

SOLUTION: Use the sinusoidal voltage source. The input and output waveforms are shown on the left screen capture, and the hysteresis curve is shown on the right screen capture.

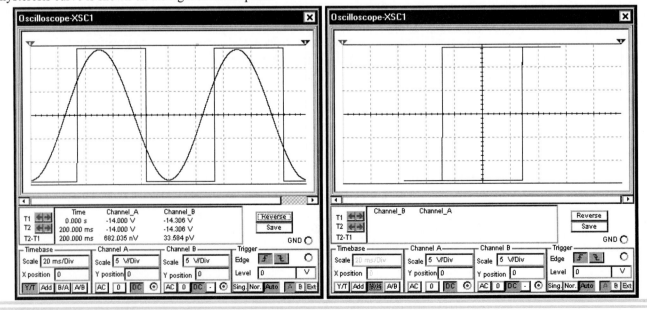

6.K. Parameter Sweep — Inverter Switching Speed

When designing digital circuits we are concerned with how different circuit elements affect the operation of the circuit. In this section we will look at switching speed. First, we will look at a basic BJT inverter and observe its operation. Wire the circuit below:

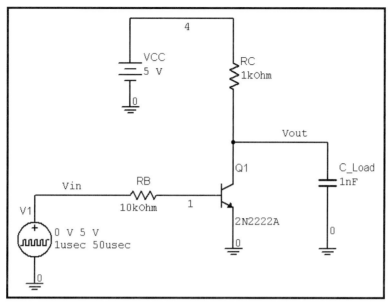

The input to the inverter will be a short 1 µs pulse. Double-click the **LEFT** mouse button on the pulse voltage source graphic to view the parameters for this source. Fill in the parameters as shown:

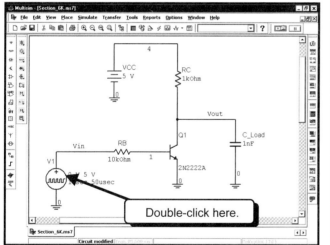
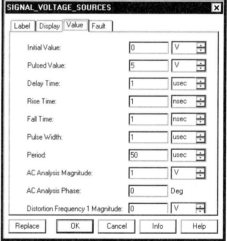

The **Pulse Width** of the pulse is 1 µs and the **Period** is 50 µs. These settings produce a short 1 µs pulse that repeats every 50 µs. We will only run the simulation for 50 µs, so we will only see a single pulse. Normally, the pulse will start at the beginning of the simulation. However, we have specified a **Delay Time** of 1 µs, so the voltage stays at the **Initial Value** of 0 V for 1 µs before the source starts the pulse. The **Rise Time** and **Fall Time** are set to 1 ns because we are looking at the switching speed of a logic gate and we want the edges of our input pulse to be much faster than the rise and fall times of the circuit we are testing.

We will use a Transient Analysis here instead of a virtual simulation because we are going to look at the response of this circuit for several different values of the collector resistor, RC. We cannot do this with a virtual simulation. We will first run a Transient Analysis to look at the response for a single value of RC. Select **Simulate**, **Analyses**, and then **Transient Analysis** from the menus and fill in the dialog box as shown:

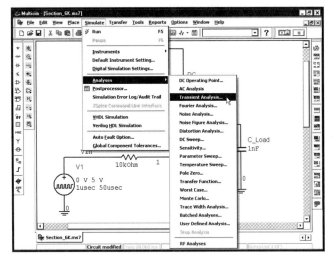 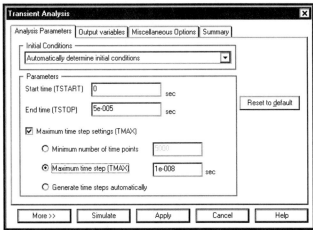

Notice that we have set the **End time** to 50 μs and have specified the **Maximum time step** to be 10 ns.

We will first look at the input voltage pulse to see that it is correct. Select the **Output variables** tab and then specify to plot **$vin** during the simulation:

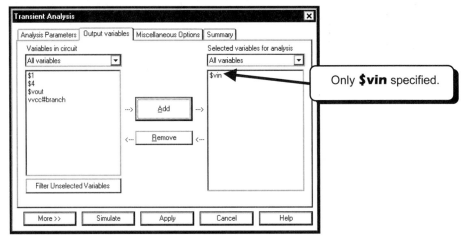

Click on the **Simulate** button to run the simulation. Your input voltage waveform should be as shown below:

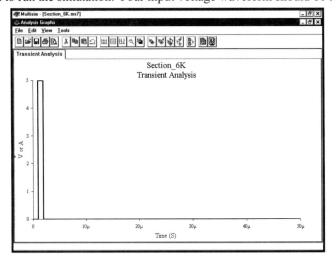

We see a 0 to 5 V pulse that lasts for 1 μs. We also note that the pulse starts after a delay of 1 us, as specified by the Delay Time in the pulse voltage source parameters. If your pulse does not look as shown above, modify the source parameters and rerun the simulation.

Now that we have verified the input pulse waveform, we need to test the operation of the circuit and plot the output. Select **Simulate**, **Analyses**, and then **Transient Analysis** from the menus and verify that your dialog box still has the same settings as shown below:

Select the **Output variables** tab and change the settings so that only **$vout** will be plotted during the simulation:

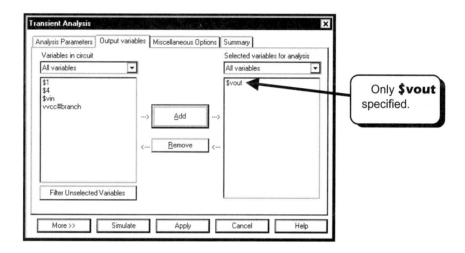

Click on the **Simulate** button to run the simulation and view the output voltage waveform:

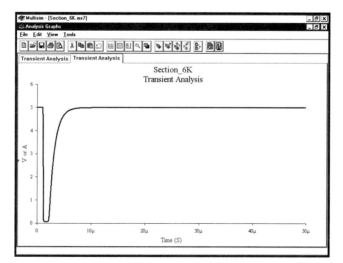

We see that the fall time of the output (the output goes from 5 V down to 0 V) is very fast, but the circuit has a slow rise time (the output goes from about 0 V up to 5 V).

We would now like to answer the question of how the rise and fall times of the output of the circuit are affected by the value of the collector resistor. We will answer this question using the Parameter Sweep. Select **Simulate**, **Analyses**, and then **Parameter Sweep** from the menus and fill in the dialog box as shown:

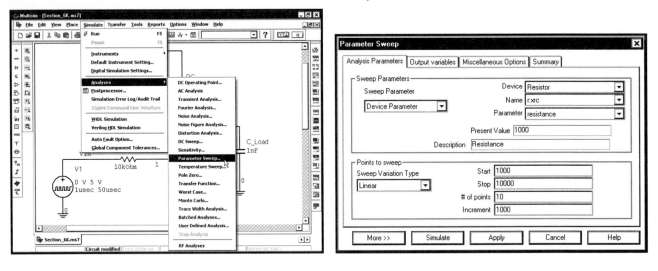

The top portion of the dialog box specifies that we are going to change the resistance of resistor RC. The bottom portion of the dialog box specifies the values that RC will have. In this case we have specified a linear sweep from 1000 Ω to 10,000 Ω in 1000 Ω steps. Thus, RC will have values of 1000 Ω, 2000 Ω, 3000 Ω, and so on, up to 10,000 Ω.

The next thing we must do is specify that the Parameter Sweep works in conjunction with a Transient Analysis. Select the **More** button:

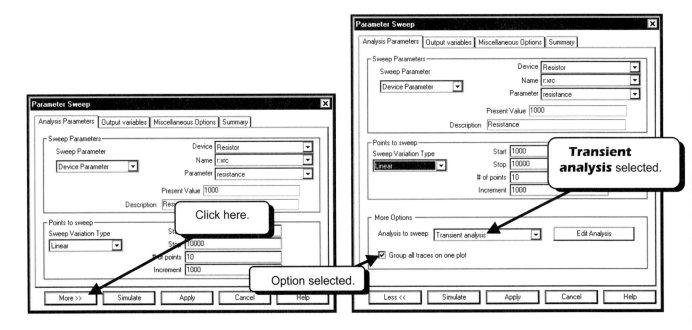

By default, the **Transient analysis** is selected. If not, select it.

We also note that the option **Group all traces on one plot** is selected. With this option selected, all runs from the Parameter Sweep will be plotted on a single graph. For our example, the resistor value takes on ten values. When we plot any trace, ten traces will be plotted, one for each value of the resistor. If we plot the input voltage, ten traces will be drawn on the graph. The ten traces will be the input voltages for all the different values of the resistor. If we plot the output voltage, ten traces will also be drawn. Each trace will be the output voltage for each value of the resistor.

The next thing we must do is edit the parameters for the Transient Analysis. Select the **Edit Analysis** button and fill in the parameters with the values we used in our analysis at the beginning of this section:

Time Domain Analyses

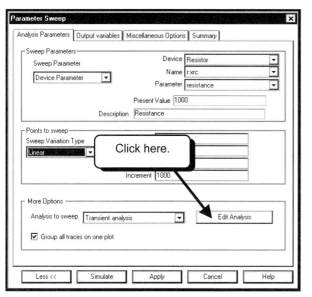
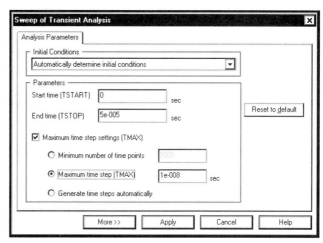

Click on the **Apply** button to return to the **Parameter Sweep** setup dialog box:

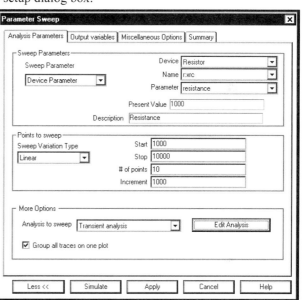
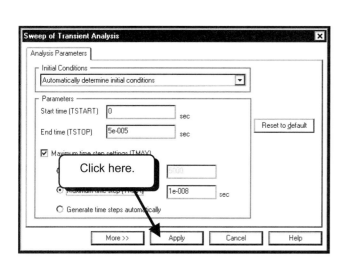

The last thing we must do is specify the **Output variables** for the run. Select the **Output variables** tab and specify that **$vout** be displayed during the simulation:

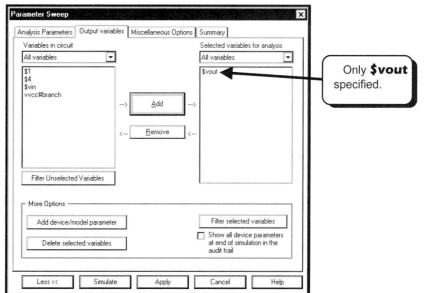

We are now ready to run the simulation, so click on the **Simulate** button:

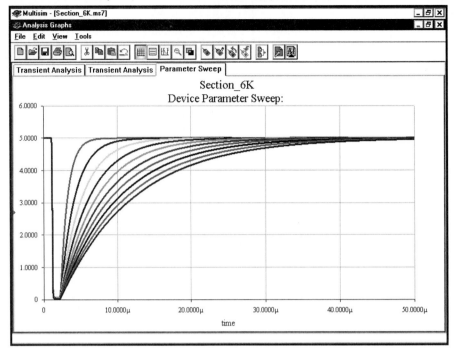

We see that the fall time is independent of our choice of RC. With the time scale shown, all of the runs go from 5 V down to 0 V so fast that it looks like a square edge. The rise time, however, appears to be a direct function of our choice of RC.

The question may arise as to how we tell which trace is for what value of RC. This can be done by displaying the legend. Click the ***RIGHT*** mouse button on the graph and then select **Toggle Legend** to display the legend:

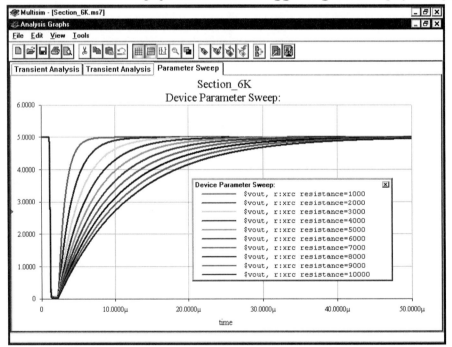

With the legend displayed, we can easily identify each trace, and we see that a smaller value of RC yields a shorter rise time. Thus, to make a logic gate of this type with a fast rise time, we need to design our circuit using a small value of RC.

EXERCISE 6-18: For the BJT inverter of this section, run the circuit to observe how RB affects switching speed. Let RC remain constant at 1 kΩ.

SOLUTION: Modify the Parameter Sweep to sweep resistor RB from 5 kΩ to 30 kΩ in 5 kΩ steps.

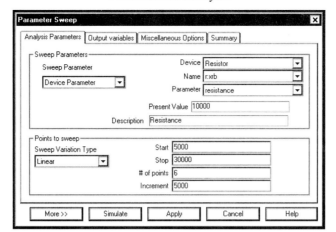

Modify the Transient Analysis setup to run the simulation for 7.5 μs and use a **Maximum step size** of 1 ns:

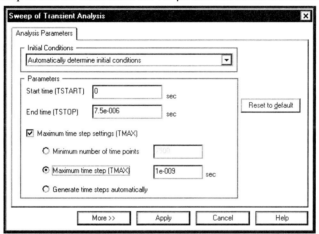

The results of the simulation are:

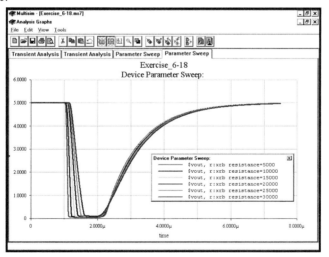

The results show that the base resistance has a direct effect on the fall time and a small effect on the rise time.

6.L. Temperature Sweep — Linear Regulator

A temperature sweep can be used in conjunction with most of the analyses, and it allows you to run the same analysis several times at different temperatures. Here we will show how temperature affects the performance of a linear voltage regulator. Create the circuit shown below:

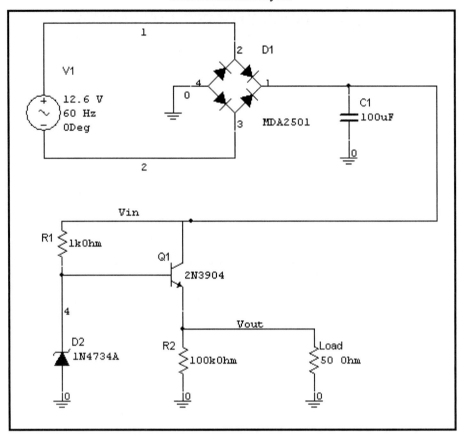

Voltage source **V1** represents the output of a 12.6 V RMS transformer when it is connected to the 115 V RMS line. This circuit is a bridge rectifier followed by a filter capacitor to produce a DC voltage with ripple at **Vin**. After **Vin** is a linear regulator made of a Zener voltage reference and an NPN pass transistor. This regulator should produce a DC output voltage of approximately 5 volts. We will first run a Transient Analysis to see the operation of the circuit at room temperature (27°C). To set up a Transient Analysis, select **Simulate, Analyses**, and then **Transient Analysis** from the menus. Fill in the parameters as shown below:

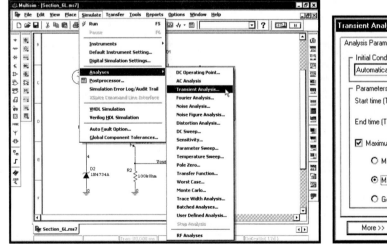
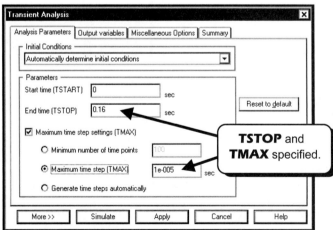

Select the **Output variables** tab and specify to display **$vin** and **$vout** during the simulation:

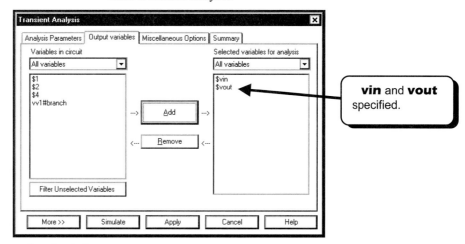

Select the **Simulate** button to run the simulation:

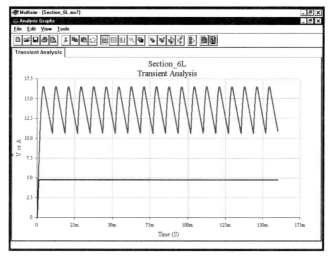

We will zoom in on the **vout** trace to see it more closely. Use the cursors to find the magnitude of the ripple.

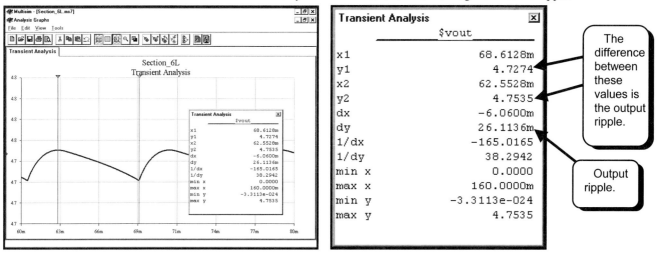

We see that the output voltage is about 4.7 V and the magnitude of the ripple on the output is about 26 mV.

Next we will see how the output changes with temperature. We will run a Transient Analysis at –25°C, 25°C, and 125°C. Return to the schematic and select **Simulate**, **Analyses**, and then **Temperature Sweep** from the menus, and fill in the **Temperature Sweep Analysis** dialog box as shown:

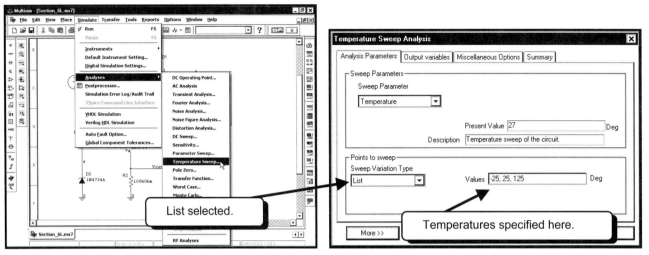

We see that we have specified the **Sweep Variable Type** as a **List**. This option allows us to specify a list of temperatures at which we want to run the simulation. There does not need to be any relationship between the temperatures in the list.

The Temperature Sweep Analysis must run in conjunction with another analysis. To specify the analysis, click on the **More** button:

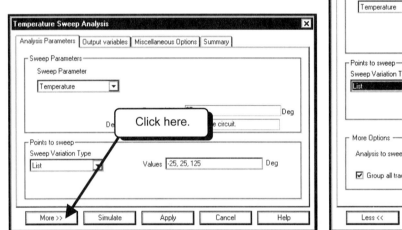

By default, a **Transient analysis** is selected. Had you wanted to run an AC Analysis or a DC Operating Point analysis, you would select that analysis here. We need to set up the parameters for the Transient Analysis, so select the **Edit Analysis** button and fill in the parameters with the values we used for the Transient Analysis at the beginning of this section:

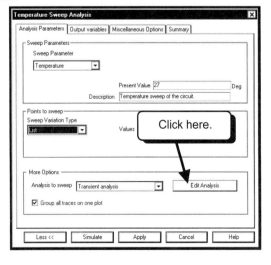
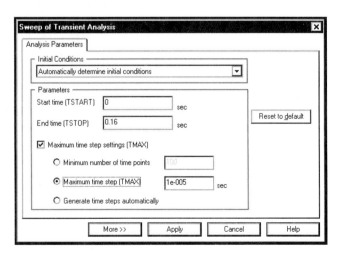

Click on the **Apply** button to return to the **Temperature Sweep Analysis** dialog box. Select the **Output variables** tab and specify to plot **$vin** and **$vout** during the simulation:

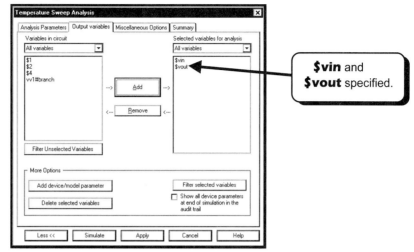

Click on the **Simulate** button to run the simulation. It will take three times as long as the previous simulation because the Transient Analysis will run three times. Logically, the Transient Analysis runs inside the Temperature Sweep. For this example, the temperature will first be set to –25°C, and then the Transient Analysis will be run. Next, the temperature will be set to 25°C, and then the Transient Analysis will be run. Finally, the temperature will be set to 125°C, and then the Transient Analysis will be run.

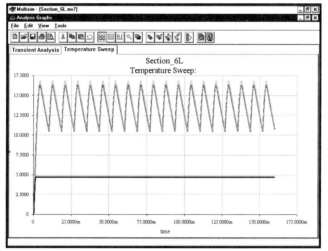

Three traces are shown for each variable, one trace for each simulation temperature. We see that both the input and the output voltage waveforms are affected by temperature. We will first look at the input voltage waveforms more closely:

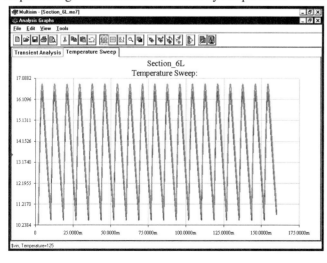
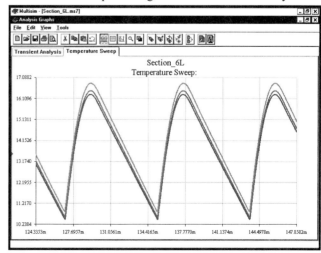

The variation in the input voltage waveform is mostly due to the temperature dependence of the diodes in the bridge rectifier. The voltage regulator also draws a slightly different current at different temperatures, and this also changes the input voltage waveform, but not nearly as much as bridge diodes.

Next, we will look at the output voltage traces. Three traces are shown, one for each temperature.

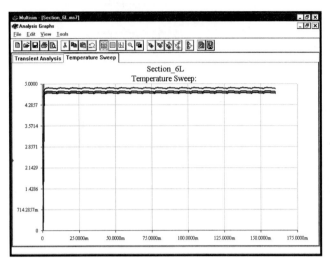 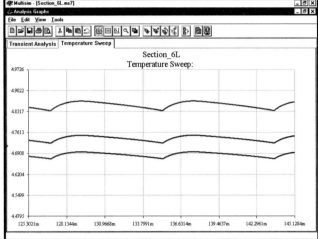

We can use the cursors to measure the ripple on each trace:

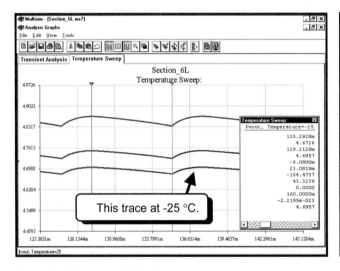 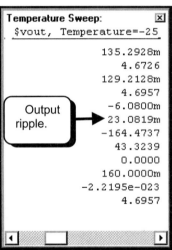

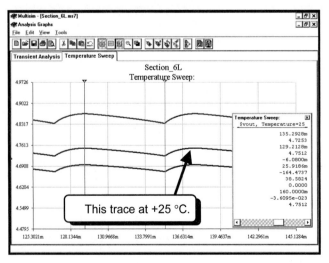 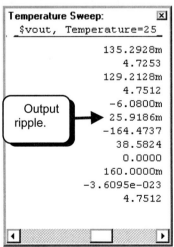

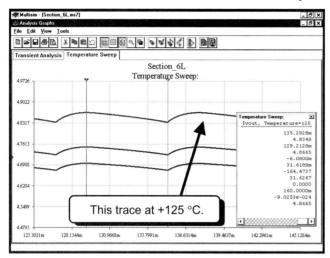
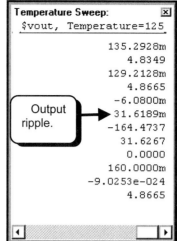

The top trace has 31 mV of ripple, the middle trace has 26 mV of ripple, and the bottom trace has 23 mV of ripple.

The question may arise as to how to tell which trace is at what temperature. To identify the traces, click the **RIGHT** mouse button on the graph and then select **Toggle Legend**:

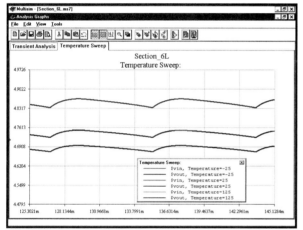
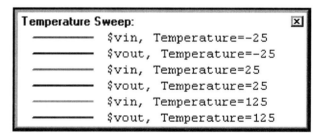

The legend tells us which trace is **vin** and **vout**, and also tells us the temperature of each trace.

6.M. Rated Components

Multisim 7 has added a feature called rated components. With some of the components, you can add rated limits to the components. If the component's limits are exceeded during a simulation, Multisim 7 will indicate in real time on the schematic that the part has been destroyed. Examples are the power rating of a resistor, the voltage ratings of a capacitor, the reverse repetitive voltage rating of a diode, the base-emitter voltage rating of a BJT, and the maximum collector current rating of a BJT. We will show a simple example to place a rated component in your schematic. You need to use the Rated Virtual Components Bar. To open this bar, click the **LEFT** mouse button on the **Show Rated Virtual Components Bar** button :

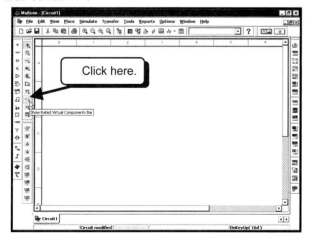
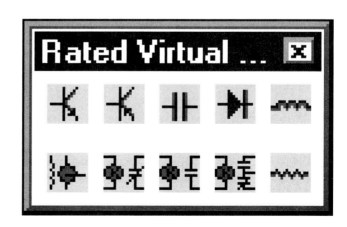

Use this button bar to create the circuit below:

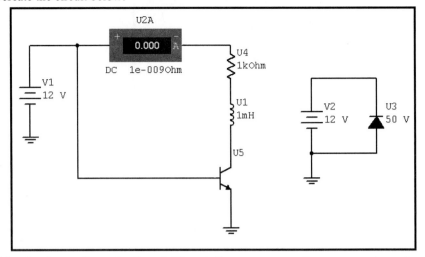

All components in this circuit are from the Rated Virtual Components toolbar except the voltage source, the current indicator, and the ground. If you double-click on the BJT, you can see its rated values. I have changed them slightly:

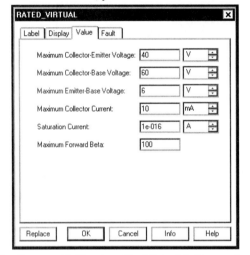

The ratings for this device are the **Maximum Collector-Emitter Voltage** (specified as **40** volts), the **Maximum Collector-Base Voltage** (specified as **60** volts), the **Maximum Emitter-Base Voltage** (specified as **6** volts), and the **Maximum Collector Current** (specified at **10 mA**; I changed this value — the default value was 200 mA). Change your settings to match this dialog box and click the **OK** button.

Next, double-click on the inductor graphic symbol to see its properties:

The only rating for an inductor is the peak current, and the default is 1 amp. Change the current rating to 10 mA and then click the **OK** button.

Next, double-click on the graphic for the resistor to see its properties:

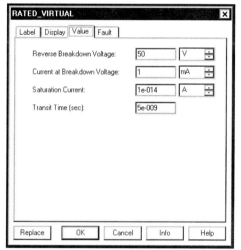

For resistors, the rating is the power limitation of the resistor and the default value is **0.25** watt. We will not change any of the settings here. Click the **Cancel** button to return to the schematic.

Lastly, double-click on the graphic for the diode:

The rating for this component is the **Reverse Breakdown Voltage** of the diode, with a default value of 50 volts. The other three parameters are actually model parameters for the diode model. If the reverse voltage exceeds 50 volts, a component failure will be indicated on the schematic. We will not change any of the values here either, so click the **Cancel** button.

We are now ready to run the circuit, so press the **F5** key to start the simulation. Multisim 7 indicates that the BJT has reached one of its specified limits and failed:

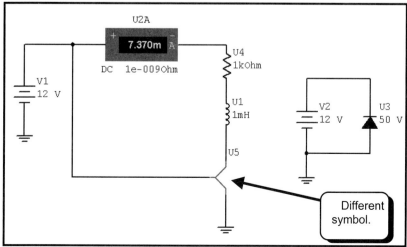

Note that the BJT symbol is a little different and it is also shown in red. Press the **F5** key to stop the simulation.

We must now ask the question why the BJT failed. Looking at the current monitor, we see that the current is 7.37 mA, well below the 10 mA collector current rating of BJT. After staring at the circuit a while, we realize that we have wired 12 V directly across the base-emitter junction of the device, and this is incorrect. A BJT is current driven, and we need a base resistor, so I will add a 1 kΩ resistor as shown below:

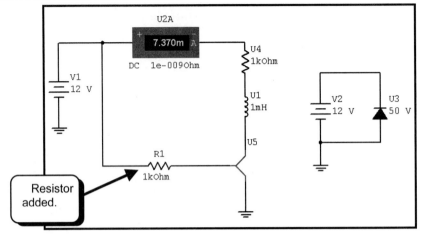

In my case, R1 was not a rated component, but it could have been, had I wanted to test this component to see if it too was exceeding its power limitations. Press the **F5** key to run the simulation again:

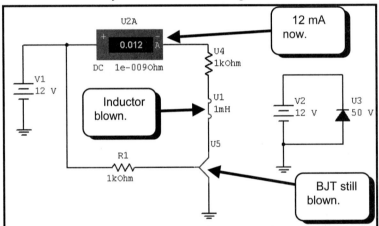

The BJT remains blown, but this time for a different reason. The current meter now indicates that we now have a collector current of 12 mA, and this exceeds the rating of the BJT. We also see that the inductor has blown because its current rating was exceeded. Press the **F5** key to stop the simulation.

We must modify the circuit to limit the current to a value below the rated limit of the device (and continue to use the same BJT), or increase the rated limit of the BJT (and find a new part with higher limitations). We will choose the latter and increase the limit of the BJT. Double-click on the graphic for the BJT and change the **Maximum Collector Current** to 20 mA:

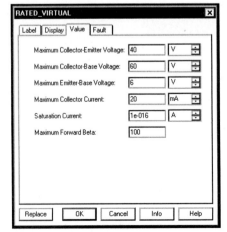

Click the **OK** button and then rerun the simulation.

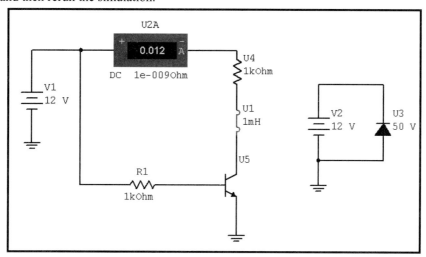

Now only the rated limit of the inductor is exceeded.

For the last example, we will change the voltage of V2 to 75 volts and observe that the diode indicates that its rated breakdown voltage limit has been exceeded:

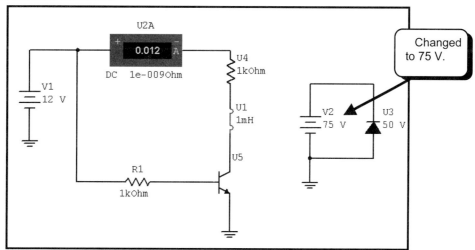

Now when we run the simulation, the diode will indicate a fault:

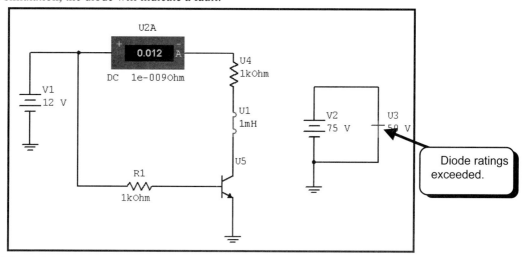

These circuits were fairly trivial and were set up to show us how the rated parts behave when conditions exceed their rated limits. Normally, we will run a much more complicated circuit and test parts in applications where it is not so obvious whether the parts' ratings are being exceeded.

6.N. Problems

Problem 6-1: Find the phase in degrees of the voltage at node V1 using the Oscilloscope instrument in the circuit below:

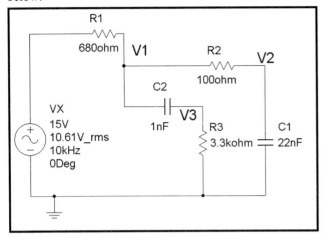

Problem 6-2: For the circuit of Problem 6-1, find the phase in degrees of the voltage at node V2 and the current through R1 using the Oscilloscope instrument.

Problem 6-3: For the circuit of Problem 6-1, find the phase in degrees of the voltage at node V3 and the current through C2 using the Oscilloscope instrument.

Problem 6-4: For the circuit of Problem 6-1, find the phase in degrees of the voltage across R1 and the current through C2 using the Oscilloscope instrument.

Problem 6-5: For the circuit of Problem 6-1, find the phase in degrees of the voltage at node V1 and the current through R2 using the Oscilloscope instrument.

Problem 6-6: For the circuit of Problem 6-1, find the phase in degrees of the voltage at node V2 and the current through C1 using the Oscilloscope instrument.

Problem 6-7: For the circuit of Problem 6-1, find the phase in degrees of the voltage at node V3 and the current through R3 using the Oscilloscope instrument.

Problem 6-8: For the circuit of Problem 6-1, find the phase in degrees of the voltage across C2 and the current through C2 using the Oscilloscope instrument.

Problem 6-9: Find the phase in degrees of the voltage at node V1 using indicators in the circuit below:

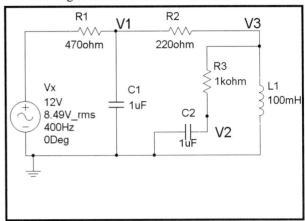

Problem 6-10: For the circuit of Problem 6-9, find the phase in degrees of the voltage at node V2 and the current through R1 using the Oscilloscope instrument.

Problem 6-11: For the circuit of Problem 6-9, find the phase in degrees of the voltage at node V3 and the current through C2 using the Oscilloscope instrument.

Problem 6-12: For the circuit of Problem 6-9, find the phase in degrees of the voltage at node V1 and the current through R2 using the Oscilloscope instrument.

Problem 6-13: For the circuit of Problem 6-9, find the phase in degrees of the voltage at node V2 and the current through C1 using the Oscilloscope instrument.

Problem 6-14: For the circuit of Problem 6-9, find the phase in degrees of the voltage across C2 and the current through C1 using the Oscilloscope instrument.

Problem 6-15: For the circuit of Problem 6-9, find the phase in degrees of the voltage at node V3 and the current through L1 using the Oscilloscope instrument.

Problem 6-16: Find the phase in degrees of the voltage at node V1 using the Oscilloscope instrument in the circuit below:

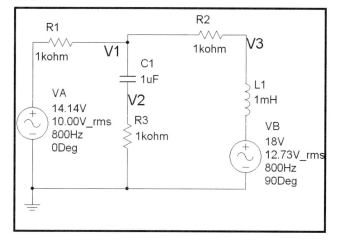

Problem 6-17: For the circuit of Problem 6-16, find the phase in degrees of the voltage at node V2 and the current through R1 using the Oscilloscope instrument.

Problem 6-18: For the circuit of Problem 6-16, find the phase in degrees of the voltage at node V3 and the current through C1 using the Oscilloscope instrument.

Problem 6-19: For the circuit of Problem 6-16, find the phase in degrees of the voltage at node V1 and the current through R2 using the Oscilloscope instrument.

Problem 6-20: For the circuit of Problem 6-16, find the phase in degrees of the voltage across L1 and the current through R2 using the Oscilloscope instrument.

Problem 6-21: For the circuit of Problem 6-16, find the phase in degrees of the voltage at node V2 and the current through C1 using the Oscilloscope instrument.

Problem 6-22: For the circuit of Problem 6-16, find the phase in degrees of the voltage at node V3 and the current through L1 using the Oscilloscope instrument.

Problem 6-23: For the full-wave power supply shown on page 353, what happens to the voltages Vin and Vout if capacitor C1 is made larger?

Problem 6-24: For the full-wave power supply shown on page 353, what happens to the voltages Vin and Vout if capacitor C1 is made smaller? What is the smallest value of C1 such that the output voltage remains constant?

Problem 6-25: For the full-wave power supply shown on page 353, what happens to the diode currents if capacitor C1 is made larger?

Problem 6-26: For the full-wave power supply shown on page 353, what happens to the diode currents if capacitor C1 is made smaller?

Problem 6-27: For the full-wave power supply shown on page 353, what happens to the voltages Vin and Vout if the load is made smaller (higher resistance)?

Problem 6-28: For the full-wave power supply shown on page 353, what happens to the voltages Vin and Vout if the load is made larger (lower resistance)? What is the largest load such that the output voltage remains constant?

Problem 6-29: For the full-wave power supply shown on page 353, what happens to the diode currents if the load is made larger (lower resistance)?

Problem 6-30: For the full-wave power supply shown on page 353, what happens to the diode currents if the load is made smaller (higher resistance)?

Problem 6-31: For the half-wave power supply shown on page 362, what happens to the voltages Vin and Vout if capacitor C1 is made larger?

Problem 6-32: For the half-wave power supply shown on page 362, what happens to the voltages Vin and Vout if capacitor C1 is made smaller? What is the smallest value of C1 such that the output voltage remains constant?

Problem 6-33: For the half-wave power supply shown on page 362, what happens to the diode currents if capacitor C1 is made larger?

Problem 6-34: For the half-wave power supply shown on page 362, what happens to the diode currents if capacitor C1 is made smaller?

Problem 6-35: For the half-wave power supply shown on page 362, what happens to the voltages Vin and Vout if the load is made smaller (higher resistance)?

Problem 6-36: For the half-wave power supply shown on page 362, what happens to the voltages Vin and Vout if the load is made larger (lower resistance)? What is the largest load such that the output voltage remains constant?

Problem 6-37: For the half-wave power supply shown on page 362, what happens to the diode currents if the load is made larger (lower resistance)?

Problem 6-38: For the half-wave power supply shown on page 362, what happens to the diode currents if the load is made smaller (higher resistance)?

Problem 6-39: For the circuit below let Vin be a triangle wave that varies between –20 and +20 volts. (a) Plot Vo(t) and Vin(t) for one cycle of the input waveform using a Transient Analysis. (b) Plot the transfer curve Vo versus Vin using the Oscilloscope instrument.

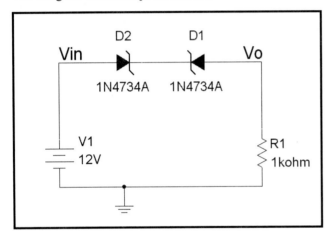

Problem 6-40: For the circuit of Problem 6-39 let Vin be a triangle wave that varies between –20 and +20 volts. (a) Plot Vo(t) and Vin(t) for about one cycle of the input waveform using the Oscilloscope instrument. (b) Plot the transfer curve Vo versus Vin using the Oscilloscope instrument.

Problem 6-41: For the circuit below let Vin be a triangle wave that varies between –20 and +20 volts. (a) Plot Vo(t) and Vin(t) for one cycle of the input waveform using a Transient Analysis. (b) Plot the transfer curve Vo versus Vin using the Oscilloscope instrument.

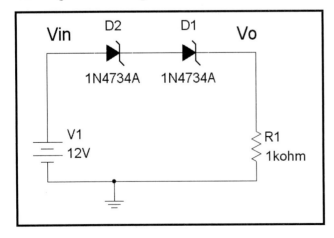

Problem 6-42: For the circuit of Problem 6-41 let Vin be a triangle wave that varies between –20 and +20 volts. (a) Plot Vo(t) and Vin(t) for about one cycle of the input waveform using the Oscilloscope instrument. (b) Plot the transfer curve Vo versus Vin using the Oscilloscope instrument.

Problem 6-43: For the circuit below let Vin be a triangle wave that varies between –20 and +20 volts. (a) Plot Vo(t) and Vin(t) for one cycle of the input waveform using a Transient Analysis. (b) Plot the transfer curve Vo versus Vin using the Oscilloscope instrument.

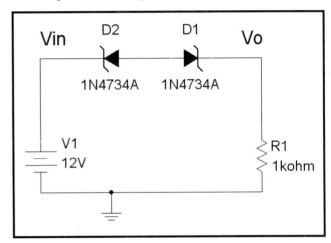

Problem 6-44: For the circuit of Problem 6-43 let Vin be a triangle wave that varies between –20 and +20 volts. (a) Plot Vo(t) and Vin(t) for about one cycle of the input waveform using the Oscilloscope instrument. (b) Plot the transfer curve Vo versus Vin using the Oscilloscope instrument.

Problem 6-45: For the circuit below let Vin be a triangle wave that varies between –20 and +20 volts. (a) Plot Vo(t) and Vin(t) for one cycle of the input waveform using a Transient Analysis. (b) Plot the transfer curve Vo versus Vin using the Oscilloscope instrument.

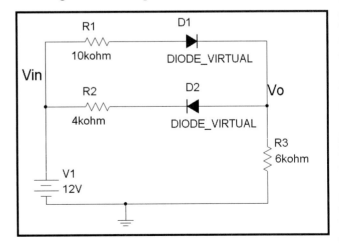

Problem 6-46: For the circuit of Problem 6-45 let Vin be a triangle wave that varies between –20 and +20 volts. (a) Plot Vo(t) and Vin(t) for about one cycle of the input waveform using the Oscilloscope instrument. (b) Plot the transfer curve Vo versus Vin using the Oscilloscope instrument.

Problem 6-47: For the circuit below let Vin be a triangle wave that varies between –20 and +20 volts. (a) Plot Vo(t) and Vin(t) for one cycle of the input waveform using a Transient Analysis. (b) Plot the transfer curve Vo versus Vin using the Oscilloscope instrument.

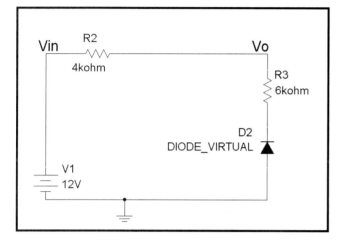

Problem 6-48: For the circuit of Problem 6-47 let Vin be a triangle wave that varies between –20 and +20 volts. (a) Plot Vo(t) and Vin(t) for about one cycle of the input waveform using the Oscilloscope instrument. (b) Plot the transfer curve Vo versus Vin using the Oscilloscope instrument.

Problem 6-49: For the circuit below let Vin be a triangle wave that varies between –20 and +20 volts. (a) Plot Vo(t) and Vin(t) for one cycle of the input waveform using a Transient Analysis. (b) Plot the transfer curve Vo versus Vin using the Oscilloscope instrument.

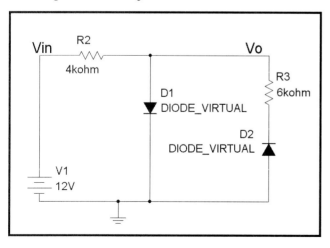

Problem 6-50: For the circuit of Problem 6-49 let Vin be a triangle wave that varies between –20 and +20 volts. (a) Plot Vo(t) and Vin(t) for about one cycle of the input waveform using the Oscilloscope instrument. (b) Plot the transfer curve Vo versus Vin using the Oscilloscope instrument.

Problem 6-51: For the circuit below let Vin be a sine wave at 1000 Hz. (a) Find the maximum voltage swing at Vout without clipping or a large amount of distortion. (b) What is the amplitude of Vin and Vout found in part a? (c) Run a Fourier Analysis with the output at its maximum found in part a. What is the total harmonic distortion contained in the output? What is the magnitude of the largest harmonic?

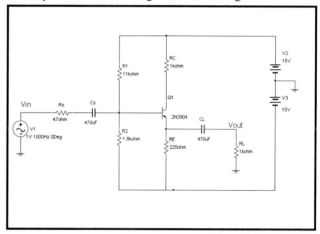

Problem 6-52: Repeat the simulations of Problem 6-51 with RL = 10 kΩ.

Problem 6-53: Repeat the simulations of Problem 6-51 with RL = 100 Ω.

Problem 6-54: Repeat the simulations of Problem 6-51 with RL = 10 Ω.

Problem 6-55: Repeat the simulations of Problem 6-51 with RC = 0 Ω. What is the effect of RC on this circuit?

Problem 6-56: For the circuit below let Vin be a sine wave at 1000 Hz. (a) Find the maximum voltage swing at Vout without clipping or a large amount of distortion. (b) What is the amplitude of Vin and Vout found in part a? (c) Run a Fourier Analysis with the output at its maximum found in part a. What is the total harmonic distortion contained in the output? What is the magnitude of the largest harmonic?

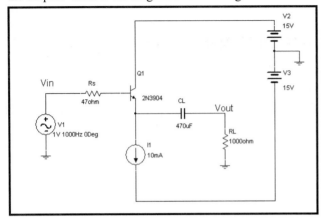

Problem 6-57: Repeat the simulations of Problem 6-56 with RL = 10 kΩ.

Problem 6-58: Repeat the simulations of Problem 6-56 with RL = 100 Ω.

Problem 6-59: Repeat the simulations of Problem 6-56 with RL = 10 Ω.

Problem 6-60: For the circuit below let Vin be a sine wave at 1000 Hz. (a) Find the maximum voltage swing at Vout without clipping or a large amount of distortion. (b) What is the amplitude of Vin and Vout found in part a? (c) Run a Fourier Analysis with the output at its maximum found in part a. What is the total harmonic distortion contained in the output? What is the magnitude of the largest harmonic?

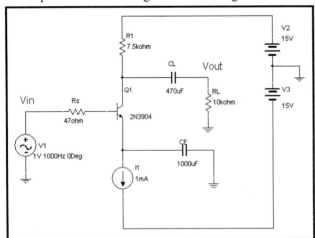

Problem 6-61: Repeat the simulations of Problem 6-60 with RL = 1 kΩ.

Problem 6-62: Repeat the simulations of Problem 6-60 with RL = 100 Ω.

Problem 6-63: Repeat the simulations of Problem 6-60 with RL = 10 Ω.

Problem 6-64: For the circuit below let Vin be a sine wave at 1000 Hz. (a) Find the maximum voltage swing at Vout without clipping or a large amount of distortion. (b) What is the amplitude of Vin and Vout found in part a? (c) Run a Fourier Analysis with the output at its maximum found in part a. What is the total harmonic distortion contained in the output? What is the magnitude of the largest harmonic?

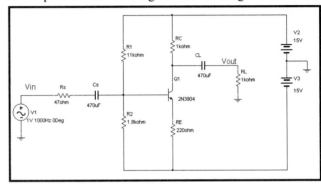

Problem 6-65: Repeat the simulations of Problem 6-64 with RL = 10 kΩ.

Problem 6-66: Repeat the simulations of Problem 6-64 with RL = 100 Ω.

Problem 6-67: Repeat the simulations of Problem 6-64 with RL = 10 Ω.

Problem 6-68: Find the transfer curve for the Schmitt Trigger shown below:

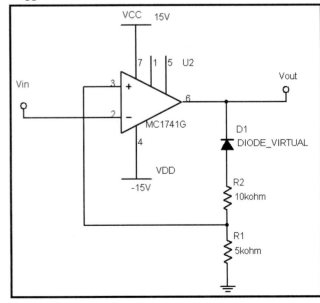

Problem 6-69: Find the transfer curve for the Schmitt Trigger shown below.

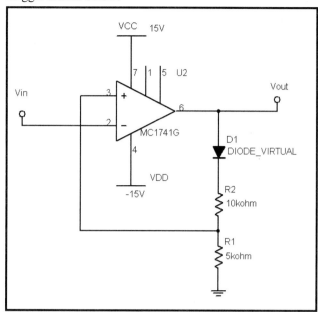

Problem 6-70: Find the transfer curve for the Schmitt Trigger shown below.

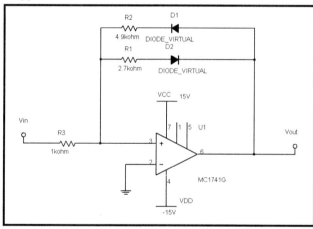

Problem 6-71: Find the transfer curve for the Schmitt Trigger shown below.

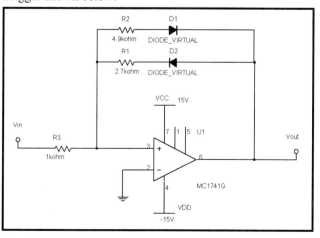

Problem 6-72: For the circuit below, run a Parameter Sweep in conjunction with a Transient Analysis to see how varying Rd from 1 kΩ to 10 kΩ in 1 kΩ steps affects the rise and fall times of the output waveform.

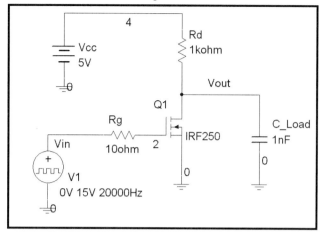

Problem 6-73: For the circuit of Problem 6-72:

(a) Run a Parameter Sweep in conjunction with a Transient Analysis to see how varying Rg from 10 Ω to 100 Ω in 10 Ω steps affects the rise and fall times of the output waveform.

(b) Run a Parameter Sweep in conjunction with a Transient Analysis to see how varying Rg from 10 Ω to 100 Ω in 10 Ω steps affects the gate-source voltage of the MOSFET.

Problem 6-74: For the circuit of Problem 6-72:

(a) Run a Parameter Sweep in conjunction with a Transient Analysis to see how varying Rg from 100 Ω to 500 Ω in 100 Ω steps affects the rise and fall times of the output waveform.

(b) Run a Parameter Sweep in conjunction with a Transient Analysis to see how varying Rg from 100 Ω to 500 Ω in 100 Ω steps affects the gate-source voltage of the MOSFET.

Problem 6-75: For the circuit below, run a Parameter Sweep in conjunction with a Transient Analysis to see how varying R from 1 kΩ to 10 kΩ in 1 kΩ steps affects the rise and fall times of the output waveform. V1 is a 0 to 15 volt square wave. You must choose the appropriate parameters for the pulse voltage source so that the output has time to fully charge and discharge.

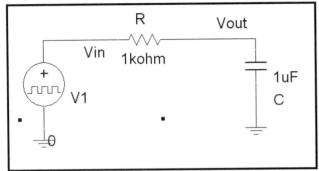

Problem 6-78: For the circuit below, run a Parameter Sweep in conjunction with a Transient Analysis to see how varying R from 1 kΩ to 10 kΩ in 1 kΩ steps affects the rise and fall times of the output waveform. V1 is a 0 to 15 volt square wave. You must choose the appropriate parameters for the pulse voltage source so that the output has time to fully charge and discharge.

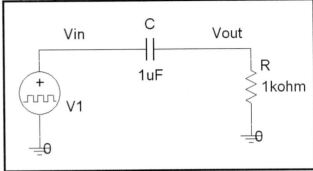

Problem 6-76: For the circuit of Problem 6-75, run a Parameter Sweep in conjunction with a Transient Analysis to see how varying R from 100 Ω to 1 kΩ in 100 Ω steps affects the rise and fall times of the output waveform. V1 is a 0 to 15 volt square wave. You must choose the appropriate parameters for the pulse voltage source so that the output has time to fully charge and discharge.

Problem 6-79: For the circuit of Problem 6-78, run a Parameter Sweep in conjunction with a Transient Analysis to see how varying R from 100 Ω to 1 kΩ in 100 Ω steps affects the rise and fall times of the output waveform. V1 is a 0 to 15 volt square wave. You must choose the appropriate parameters for the pulse voltage source so that the output has time to fully charge and discharge.

Problem 6-77: For the circuit of Problem 6-75, run a Parameter Sweep in conjunction with a Transient Analysis to see how varying C from 1 μF to 10 μF in 1 μF steps affects the rise and fall times of the output waveform. V1 is a 0 to 15 volt square wave. You must choose the appropriate parameters for the pulse voltage source so that the output has time to fully charge and discharge.

Problem 6-80: For the circuit of Problem 6-78, run a Parameter Sweep in conjunction with a Transient Analysis to see how varying C from 1 μF to 10 μF in 1 μF steps affects the rise and fall times of the output waveform. V1 is a 0 to 15 volt square wave. You must choose the appropriate parameters for the pulse voltage source so that the output has time to fully charge and discharge.

Problem 6-81: In the linear regulator example of section 6.L, we ran a Temperature Sweep to see how the temperature dependence of the semiconductor devices affected the output voltage of the regulator. In this problem we will add the temperature dependence of R1 and R2 to the simulation. We can add temperature dependence to a virtual resistor. A virtual resistor is one you place in your circuit and then change its resistance value. A nonvirtual resistor is one of those you pick from the list of available resistors such as 1.0, 1.5, 1.8, 2.2, 2.7, and so on. R1 and R2 must be virtual resistors. If you double-click on the graphic of the resistor, you will see that there are two parameters you can set called TC1 and TC2. These parameters are the temperature coefficients of the individual resistors. Resistance as a function of temperature is given as:

$$\text{Resistance} = R\left(1 + TC1(T - Tnom) + TC2(T - Tnom)^2\right)$$

R is the value of the resistor specified on the schematic. TC1 is the linear temperature coefficient and TC2 is the quadratic temperature coefficient. If you do not specify values for TC1 and TC2, their values are set to zero, which means no temperature dependence. Tnom is the nominal temperature of the simulation and has a default value of 27°C.

A typical resistor may have a linear temperature dependence of 100 parts per million (TC1 = 100e-6, TC2 = 0). Add this temperature dependence to R1 and R2 and simulate the circuit with the same analyses used in section 6.L.

PART 7
Digital Simulations

Multisim can simulate pure analog, mixed analog/digital, and pure digital circuits, and also has VHDL capabilities. This part describes how to run pure digital and mixed analog/digital simulations. The circuits given are fairly simple, but the examples can be applied to larger systems. We can either use the Transient Analysis simulation or run a Virtual Lab simulation.

For this section, we will assume that you have studied Parts 1 and 2, which show how to draw circuits and use the Postprocessor. We will also assume that you have done many of the exercises in the other sections so that you are familiar with the use of the oscilloscope and know the location of most of the parts we will be using in this section.

7.A. Digital Indicators, Signal Generators, and Instruments

To demonstrate the different instruments and indicators available for use in digital simulations, we will simulate the circuit below:

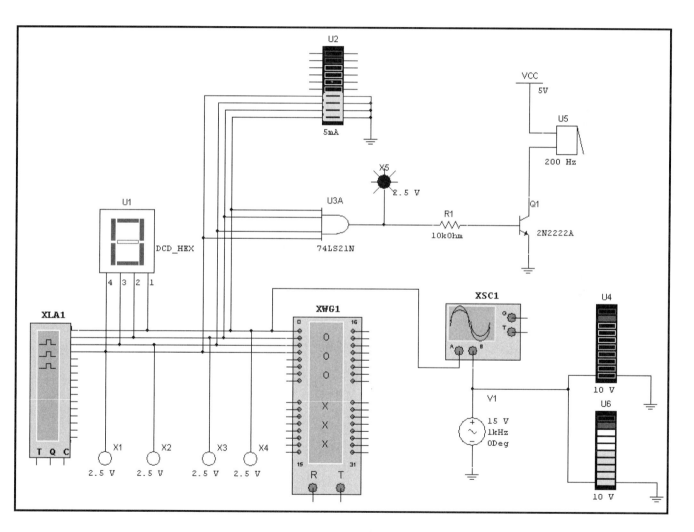

This is a fairly large circuit, so we will build it slowly and test it piece by piece. It contains the Word Generator, Logic Analyzer, and Oscilloscope instruments, some indicator lights, a buzzer, a logic gate, and some bar graphs. These are the items you will use to simulate and observe the behavior of digital circuits.

7.A.1. Word Generator and the Logic Analyzer

We will start by creating the portion of the circuit shown below:

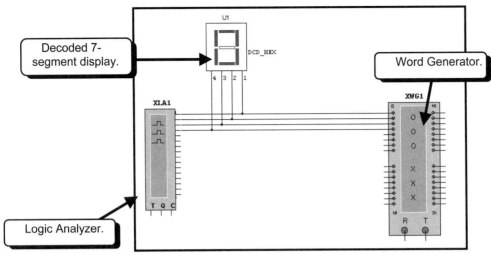

The Word Generator and the Logic Analyzer are located in the Instruments toolbar. To place the Word Generator, click the **LEFT** mouse button on the **Word Generator** button as shown below. The instrument will become attached to the mouse pointer:

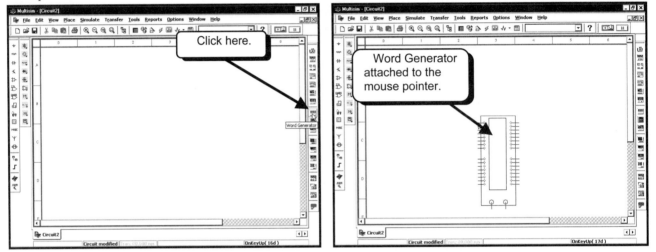

To place the part, use the mouse to move the part to the desired location and click the **LEFT** mouse button. Place the part in a convenient location:

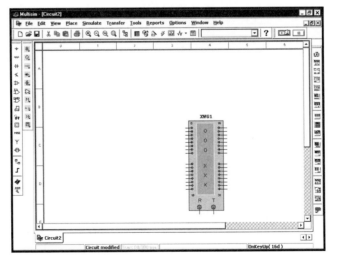

To place the Logic Analyzer, click the **LEFT** mouse button on the **Logic Analyzer** button as shown below. The instrument will become attached to the mouse pointer:

Multisim 7 — Digital Simulations

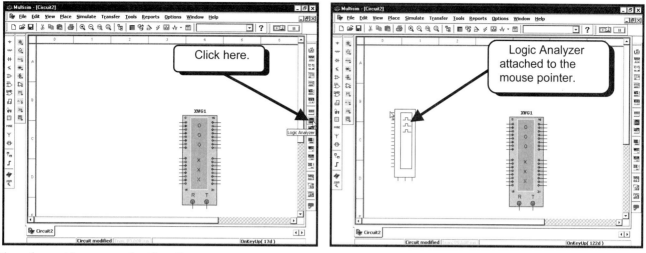

Place the part in a convenient location:

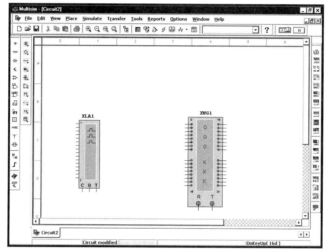

We will now take a closer look at the Logic Analyzer and the Word Generator graphics. The Logic Analyzer is shown below.

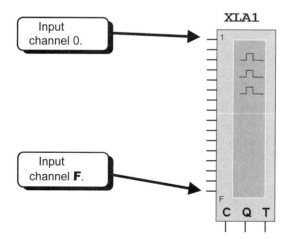

The Logic Analyzer has sixteen inputs labeled **1** through **F** (although they really should be 0 through F). Remember that when dealing with logic circuits, F is a hexadecimal number and represents the number 15 in base 10, or the number 1111 in binary. Either way, the Logic Analyzer has sixteen inputs that can be displayed simultaneously. The labels **1** and **F** tell us which are the first and last inputs.

An enlargement of the Word Generator is:

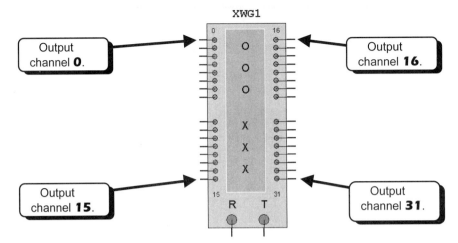

This instrument has 32 outputs, labeled 0 through 31. Each output generates a logic signal that can be used as an input to a logic circuit. The output of each channel is independent of the other channels.

For our example, we will be using the first four outputs of the Word Generator and the first four inputs of the Logic Analyzer. To make wiring the two instruments together easier, we will need to flip the Logic Analyzer horizontally so that the input terminals face the Word Generator. To flip the Logic Analyzer, click the **RIGHT** mouse button on the Logic Analyzer graphic. A menu will appear:

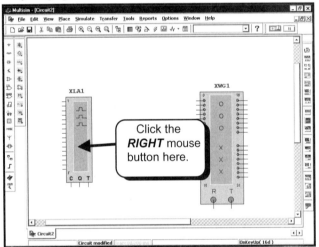
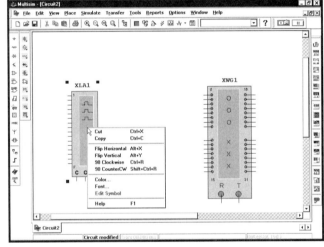

Select **Flip Horizontal**:

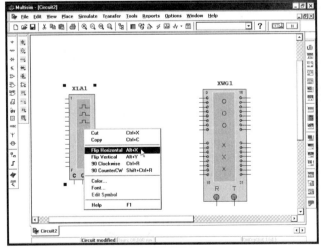
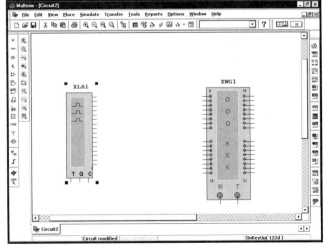

The input terminals for the Logic Analyzer are now on the right side of the instrument.

We can now easily wire the Logic Analyzer and the Word Generator together. If you are unfamiliar with wiring components together, see section 1.D. You may need to shift one of the instruments up or down so that the pins line up nicely:

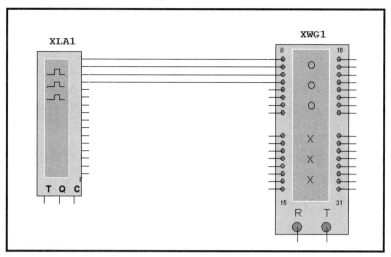

The next thing we want to add is the decoded 7-segment display. This part is located in the **Indicator** group. Click the **LEFT** mouse button on the **Indicator** button as shown below:

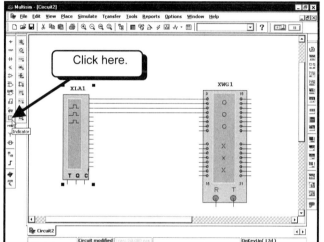
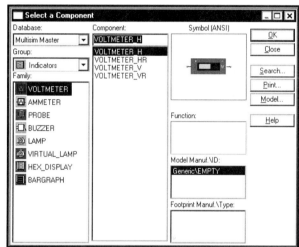

Select the text **HEX_DISPLAY** to select the **Family**:

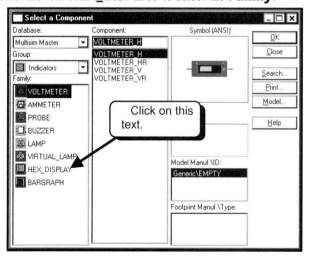
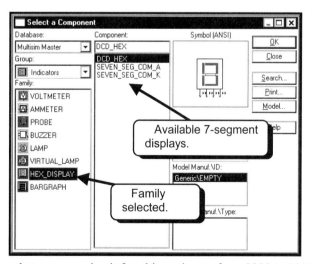

The **DCD_HEX** part is a decoded 7-segment display. This means that you can give it four binary inputs from 0000 to 1111, and it will display an alphanumeric character of 0 through F. No other logic is needed to convert between the binary inputs and the LED segments because the conversion logic is built in to the part. The other two parts are basic 7-segment displays that have inputs that directly control a single segment. You need to add the conversion logic to control each segment. If you are studying 7-segment displays and their driver circuits, you will want to use either the **SEVEN_SEG_COM_K** or the **SEVEN_SEG_COM_A** because you have to add all of the required circuitry to get these parts to function. If you want a quick display of four binary inputs, use the **DCD_HEX** part because all you need to do is place it in your circuit and connect the four inputs.

Because the part DCD_HEX is already selected, all we have to do is click on the **OK** button to place the part in our circuit. Place the part and then wire it as shown:

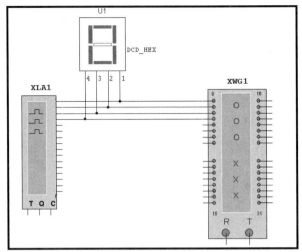

The last thing we must do before running a simulation is to program the Word Generator. Double-click the *LEFT* mouse button on the graphic of the Word Generator:

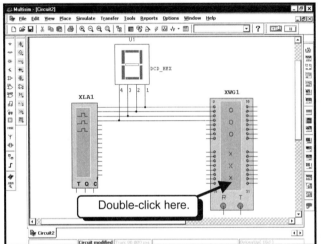
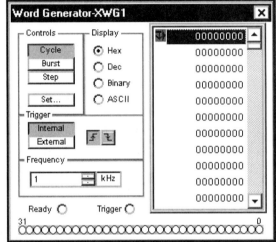

We would like the Word Generator to count from 0 to 15 (F) and then back down to zero. This is a total of 30 values (1E) that will use 30 memory locations in the Word Generator. Only 30 memory locations are required because we will loop back to the memory location containing the value 0.

We can generate the increasing count automatically. Click on the **Set** button:

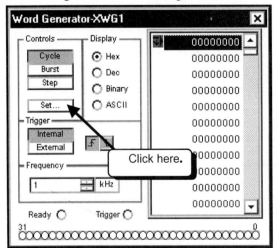
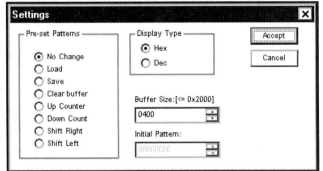

This dialog box allows us to load and save data patterns to a file, as well as easily generate some standard counting patterns. Here we want to initially generate numbers that increase, so select the **Up Counter** option and click on the **Accept** button:

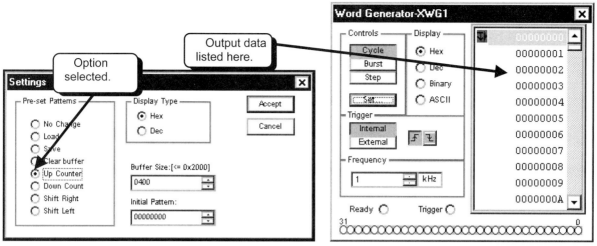

If you look at the output data, you will notice that the output counts from 0 to 3FF in hexadecimal:

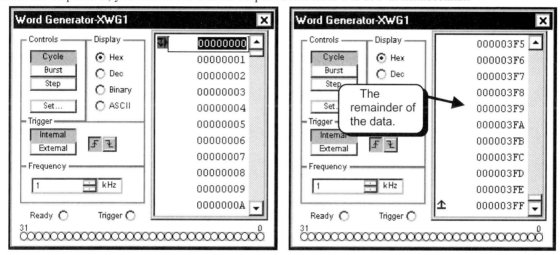

We see that we can easily create data that counts up. To create the portion of the data that counts down, we have to enter it manually. Locate address location 10(hexadecimal) and select it. To select it, click on the data:

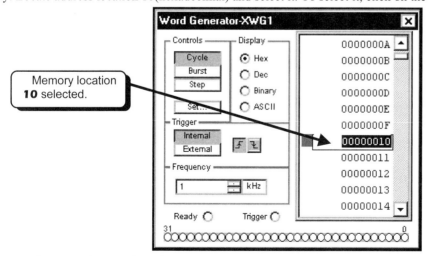

We can enter the new value as a hexadecimal number, an ACSII code (which doesn't make much sense in this example), or a binary number Word Generator to output in this memory location. For this example, we want the output to be the hex value E. Type the letter **E** and press the **ENTER** key. The value is entered and the cursor moves to the next memory location, ready to modify its contents:

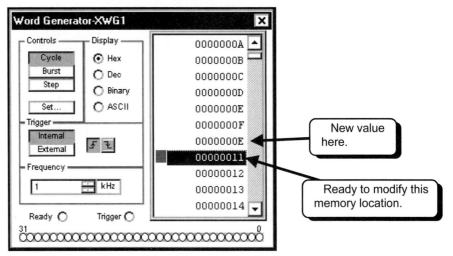

We can now modify the next memory location.

To see how the **Display** function works, we will change the display to Decimal. Select the **Dec** option as shown below:

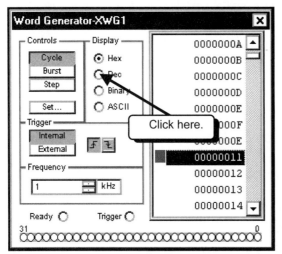 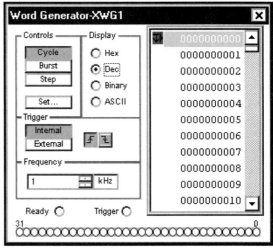

The data address was reset to the beginning address of zero. If we scroll through the data, we notice that the data is now displayed in decimal, which is nice for humans:

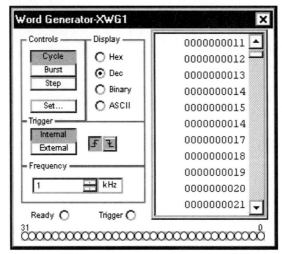

We see that the counter now counts from **0** to **15**, then down to **14** because we started counting down, then jumps up to **17** and continues counting up again because we have yet to modify any more of the data. Click on the text **17** to modify the data:

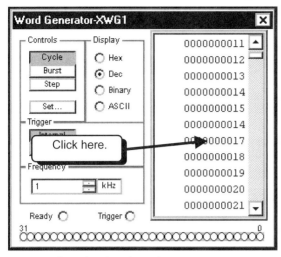 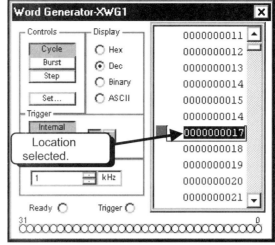

When the memory location is selected, you can enter a new value to be saved in the location. We want to count down, so type the text **13** and press the **ENTER** key:

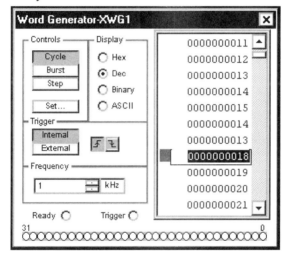

The value is entered and we are ready to modify the next memory location. Type **12** and press the **ENTER** key:

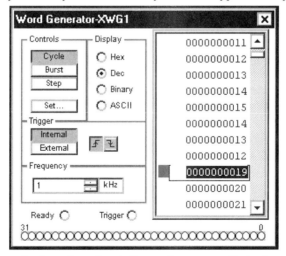

You can use this method until you get back to the value 1. Presently, we are displaying values in Decimal. You can display the values in hexadecimal, decimal, binary, or ASCII, whichever method makes sense for your application. When you fill in all the data for this example, your Word Generator should look as shown below if you have set your display to show values in decimal:

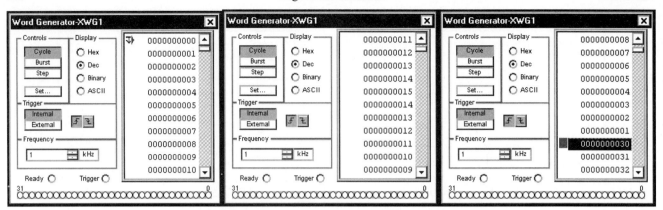

Next, we need to tell the Word Generator where the initial and final memory positions are for the data. The initial memory position is where the simulation starts running, and the final memory position is where it stops. By default, the initial memory position is memory location 0 and is indicated by a blue arrow that points down ⇩:

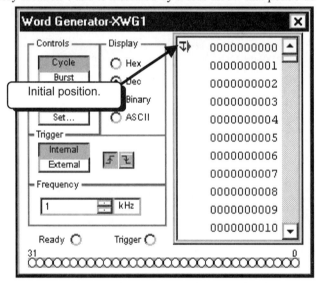

When the simulation starts a new memory cycle, it will start at the initial position indicator.

By default, the final position is at the end of the Word Generator, which is after all 1024 memory locations. For this example, we are using only 30 memory locations (which have addresses 0 to 29), so we need to place the final position indicator where we need it. Locate the last memory cell in our sequence and click the **RIGHT** mouse button on it:

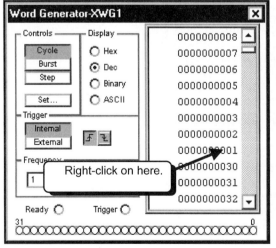
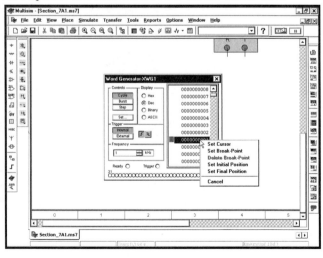

Select **Set Final Position**:

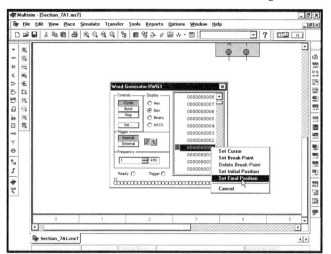 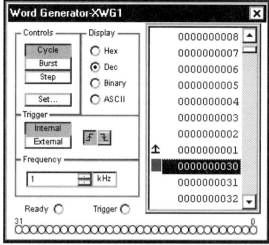

You will notice that the final position indicator ⬆ appears next to the memory location. When the simulation runs, the Word Generator will only cycle through the data contained within the initial and final position indicators.

Now that we have programmed the Word Generator, we need to tell it how to run. **Cycle** mode means that the Word Generator starts at the initial position indicator and outputs the contents of each memory location sequentially. After the Word Generator reaches the final position indicator, it goes back to the initial position indicator and starts over. In **Cycle** mode, the Word Generator will cycle through the data over and over until you stop the simulation. In **Burst** mode, the Word Generator will output the contents of each memory location once and then stop. In **Step** mode, the Word Generator will only output the contents of one memory location at a time. Each time you click on the **Step** button, the Word Generator will go to the next memory location and output its contents. It will remain at that memory location until you click on the **Step** button again. This mode of operation is good for debugging circuits. When programming the Word Generator, if you right-click on a memory location and then select **Set Break-Point**, that memory location will be specified as a breakpoint. When you are in Cycle mode or Burst mode, the Word Generator will stop and hold the output value at a breakpoint. The Word Generator will not resume stepping through the memory locations until you click on the **Cycle**, **Burst**, or **Step** buttons again. To remove a breakpoint, right-click on a memory location and then select **Remove Break-Point**.

As you are running a simulation, a cursor ▶ in the Word Generator window indicates the memory location that is currently being output by the Word Generator:

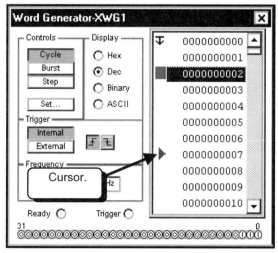

When you start a simulation, the Word Generator will start its sequence beginning at the present location of the cursor. As the simulation runs, you will notice that the location of the cursor changes as the Word Generator steps through each memory location. If you stop a simulation, you can change the location of the cursor by clicking the *RIGHT* mouse button on a memory location and then selecting **Set Cursor**. Make sure that your cursor is set to memory location 0, as it should be if you have not yet run a simulation with the present circuit:

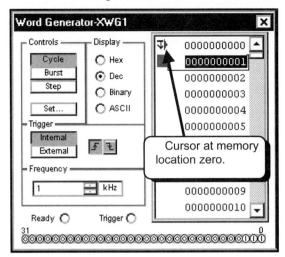

We are now ready to start the simulation. Make sure that the Word Generator window is not covering the alphanumeric display and click on the **Cycle** button:

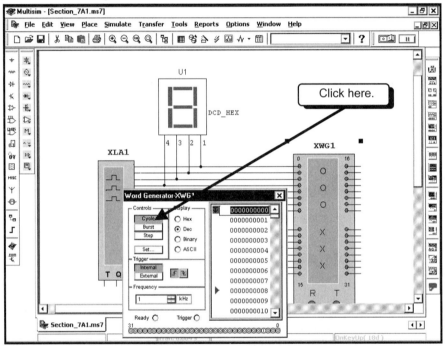

Clicking on the **Cycle** button starts the simulation. You should see the alphanumeric display go from 0 to F and then back down to zero. You will also see the current memory address that the Word Generator is outputting:

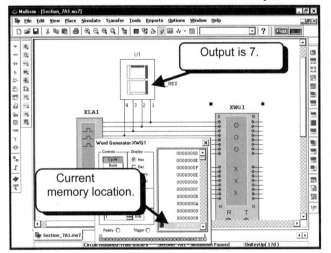
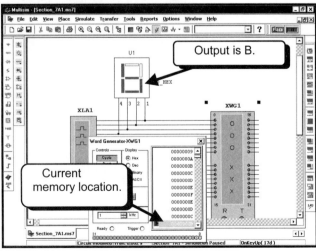

To stop the simulation we need to click on the **Run / stop simulation** button as shown:

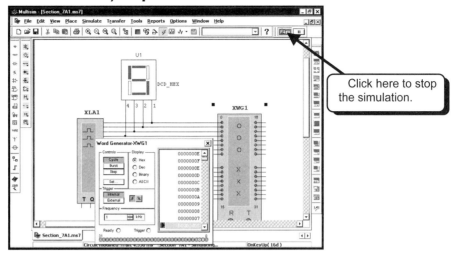

You can observe the behavior of the Word Generator by using the **Burst** and **Step** buttons if you wish. Remember that with the **Step** button, you must click on the button each time you want the Word Generator to go to the next memory location.

The last thing we will do is observe the waveforms using the Logic Analyzer. Double-click the **LEFT** mouse button on the Logic Analyzer to open its window:

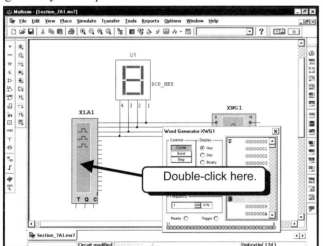
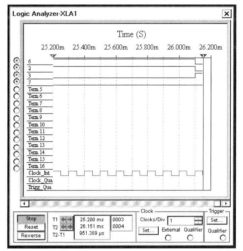

The window displays a few incorrect waveforms because the Logic Analyzer is not set up correctly. The only thing we need to do is change the clock section. Click the **LEFT** mouse button on the **Set** button in the **Clock** section to adjust the sampling frequency:

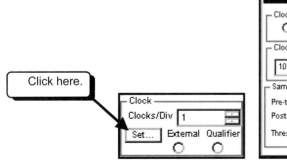
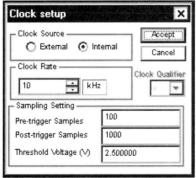

The clock rate is the sample rate of this instrument. Once every clock period, the Logic Analyzer measures the input channels and displays those values on the screen. The displayed values will remain unchanged until the next sample is taken, even if the inputs change. Thus, we need to make the sample frequency much higher than the frequency of the waveform we are measuring. Our Word Generator was set at 1 kHz, so we need to make the **Clock Rate** of the Logic Analyzer much higher than 1 kHz. Choose a frequency that is a factor of 2^n higher than 1 kHz. The reason for the 2^n will become apparent later. I will choose 16 kHz for the **Clock Rate**:

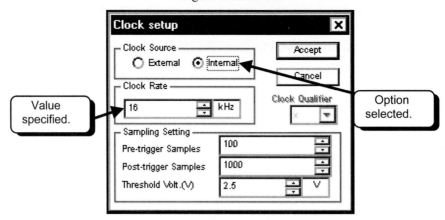

Also make sure that the **Clock Source** is specified as **Internal**. We are not providing a clock for the Logic Analyzer, so it will need to generate its own clock signal. Click on the **Accept** button when done.

The next thing we need to do is to set up the display of the Logic Analyzer. The display will be broken up into divisions, and we need to specify the number of Logic Analyzer clock pulses per division (**Clocks/Div**). The number of clocks per division can be set to 1, 2, 4, 8, 16, and so on, up to 128. This is why I chose the clock rate to be a factor of 2^n times as fast as the Word Generator frequency. I will choose the **Clocks/Div** to be 16. Because the Logic Analyzer clock frequency is 16 kHz, this will make the time per division equal to 1 ms, which happens to be the period of our Word Generator frequency:

We can now run the simulation. Click on the **Cycle** button of the Word Generator to start the simulation:

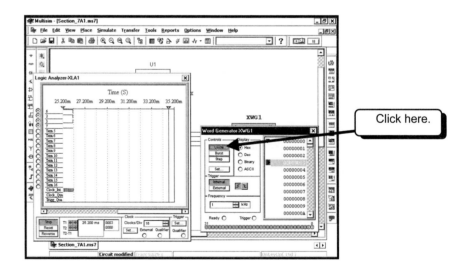

The simulation will run and the data will be displayed on the Logic Analyzer screen:

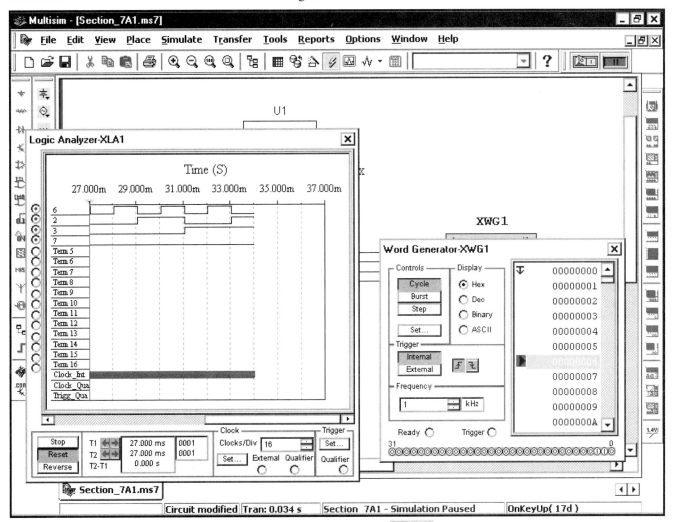

If you stop the simulation by clicking on the **Run/Deactivate** button, you can use the cursors to measure values on the Logic Analyzer screen. Cursor 1 is red and cursor 2 is blue. To move a cursor, drag the triangle at the top of the cursor. The procedure for moving the cursors is the same as it was for the oscilloscope in section 6.A.4. Shown below is a screen capture with the cursors:

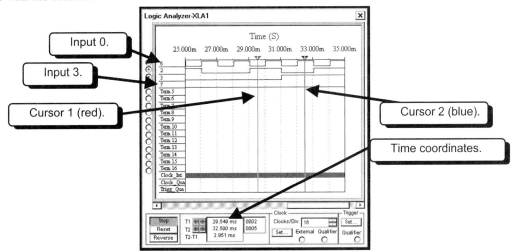

The cursors show the time coordinates of each cursor and the time difference between the cursors. The cursors also give the value of each input at the cursors, although this is not as obvious. If you treat input 0 as the least significant bit, and input 15 as the most significant bit, the inputs to the Logic Analyzer can be treated as a 16-bit binary number. The value of this number is displayed in hexadecimal in the cursor information:

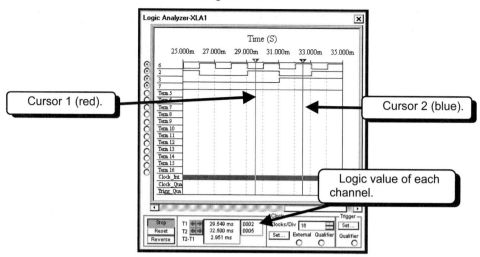

If we look at cursor 1, we see that inputs 0, 2, and 3 are low, and input 1 is high. The other inputs are not used and are treated as a logic zero. Thus, the inputs at cursor 1 are 0010, which corresponds to 2. If we look at cursor 2, we see that inputs 1 and 3 are high, and inputs 2 and 4 are low. Thus, the inputs at cursor 2 are 0101, which corresponds to 5. This information is displayed as part of the cursor display:

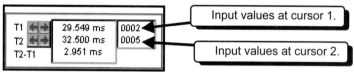

We are now done with these instruments for the moment, so we need to close their windows. Click on the ⊠ buttons as shown to close the windows:

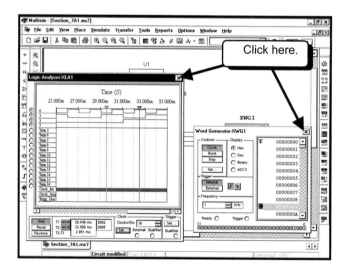

We will continue with this circuit in the next section.

7.A.2. Digital Probe and Undecoded Bar Graph Lights

We will next look at two indicators that can be used to view the status of a single bit. The digital Probe part can be directly connected to a wire, and it will indicate if the voltage on the wire is high or low. An undecoded bar graph is a package of ten lights arranged similar to an LED bar graph. You must drive this bar graph using a method similar to what you would use with an LED bar graph, but it is not an LED bar graph and should not be confused with one. The lights have a built-in resistance and require 5 mA of current to illuminate. We will describe the properties of this bar graph later.

The digital Probe is located in the Indicators group. Click the **LEFT** mouse button on the Indicator button as shown:

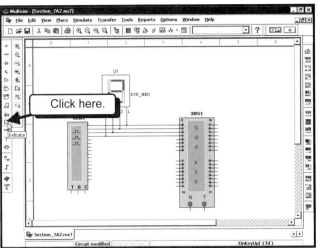
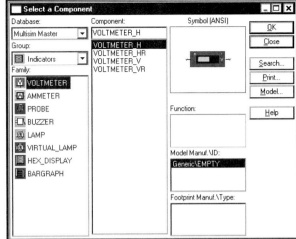

Click on the **PROBE** text to select the family:

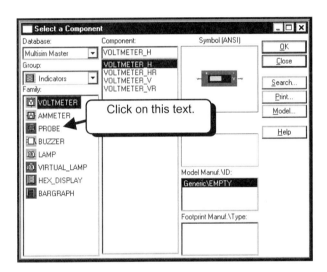
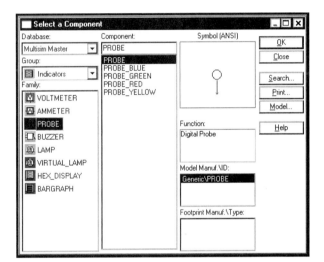

Five Probes are available. One is plain or not colored, and the others are colored. Select one of the Probes and click the **OK** button. The graphic will become attached to your mouse pointer and move with the mouse:

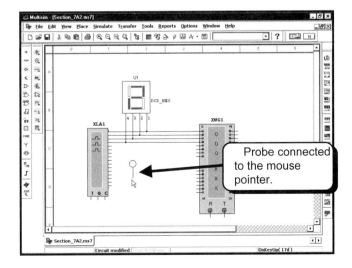

Place four **PROBE**s in your circuit as shown below. To rotate a part, type **CTRL-R** after you have placed the part on your page. I will place Probes of four different colors in my circuit:

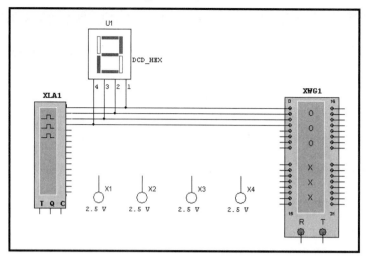

Notice that the text **2.5 V** is displayed on the graphic for the Probe:

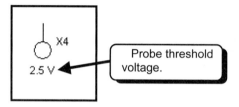

2.5 V is the threshold for the Probe. If the Probe measures a voltage less than the threshold, the Probe will be off. If the Probe measures a voltage greater than the threshold, the Probe will illuminate. The threshold is displayed on the schematic so that you can easily change it if you wish. A threshold of 2.5 V works well for our example, so we will not change it. A Probe draws no current, so you can connect it to your circuit and not worry that it will affect the operation of your circuit. Wire up the Probes as shown:

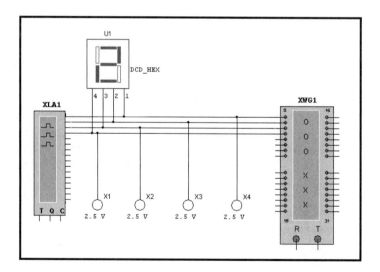

The Probes will now display the status of each output of the Word Generator. The rightmost Probe (**X4** in my circuit) is the least significant bit and the leftmost Probe (**X1**) is the most significant bit.

Next, we will add the undecoded LED bar graph to our circuit. This part is also located in the **Indicators Group**. From the **Indicators** group, click on the text **BARGRAPH** to select the bargraph **Family**:

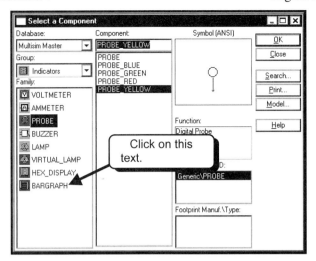 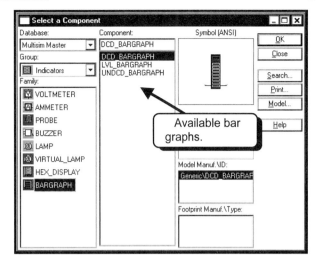

The dialog box displays all of the available bar graph components. We want an undecoded bar graph, which is called **UNDCD_BARGRAPH**. Select this part and then click on the **OK** button. The part will become attached to the mouse pointer:

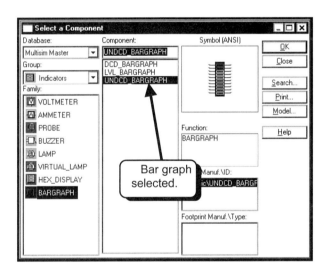 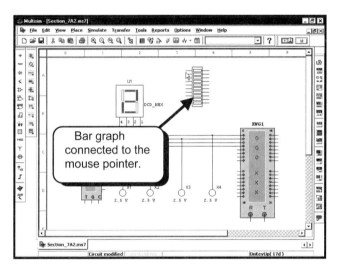

This bar graph is just ten individual lights mounted in a single package and arranged into a bar graph display. The part is similar to an LED bar graph display, but the lights are not actually LEDs. Current can flow through the lights in both directions, and the lights do not have the 1.5 V LED diode voltage drop. Instead, each light has a built-in 500-ohm resistance and a current threshold. As shown above, for an individual light in the array, if the current enters the left pin and exits the right pin (the current flows from left to right) and the current exceeds 5 mA, the light will illuminate. If the current is less than the threshold or if the current is the wrong direction, the light will not illuminate. This behavior is similar to an LED but is not the same. Both the threshold and the resistance of a part can be changed. Note that if you rotate the part, the pin designations of left and right just mentioned do not apply. Thus, when you first place the part, you know which terminal is which. When you rotate the part, you must keep track of the pins.

Place the bar graph in a convenient location and wire the remainder of the circuit as shown. Note that we do not need to add external series resistors to limit the current as we would for an LED bar graph because this part has internal resistance.

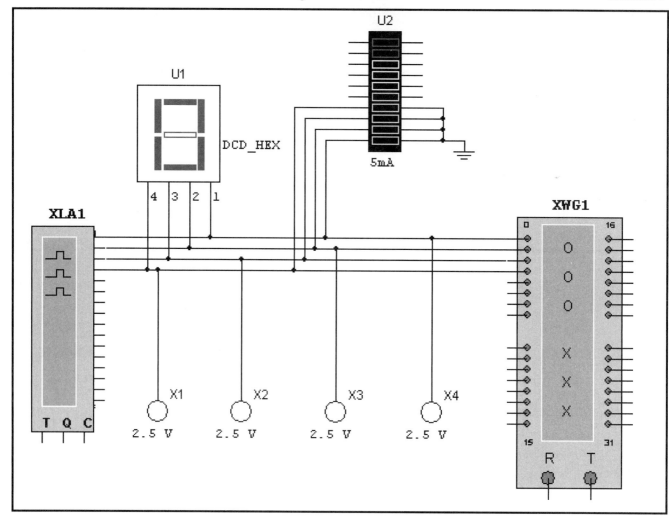

We are not using any driver circuits for the bar graph because the Word Generator can source enough current to illuminate the lights. If you double-click on the bar graph you will see that the threshold for the lights is 5 mA and the internal resistance is 500 ohms:

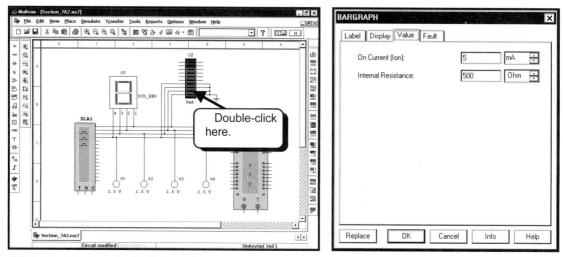

Click the **OK** button to return to the Schematic.

We can now run the simulation. Click on the **Cycle** button in the Word Generator to start the simulation. The bar graph and the Probes should display binary codes that agree with the alphanumeric display. A few screen captures are shown below:

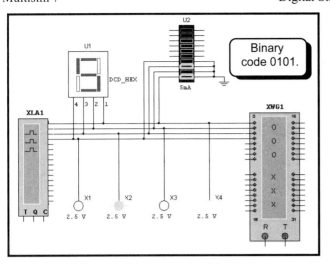
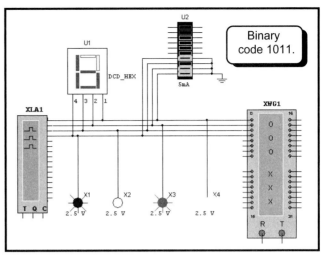

7.A.3. Mixed Signal Indicators

We will now add a buzzer to the circuit and test the other types of bar graph displays. We will also look at two of the waveforms using the oscilloscope. Finish the remainder of the circuit as shown below. This section contains a number of analog parts that have been covered in Parts 3 through 6 of this book.

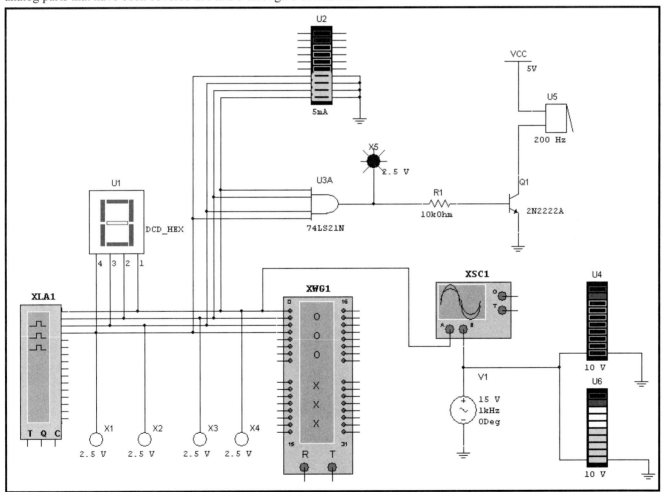

The buzzer is located in the **Indicators** group, and the two other bar graph displays are placed the same way as the bar graph of the previous section. The U4 and U6 are called LVL_BARGRAPH and DCD_BARGRAPH and are located in the Bargraph family of the Indicators group. The 4-input AND gate is located in the TTL group. To place the AND gate, click on the TTL button as shown below:

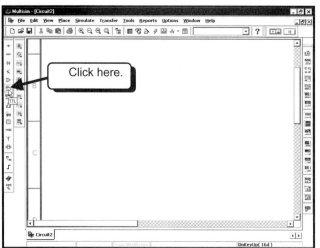
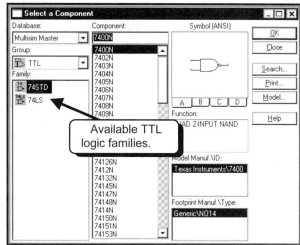

The left pane shows the available TTL logic families. We see that the standard TTL and low-power Schottky TTL logic families are available. We will use a low-power Schottky part, so select the text **74LS** to select the low-power Schottky family:

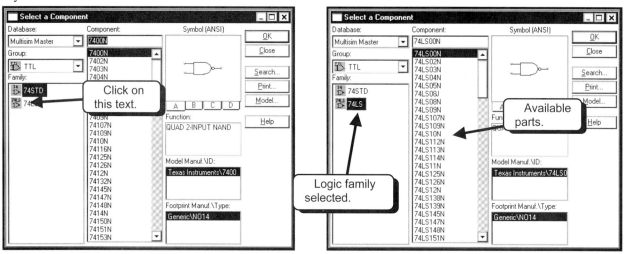

The dialog box lists all of the low-power Schottky parts available to us. Locate the 74LS21N and select it:

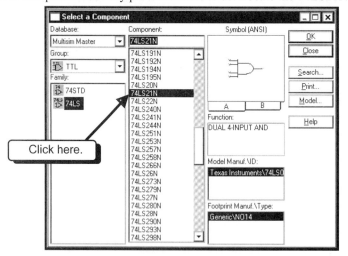

Click on the **OK** button to place the part.

To place the AC voltage source, click the *LEFT* mouse button on the **Show Signal Source Components Bar** button and then click on the **AC_VOLTAGE_SOURCE** button :

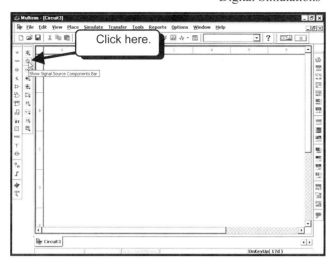
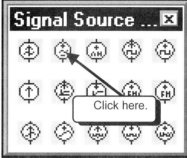

To place the 2N2222A transistor, click on the **Transistor** button :

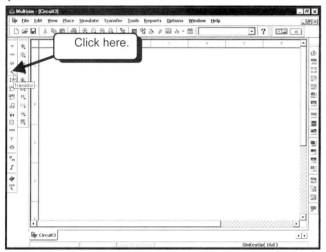
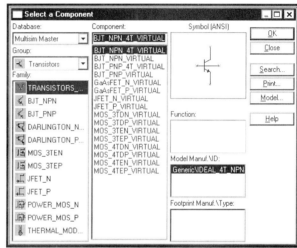

The 2N2222A is an NPN transistor, so click on the text BJT_NPN to select the **Family**:

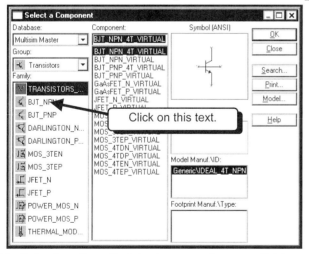
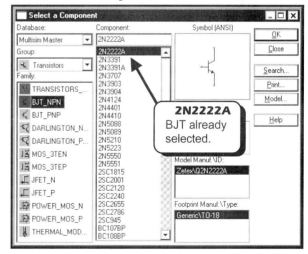

This dialog box lists all of the NPN transistors available to us. The BJT selected happens to be the one we want to use. To place the part in your circuit, just click on the **OK** button.

You should now be able to place all of the parts in the circuit and wire it up. A few of the components require some discussion. We must modify the settings of the buzzer. Double-click the ***LEFT*** mouse button on the buzzer graphic to edit its properties:

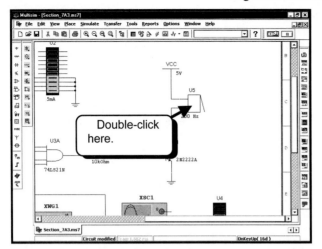 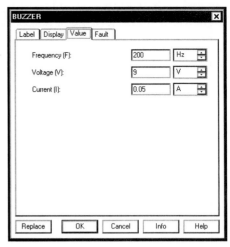

The frequency of the buzzer is 200 Hz, it requires 9 volts to operate, and it draws 50 mA of current. Our driver circuit was designed to drive a 4-volt buzzer, so change the voltage to 4 volts as shown:

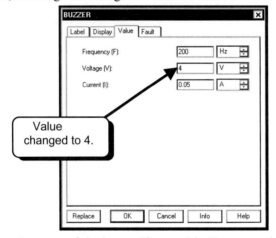

You can change the frequency and current of the buzzer if you wish. However, choosing too large a current for the buzzer may cause problems with the 2N2222A transistor. Click on the **OK** button to accept your modifications. The 4-input AND gate detects when the Word Generator emits the code 1111. When this occurs, the buzzer should make a noise.

The **LVL_BARGRAPH** and the **DCD_BARGRAPH** are actually analog indicators. These indicators take a single analog input and then display the value as a level on the bar graph. These parts contain the logic circuits to convert the analog signal to 10 digital signals, and also contain the drive circuits so that those 10 logic signals can be displayed on the LED bar graph. To use them, all you have to do is place them in your circuit and then connect the two inputs to ground and the signal you want to measure. For our circuit, these bar graphs are connected to a 15-volt sinusoidal voltage source, so the display should go up and down with the sinusoidal voltage.

When you have completed wiring the circuit, you can start the simulation. Click on the **Cycle** button in the Word Generator to start the simulation. The buzzer should sound every time the Word Generator hits the code 1111. The two decoded bar graphs should rise up and down as they follow the sinusoidal voltage source. A few screen captures are shown below:

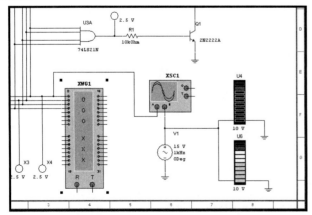 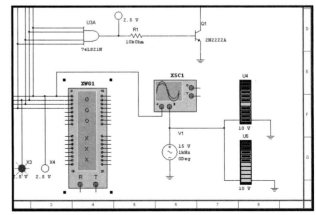

The buzzer should sound when you see the following displayed on your screen:

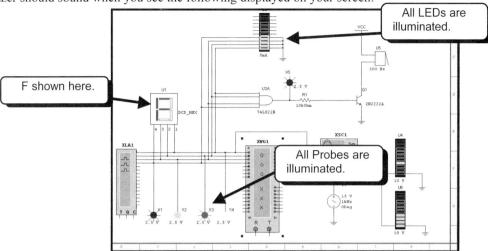

The last thing we will look at is the waveforms displayed by the oscilloscope. See section 6.A for instructions on displaying waveforms with the oscilloscope. The waveforms are shown below:

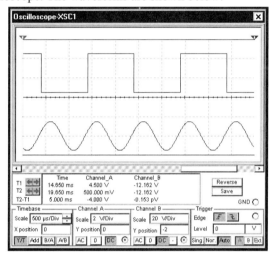

The purpose of showing these waveforms is twofold. The first point is that we can use the oscilloscope to measure waveforms in a digital circuit. This is useful if we have a digital circuit that is connected to an analog waveform somewhere in our circuit. The second point we want to emphasize is that the AC voltage source is not synchronized to the Word Generator. If we blow up the time axis and use the cursors, we see that the edges of the Word Generator do not match up with the zero crossings of the AC voltage source:

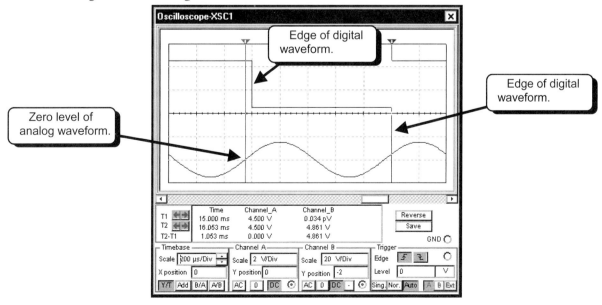

Cursor 1 (red) is positioned at the zero crossing of the analog waveform. Cursor 2 (blue) is positioned at the positive edge of the digital waveform. If we want to synchronize the Word Generator to the zero crossings of the AC voltage source, we need to add some circuitry.

7.B. Mixed Analog and Digital Simulations

The first circuit we will look at is an OPAMP circuit that drives the clock of a J-K flip-flop. Wire the circuit shown below:

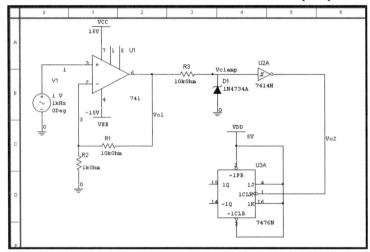

The circuit is drawn as though we were going to wire the circuit in the lab. No special circuits are required between the analog circuitry and the digital logic gates. Note that the J and K inputs of the flip-flop are held high so that the flip-flop toggles at each negative clock edge.

The sinusoidal voltage waveform produces a 1 kHz sine wave with voltages between –1 and 1 volts. The **741** OPAMP circuit has a gain of 11 and produces a ±11 volt sine wave of 1 kHz at node **Vo1**. This waveform goes into a Zener clipping circuit that limits the voltage to approximately +5.6 and –0.7 volts at node **Vclamp**. This voltage is TTL compatible and can be connected to the Schmitt Trigger (7414) input. The output of the Schmitt Trigger (**Vo2**) should be a 0- to 5-volt square wave at 1 kHz. The J-K flip-flop is wired as a divide-by-two counter, so Q and $\overline{Q}$ should be 0- to 5-volt square waves at 500 Hz. They should also be 180 degrees out of phase.

We will now run the simulation and use the oscilloscope to view the waveforms. Place the Oscilloscope instrument in your circuit and connect it as shown:

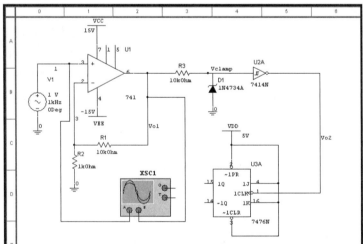

We will first test the input and output of the 741 OPAMP circuit. This circuit has a gain of 11, so the output should be in phase with the input, but 11 times larger. Click the **LEFT** mouse button on the **Run** button to start the simulation. Double-click the **LEFT** mouse button on the oscilloscope to open the oscilloscope window. Adjust the settings of the oscilloscope as shown below. If you are unfamiliar with using the oscilloscope, see section 6.A for detailed instructions:

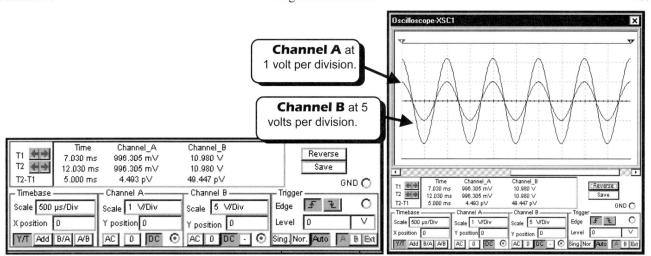

We see that the input signal (**Channel A**) is a 1-volt peak signal, and the output (**Channel B**) looks to be about an 11-volt peak signal. We could use the cursors to measure the peak of the output waveform, but we will just eyeball it. Thus, it appears that our OPAMP circuit is working properly.

Next, we will change **Channel A** of the oscilloscope to measure the voltage across our Zener clamp (**Vclamp**):

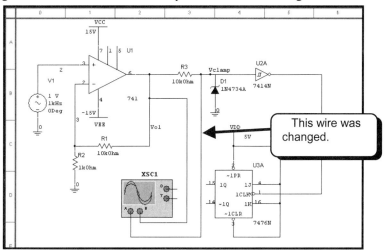

Restart the simulation and display the waveforms:

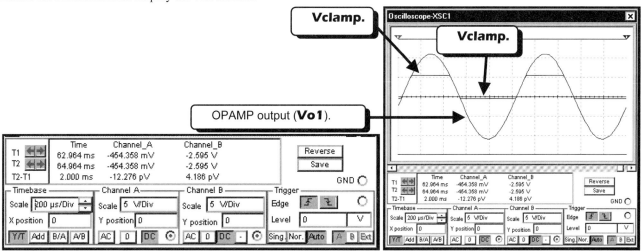

Both channels are set to 5 volts per division. We see that the Zener clipping circuit limits the voltage to a little over 5 volts and about –0.7 volts. This is the expected operation of this clipping circuit.

Next, we will change **Channel B** to measure the output of the 7414 Schmitt trigger (**Vo2**):

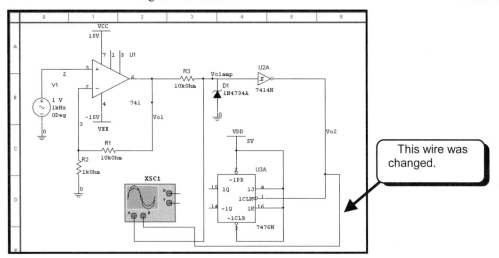

The waveforms at this point are:

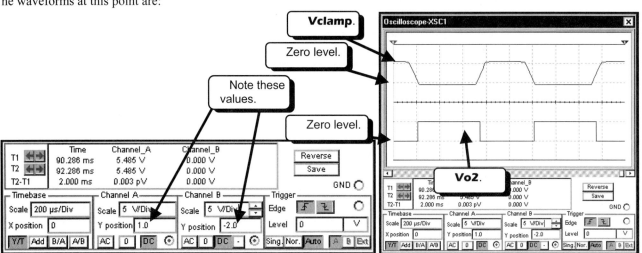

Note that we have changed the **Y position** fields in the Oscilloscope settings to separate the two traces. We see that the Schmitt trigger takes an input pulse with slow rise and fall times and converts it into a pulse with very sharp edges. This is the purpose of using the Schmitt trigger in this application. The output of the Schmitt trigger is now a suitable signal that can be used for a logic input to a digital circuit, including the clock input of a flip-flop.

The last thing we will do is observe the output of the J-K flip-flop. The flip-flop is wired as a divide-by-two counter, so the Q output of the flip-flop should be a square wave of half the frequency of the clock input. Change the oscilloscope connections as shown and then observe the waveforms:

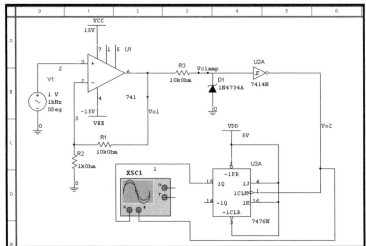

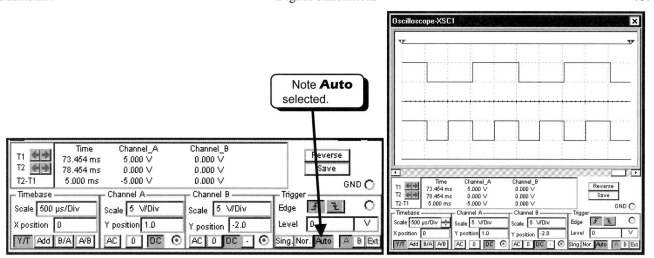

If you have the oscilloscope set to normal triggering, you will need to change the trigger **Level** of the oscilloscope to a voltage greater than 0 and less than about 4 volts. If we do not make this change, the oscilloscope will never trigger because neither of the measured waveforms ever goes below zero, which is the default trigger level. I set my scope to Auto triggering to make it easier to generate screen captures. However, the waveforms do dance across the screen when you do this. You should set your oscilloscope to normal triggering and change the level to 1 volt or so.

For a second example, we will simulate the circuit below:

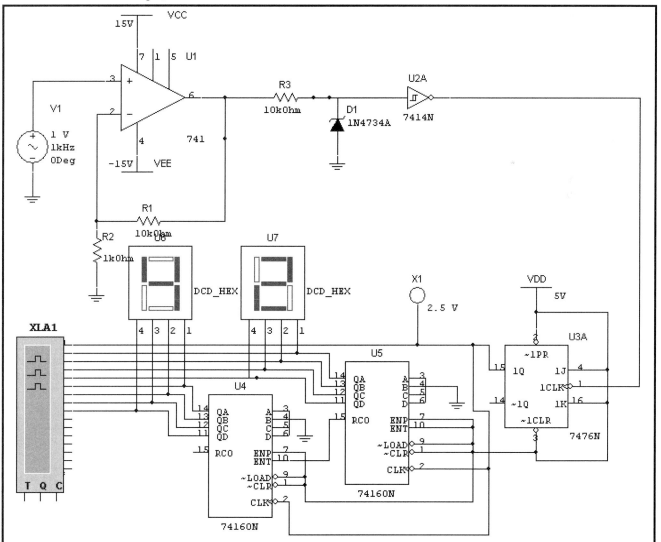

In this circuit, we are using the Q output of the 7476 as the clock input of the two 74160 counters. The 74160s are decade counters, which means that they count from 0 to 9. The two counters are cascaded so that they count from 0 to 99. If you click on the **Run / stop simulation** button, you should observe that the counters count from 0 to 99.

We have also connected the Logic Analyzer so that we can view the waveforms. Some sample waveforms are shown below:

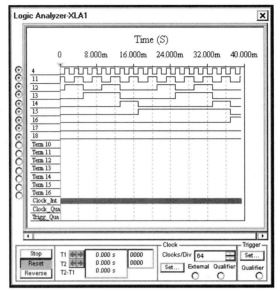

Note that the Logic Analyzer clock frequency is set to 16 kHz (click on the **Set** button to change) and the **Clocks/Div** is set to 64.

EXERCISE 7-1: Design a 60 Hz sync circuit. The input to the circuit is a 12 V amplitude, 60 Hz sine wave. The output of the circuit should be a 1 ms 5 V pulse that occurs when the sine wave crosses zero with a positive slope.

SOLUTION: Multisim does not have a 74xx123 one-shot. However, in the **Mixed** toolbar there is a monostable one-shot that appears to be nearly equivalent to a 74xx123 one-shot. We will use this monostable one-shot instead of a 74xx123.

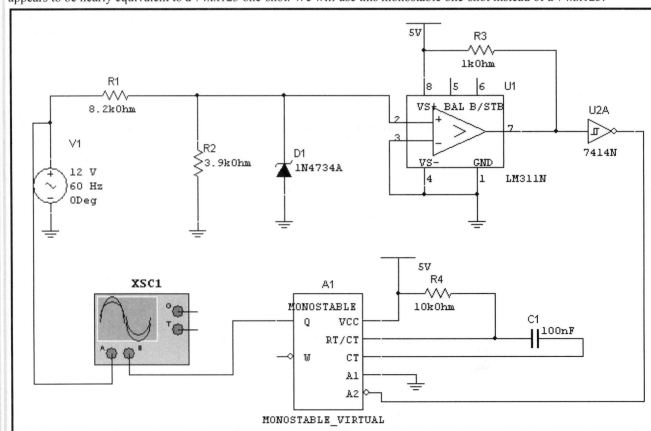

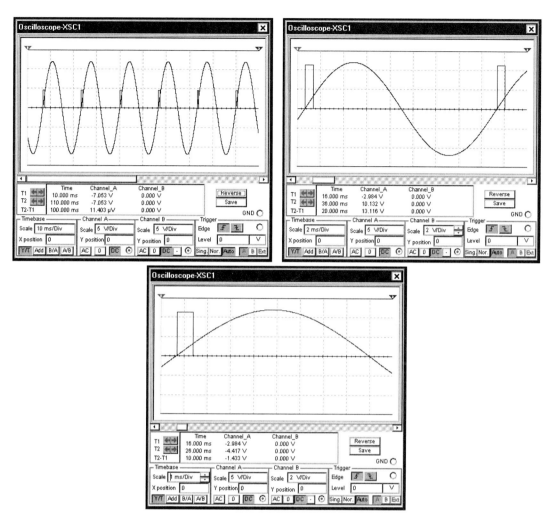

The screen capture above shows that the pulse width is not quite 1 ms. We can increase the pulse width by making **R3** or **C1** larger.

7.C. Pure Digital Simulations

Multisim can be used as a logic simulator as well as a mixed analog/digital simulator. We will now simulate a switch-tail counter. Wire the circuit as shown:

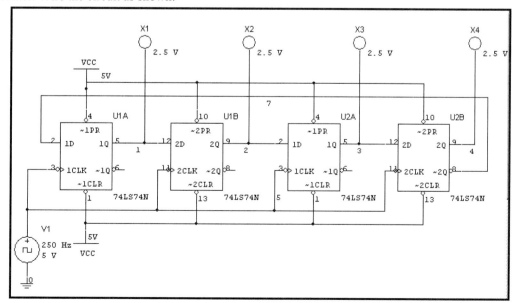

We note that there are no power and ground connections in the circuit. Multisim automatically connects the power and ground for us. Multisim has two modes for simulating circuits with digital components, ideal and real. Ideal mode is faster, but it does not take into account power supply variations, device tolerances, and rise and fall times, and it does not require you to specify power and ground connections. Note that ideal mode does include gate propagation delay times. Real mode produces a more accurate simulation in terms of switching speed and signal levels, but it requires you to use the digital ground ( ) and specify a supply ( ). The default mode of operation is ideal mode. To change to real mode, you must specifically select the option, which we will not show here. We will discuss the difference between these two modes in section 7.E.

If you are familiar with a switch-tail counter, you can click on the **Run** button and observe the operation with the Probes. We cannot demonstrate the operation of the circuit in screen captures, so we will add the Logic Analyzer and display the waveforms you should see with this circuit:

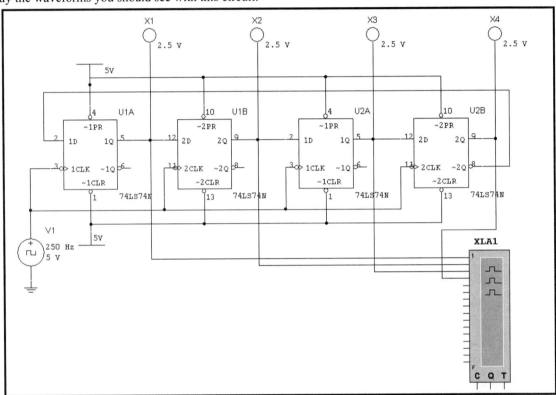

With the Logic Analyzer, the pattern produced by this counter is quite distinctive and easy to recognize when it is correct. The clock rate was set to 2.5 kHz for the Logic Analyzer waveforms below:

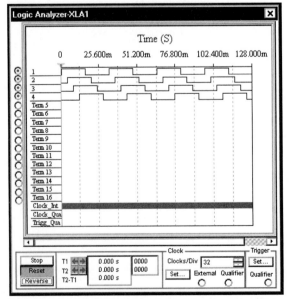

EXERCISE 7-2: Design and simulate a 4-bit decade counter.

SOLUTION: Wire the circuit as shown. A 250 Hz clock is used.

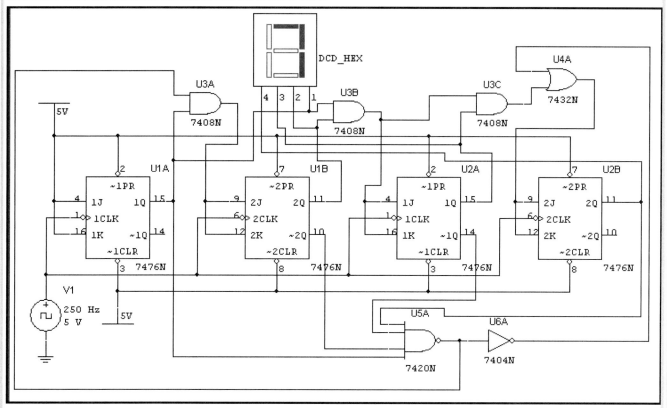

When you click on the **Run / stop simulation** button, the alphanumeric display will cycle through the digits 0 through 9. Note that this realization is not a minimal design and there are several different ways of realizing a counter. Your design may be different.

EXERCISE 7-3: Construct and simulate a circuit that uses a 555 astable multivibrator as the clock and a decade counter to drive a decoder.

SOLUTION: Wire the circuit as shown.

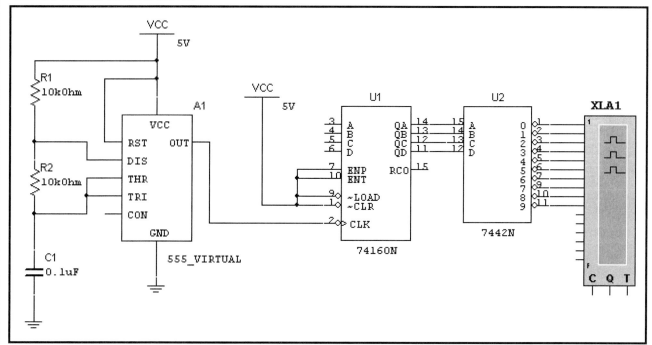

Because of the large number of signals, we will display the results with the Logic Analyzer.

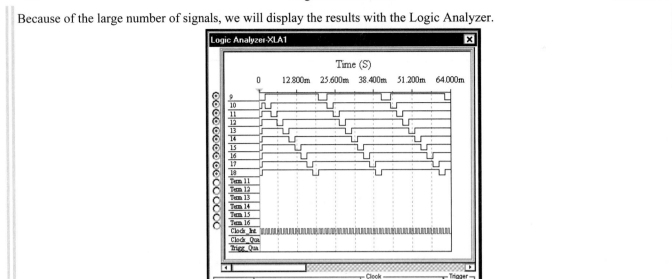

7.D. Startup Clear Circuit

The initial state of a flip-flop in a Multisim simulation is not always known. We will now present a technique that can be used to initialize flip-flops in both simulations and in circuits that you build. This technique is relevant because real flip-flops do not automatically set themselves to a specific state at startup. The circuit designer must add a circuit that presets or clears all flip-flops so that the circuit starts in a known state. The circuit below can be used to preset or clear a flip-flop at power-up as long as the preset or clear input is active low. Here we show how to clear two flip-flops at the start of the simulation.

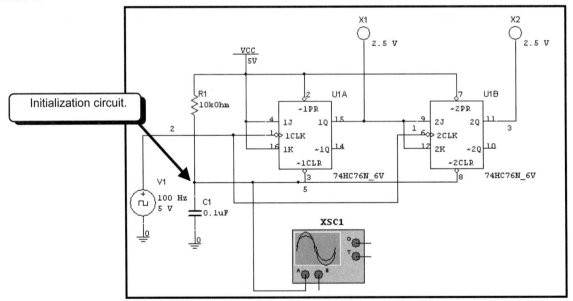

Resistor **R1** and capacitor **C1** form the initialization circuit. The initial condition of the capacitor is specified as 0 volts. When the simulation runs, the capacitor starts at 0 volts (logic zero) and charges to 5 volts. While the capacitor is close to 0 volts, it provides a logic zero to the clear input of the flip-flop. For the remainder of the simulation the capacitor is charged to 5 volts, providing a logic 1 to the flip-flop clear input, which has no effect.

To specify an initial condition for a virtual simulation, double-click on the wire as shown:

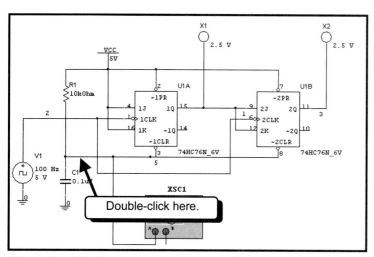
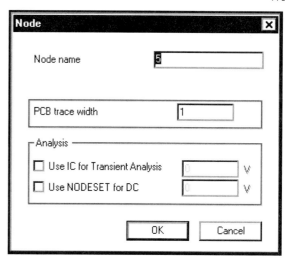

Change the **Node name** to `Clear`, enable the option **Use IC for Transient Analysis**, and make sure that the initial condition is zero:

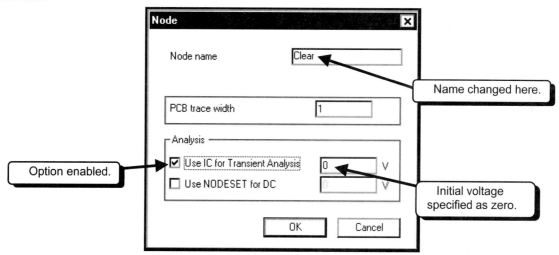

Click on the **OK** button to accept the changes. When you run the simulation, no matter what the initial states of the flip-flops are, they will be cleared. For this circuit, the flip-flops are initially cleared, and then they start counting since they are wired as a two-bit counter:

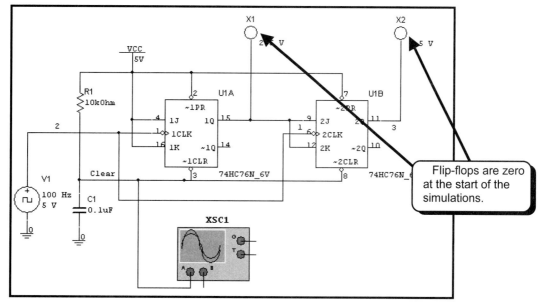

If you start and stop the simulation several times, you will see that the flip-flops always start at zero. If you disable the option **Use IC for Transient Analysis** and start and stop the simulation several times, you will notice that the Probes initially start up in some random pattern.

Next, we will look at the capacitor voltage with the oscilloscope with the **Use IC for Transient Analysis** option enabled. When we first start the simulation, the capacitor voltage looks as shown below:

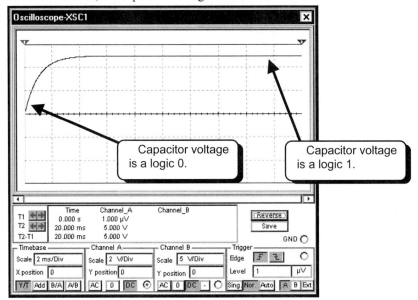

We see that at the very beginning of the simulation, the capacitor voltage is close to zero volts. It is at a low enough voltage for a long enough period of time that the flip-flops respond to the low voltage on the clear input and clear the flip-flops. For the remainder of the simulation, the capacitor voltage is a logic high, and this has no effect on the flip-flops because the clear inputs are active low.

The trick to this circuit is that the capacitor and resistor have to be large enough that the capacitor takes a long enough time to charge up. The larger **R1** and **C1** are, the longer the capacitor stays at a low voltage, and the longer the flip-flops have to react to the low voltage on their clear inputs. If **R1** and **C1** are too small, the RC transient will be too fast and not all of the flip-flops will be fast enough to react to the low input. A better procedure is to use this RC startup circuit to trigger a one-shot. The one-shot then emits a pulse of fixed and known duration that clears or presets all flip-flops.

This startup clear circuit only works at power-up because before power is turned on, the capacitor discharges to zero. When the circuit is running, there is no way to discharge the capacitor unless you add more circuitry. Thus, with the shown implementation, the only time this circuit has an effect is at power-up or at the start of the simulation.

EXERCISE 7-4: Design and simulate a 4-bit ring counter. The counter should be initialized to 1000.

SOLUTION: Wire the circuit as shown. A 250 Hz clock is used. Use the capacitor startup circuit to preset the first flip-flop to 1 and the remaining flip-flops to 0. Note that the initial condition of the capacitor is set to zero.

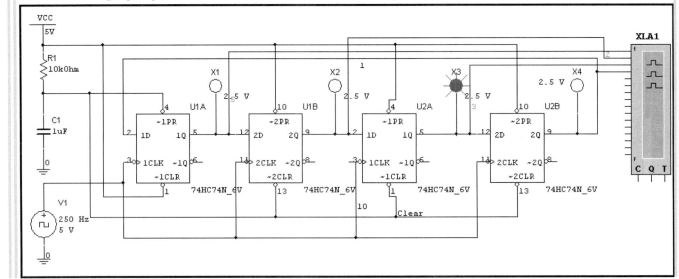

When you run the simulation, you will see a single illuminated Probe circulate around the counter from left to right. We can also view the states of the flip-flops with the Logic Analyzer:

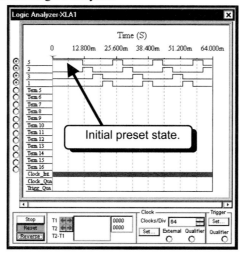

We see that only one flip-flop at a time has a logic 1 output, and this logic 1 is passed from flip-flop to flip-flop. We also see that at the start of the simulation, the initial preset state lasts longer than any of the other states.

7.E. Real and Ideal Mode Digital Simulations and Gate Delays

Multisim includes gate delays in its digital simulations. To illustrate gate delays, we will create a two-phase, non-overlapping clock using gate delays. We will show this simulation using Multisim's ideal and real digital simulation modes of operation. Both modes include gate delays. In ideal mode, gate outputs produce square edges. Only the delay time between input and output is modeled. In real mode, gate delays as well as rise and fall times are modeled, and gate outputs are not square. Real mode requires more simulation time to model the gate properties, and thus simulations take longer to complete.

7.E.1. Ideal Mode Digital Simulations

The default mode for digital simulations is the ideal mode. If we proceed as we did in previous simulations, we will be using ideal mode, so we do not need to change anything. In ideal mode you can use either ground for your drawings (⏚ or ⏚). Wire the two-phase clock circuit below:

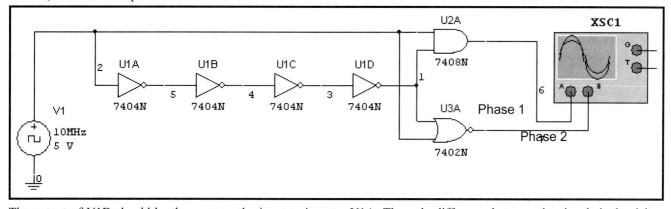

The output of U1D should be the same as the input to inverter U1A. The only difference between the signals is the delay time through the four inverters. Because gate delays are on the order of nanoseconds, we will set the clock frequency to 10 MHz. We will first observe the output of the circuit, Phase 1, and Phase 2. Two oscilloscope windows displaying the same waveform at different time scales are shown:

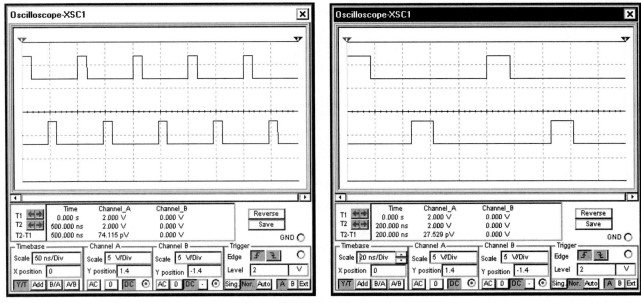

We see that the circuit produces two out-of-phase clock pulses from a single 10 MHz clock input. The circuit functions because of the gate delays of the four inverters. We can measure the delays caused by the inverters with the oscilloscope connected as shown:

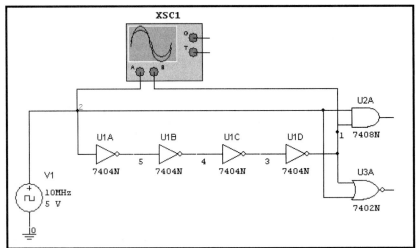

We can use the oscilloscope cursors to measure the delay through the four gates. See section 6.A.4 for using the oscilloscope cursors. Two screen captures with different time scales are shown:

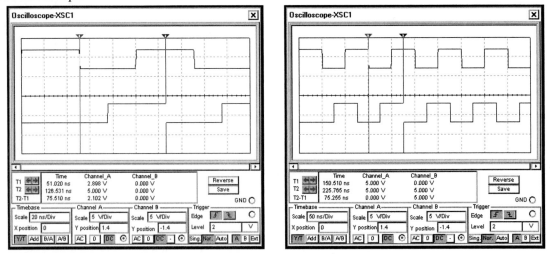

The delay for four inverters is about 75 ns, so the delay per inverter gate is approximately 19 ns.

7.E.2. Real Mode Digital Simulations

The one thing we notice about all of the previous simulations is that the edges of the digital waveforms are square. The logic and delay times of the devices are modeled correctly, but in real devices the edges of output waveforms are not square. We will simulate the same circuit as in the previous section, but we must make a few modifications so that it will work in real mode. Before we change the circuit, we will look at the analog ground symbol. Double-click the *LEFT* mouse button on the ground wire, and you will see that using the ⏚ part names the ground node zero (0).

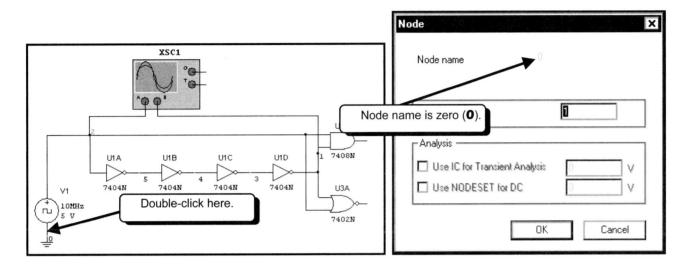

The node name is shown in green and a little hard to see in the screen capture. A ground name of 0 is the standard for SPICE simulations. For real mode simulations, we have to use the digital ground. Modify the circuit as shown below:

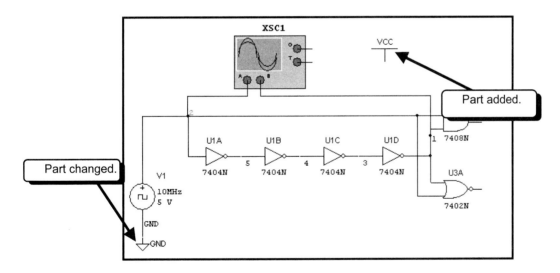

We have added a Vcc part to the diagram. In analog circuits, this part was used to connect other circuit elements to a power supply of the specified voltage. Here, it is not connected to anything, but it does set a node called **VCC** to **5** volts. All digital circuits are connected to this node even though we do not specifically wire the gates to the node. Thus, we only need to place the part in our circuit. Because the power pin of logic gates is connected to node VCC, this part is connected to the gates even though we do not draw a wire.

We have also replaced the analog ground by the digital ground symbol (⏛). If we now click on the ground wire, it will have a different label:

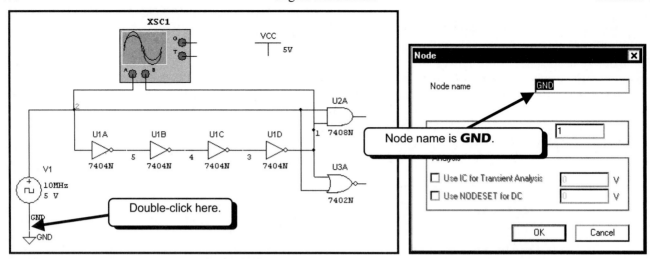

We see that the ground node is now named **GND** rather than 0. In real mode, the default power and ground nodes are **VCC** and **GND**. It is possible to change the power and ground names, but we will not do this here.

Next, double-click on a digital part such as an inverter:

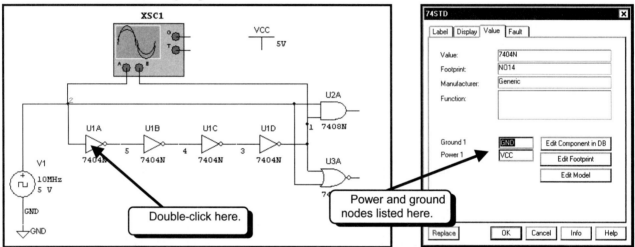

We see that the power and ground nodes are listed as **VCC** and **GND**. If you want to change them, you can use this dialog box. We do not need to change them, so click on the **Cancel** button.

Before we run the simulation, we need to switch to real mode. Select **Simulate** and then **Digital Simulation Settings** from the menus:

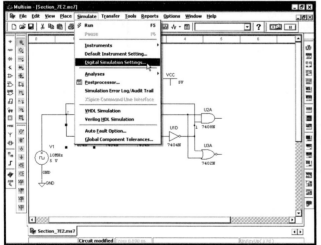

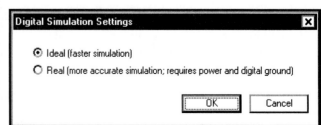

This dialog box allows us to select between real and ideal modes. We see that the default mode for digital simulations is ideal mode. To switch to real mode simulations we need to select the option here. Select the **Real** option and then click on the **OK** button:

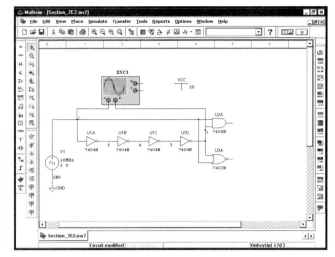

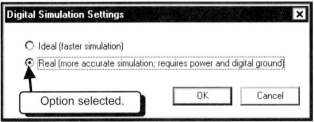

We are now ready to run the simulation. We will first look at the delay through the four inverters:

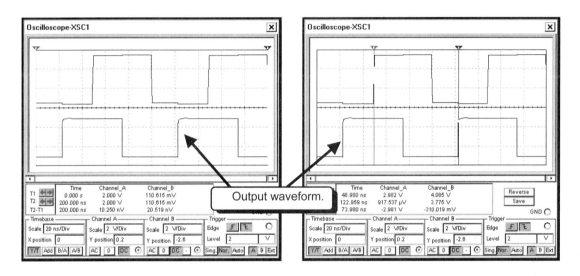

We see that the output of the digital gate has a slight rise time, a small amount of overshoot, and the rising edge corners are now rounded. The total delay through the four inverters is measured as approximately 74 ns, nearly the same as in the ideal mode simulations. The fall time of the inverter output is very fast and does not have an RC decay or undershoot.

Next, we will look at the output of the circuit:

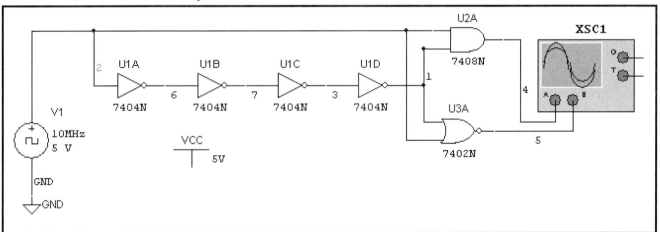

The waveforms are:

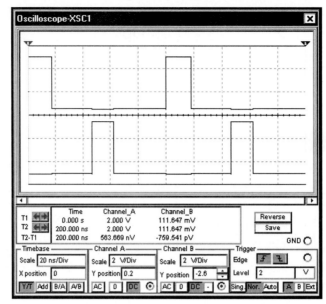

We see that the circuit still develops a two-phase clock and the output looks nearly identical to the output in ideal mode. In larger circuits with additional cascaded gate delays, you should expect the real-mode simulations to have more of an effect.

7.F. Problems

Problem 7-1: Create a 1 kHz square wave with a 50% duty cycle using a 555 timer and a J-K flip-flop.

Problem 7-2: Create a 10 kHz square wave with a 50% duty cycle using a 555 timer and a J-K flip-flop.

Problem 7-3: Using a 74xx160 counter, create a counter that counts from 0 to 5. Use the alphanumeric display to display and verify the operation of your circuit. Use the startup clear circuit covered in section 7.D to initially clear the counter.

Problem 7-4: Using a 74xx160 counter, create a counter that counts from 3 to 9. Use four Probes to display and verify the operation of your circuit.

Problem 7-5: Create a 10-bit ring counter and display the outputs on a bar graph. Use the startup clear circuit covered in section 7.D to initialize the counter.

Problem 7-6: Create a binary counter that counts from 0 to 255. Connect the 8 outputs of this counter to a DAC and add the necessary extra components so that the analog output is a voltage from 0 to 5 volts. Show that if you let the counter continuously cycle from 0 to 255, the analog output will be a ramp from 0 to 5 volts at a frequency that is equal to the clock frequency divided by 256.

Problem 7-7: Create a binary up/down counter that counts from 0 to 255 and then down to 0. Connect the eight outputs of this counter to a DAC and add the necessary extra components so that the analog output is a voltage from 0 to 5 volts. Show that if you let the counter continuously cycle, the analog output will be a triangle wave from 0 to 5 volts at a frequency that is equal to the clock frequency divided by 510.

INDEX

#

η 150
$\mu A/V^2$ 207
μV 13

@

@qq1[ic] 212
@rr1[i] 195

0

0 - ground 31

1

10:1 probes 318
1N4001GP 197
1N4734 379

2

27°C 155
2N3904 .... 162, 213

3

–3 dB frequency 279, 286
3D Components bar 38
3D parts 38
3-terminal plug 318

5

5% resistor 19
555 timer 471

6

60 Hz ripple 358

7

7414 464
74160 468
74HC21N 460

7815 linear voltage regulator
.... 355
7-segment display 443

9

90° Clockwise 22

A

A and B terminals 318
A button 136
absorbed power 140
AC Analysis 240
   Bode plots 257
   Decade Sweep type 277
   Decades 258
   Decibel 259, 277
   Linear 259
   Linear Sweep Type .. 242, 258
   Logarithmic 259
   Number of Points 242
   Octave 259
   Octave sweep type 258
   Output variables tab 242
   Plot during simulation button
   243
   points per decade 258, 277
   Simulate button 243
   Start frequency 242
   Stop frequency 242
   Sweep Type 242
   Vertical scale 242, 277
AC button 137
AC coupling 323, 360
AC Mode 236
AC portion of signal 358
AC power 287
AC Sweep 240
   Vertical scale 258
AC voltage source 8
Add device/model parameter
   button 166, 187
adding a trace 52
admittance 290
amplifier gain 264, 276

amplifier voltage swing 392
amplitude 315, 393
analog oscilloscope 316
Analysis Results 179
Analysis to sweep 284
AND gate 459
angle of load 288
Apply button 98
asterisk 152
audit trail 162
auto triggering 329
auto-hide 2
Autorange 102
axis label 95

B

bandwidth 281
bar graph 454
base current 165
Basic button 15
bins 6
bipolar junction transistor 161
BJT
   characteristic curves 211
   DC current gain 218
   emitter current 218
   HFE 218
   I-V characteristic 211
   operating point 161
   PNP 224
   VCE 225
BJT inverter 420
Bode plots 252, 257
   examples 260, 263
   phase 252
Bode Plotter
   Grapher 255
   instrument 252
   Magnitude button 254
   Vertical axis 267
   x-axis range 254
   y-axis range 254
bottom axis 95
Bottom Axis tab 97, 222
branch current 180

buzzer 459, 461

C

capacitor 15
   initial condition................ 406
change name 14
Change Part Model button.. 152, 208
Channel A minus Channel B 344
characteristic curves
   BJT 211
   jFET 213, 217
checkmark............................ 145
Circuit tab 145
clipping circuit.............. 385, 391
clock
   pulses 476
   two-phase......................... 475
Clock Rate 451
Clock Source 452
Clocks/Div............................ 452
clockwise.............................. 22
close toolbar 25
closed-loop gain 281
CMOS inverter 210
collector current... 165, 168, 212
Collector Current 428
collector current rating 427
Collector-Base Voltage 428
collector-emitter voltage...... 212
color code 39
common-base amplifier 268
Component Bin...................... 5
Component Name List......... 270
connecting components 26
connection 28, 31
correcting mistakes................ 25
correcting wiring mistakes 32
counter 464, 468
coupling 323
creating waveforms 380
crosshairs 26, 27, 30
CRT screen 326
CTRL-ESC 2
CTRL-G........................ 80, 119
CTRL-R 22
CTRL-S 33
CTRL-W 23
current................................. 340
current convention 125
   Multimeter 134
current direction 125
current indicator 123, 124, 159, 234

current direction 125
   ideal 131
   Mode................................ 125
   series resistance 126
current measurement 135
current monitor... 180, 183, 339, 430
current through resistor 186
current-controlled current
   source 149
current-controlled voltage
   source 180, 184, 185, 240, 339
current-sensing resistor 340
cursor
   Go to next Y_MAX........ 108
   Go to next Y_MIN 108
   search commands 107
   Set X_Value 108
   Set Y_Value 108
cursor information 106, 334
cursor search commands 107
cursors 103, 278, 350
   moving..................... 106, 331
   oscilloscope 330
Cursors On 105
cut button........................ 61, 66

D

damped sine wave 411
Damping Factor................... 411
dashed line..................... 26, 27
data file 380
DC coupling 323
DC current gain 218
DC Mode............................. 236
DC node voltages 143
DC Operating Point..... 161, 166
DC Operating Point Analysis
   .... 123, 143, 147
DC power supply. 352, 356, 375
DC supply ripple 371
DC Sweep............................ 176
   nested 211
DC transfer curves............... 204
DCD_BARGRAPH............ 462
DCD_HEX 443
dead-zone clipping circuit .. 385, 391
decade counter............. 468, 471
Decade Sweep type 277
Decades 258
Decibel 259, 277
decoded 7-segment display . 443

decoder................................471
default instrument settings ...154
default setting of indicator ...236
Delay Time...387, 406, 411, 415
delays 476
delete a page..........................61
Delete Diagram button 85
delete graph 83
delete Grapher page 86
Delete Page button 86
Delete Trace button................82
dependent sources 147
depletion MOSFET 207
Description field...........187, 283
device model parameters......167
Device Parameter 281
Device Type 187
digital ground ...28, 31, 470, 477
digital Probe 454
digital simulations 439, 469
digital waveforms.................477
diode
   current 150
   emission coefficient 150
   Is......................................151
   I-V characteristic..............197
   saturation current......150, 151
   toolbar 197
   virtual 150
   Zener 379
Diode button........................197
Diode family 197
diode reverse breakdown
   voltage.............................429
Diode_Virtual 150
Diodes Incorporated............353
display node names 32
display title block...................34
distortion 395
distortion-free amplifier 396
divide-by-two counter..........464
divisions 102
dot at connection 28
drain current........................171
Draw button 56, 60
drawing a circuit......................1
drawing a wire......................26

E

edit
   clear menu selection..........81
   Model button............151, 208
   properties....................10, 12
emission coefficient 150

emitter current 218
Emitter-Base Voltage 428
End time 366
enhancement MOSFET 207
ESC key 26
Excel 381
exponentially damped 411
extension
    gra 119
external trigger 318

F

F5 key 158
F8 key 10
F9 key 10
Fall Time 387, 415
File
    Recent Files 77
    Save 33
Filter Unselected Variables . 171
flip 442
flip component 22
Flip Horizontal 442
flip-flop 466
Fourier Analysis 396, 399
    Edit transient analysis 396
    Fundamental Frequency .. 396
frequency 14, 411
    components 398
    fundamental 396, 398
    response 252

G

GΩ 130
gain 264
gain bandwidth product 279
gate delays 476
GBPC2502 353
GBPC2502A 353
GBW 279
general ground symbol 28, 31
General Semiconductor 353
General tab 89, 96, 105
Generate time steps
    automatically 366
Go to next Y_MAX 108
Go to next Y_MIN 108
gra extension 119
Grapher 47, 56, 57, 87
    Apply button 98
    Autorange 102
    axis label 95

bottom axis 95
Bottom Axis tab 97
cursor information 106
cursors 103
Cursors On 105
cut button 61, 66
delete page 61
Divisions 102
Draw button 60
Edit Clear 81
File and Save 118
General tab 89, 96, 105
Graph Properties 105
Grid 90
Label 94, 99
Label text field 92
left axis 95
Left Axis tab 96
Legend 88, 90
move cursors 106
numerical values 103
opening 80
page 50
Pen Size 91, 94, 99
Properties 89, 105
Range 102
Restore Graph 112
Right Axis 100
Right Axis tab 101
save and load graphs 118
second y-axis 98
tabs 60
Title 96
Toggle Cursors 104
trace color 90
Trace field 93
trace labels 90
trace width 90
Traces tab 91
two y-axes 98
x-axis label 95, 97
x-axis scale 112
y-axis label 95, 97
y-axis scale 112
y-axis settings 96
zoom in 113
zoom out 112
zoom rectangle 111
zooming 110, 394
Grid 90
ground clips 318
ground lug 318
ground name 477

ground symbols 28
ground terminal (G) 318
grounding 28
Group all traces on one plot 284, 418

H

half-wave rectifier 362, 375
harmonic distortion 398
harmonics 398, 399
hexadecimal 441
HFE 218, 224
hover 8
hybrid-π 165
hysteresis curve 413

I

ideal mode 475
ideal mode digital simulation
    470
ideal operational amplifier ... 404
ideal parts 151
impedance measurement 292, 294
    active circuit 297
    passive circuit 294
indicator 123, 454
    AC Mode 236
    DC Mode 236
    default setting 236
    upper frequency limitation
        234
Indicators button 235
Indicators toolbar 123
inductive circuit 339
inductor current 340
inductor maximum current ... 428
inductors 15
initial condition 472
    capacitor 406
Initial Value 387, 406
initialization circuit 472
input terminals of oscilloscope
    317
Instruments toolbar 237, 288
integrator 407
Internal 452
International Rectifier 353
inverter 207, 420
I-V characteristic 197, 199
    BJT 211
    jFET 213, 217

J

jFET 297
    characteristic curves 213, 217
    I-V characteristic 213, 217
J-K flip-flop 464

K

kΩ 17
knurled knob 318
kOhm 17
Kp 207
kV 13

L

Label 94, 99
Label tab 14, 17, 21
Label text field 92
labeling nodes 32
lagging 337, 342
lambda 207
leading 337, 342
LED
    display 443
    I-V characteristic 200, 204
left axis 95
Left Axis tab 96
Legend 88, 90
Level 327
limits 427
linear
    load 287
    model 276
    network 252
    sweep 418
    voltage regulator 355, 421
Linear 222, 259
Linear Sweep Type 242, 258
linearized model 264
List 424
list of temperatures 424
List Sweep Variation Type .. 283
LM124J 269, 281
load graphs 118
loading Postprocessor pages .. 76
Logarithmic 222, 259
logarithmic x-axis 222
Logic Analyzer 440, 451, 470
    Clock Rate 451
    Clock Source 452
    Clocks/Div 452
    Internal 452
logic signal 442

logic simulations 469
lower –3 dB frequency 268
LVL_BARGRAPH 462

M

MΩ 132
magnitude 315, 340
magnitude and phase ... 233, 287
Magnitude button in Bode
    Plotter 254
magnitude versus frequency 252
maximize icon 56
maximum time between
    simulation points 369
Maximum time step 366, 393
Maximum time step settings 366
MDA2502 353
MDA25021 .... 355, 356
mean 69
measure
    current 135
    impedance 292, 298
        active circuit 297
        passive circuit 294
    phase 335, 337
    resistance 135
    voltage 135
measured data 380
measurement error 131
midband gain 279
Miller integrator 407
Minimum number of time points
    366
minus – oscilloscope 344
Miscellaneous Options tab .. 154
mixed analog/digital
    simulations 439, 464
Mode 125, 127
model
    MOSFET 207
    parameter 281
Modify Title Block 36
monostable one-shot 468
More>> button 162, 166, 186
MOS_3TDN_VIRTUAL 207
MOS_3TEN_VIRTUAL 207
MOSFET
    depletion 207
    Kp 207
    lambda 207
    load 207
    model 207, 210
    threshold 207
    VTO 207

W/L 210
move cursors 106
Multimeter 123, 133, 234
    AC measurements 237
    button 134
    current convention 134
    resistance measurement 292
    Set button 135
    settings 136
Multisim 2001 3
mV 13
MV 13

N

nΩ 132, 159
negative feedback 269
nested DC sweep 211
New Chart button 245
New Graph button 51, 62, 64,
    67
new Grapher page 51
New Page button 51, 66
NMOS inverter 207
nodal analysis 123
    bias point 161
    dependent sources 147
node
    0 31
    display names 32
    naming 32
node numbering 32
node numbers 143
node voltage
    analysis 123
Normal triggering 327, 329
Norton Equivalent 156
NPN bipolar junction transistor
    162
NPN pass transistor 422
NPN transistors 265
N-type depletion MOSFET .. 207
N-type enhancement MOSFET
    207
Number of Points 242
numerical values 103

O

Octave 259
Octave sweep type 258
ohmmeter 292
one-shot 468
OPAMP 269, 281, 404, 410
    frequency response ... 281, 286

quad package 270
    upper frequency limit 279
open circuit voltage 157
operating point..................... 161
Operating Point Analysis..... 143
Option Value 155
Options
    Modify Title Block 36
    Preferences 4, 32, 34, 144
oscilloscope
    0 323
    10:1 probes 318
    A and B terminals 318
    AC coupling.............. 323, 360
    auto triggering 329
    coupling 323, 358
    CRT screen 326
    cursor information 334
    cursors............................... 330
    cursors, moving 331
    DC coupling...................... 323
    external trigger................. 318
    ground clips 318
    ground input 323
    ground terminal (G).......... 318
    horizontal axis 316
    input terminals 317
    instrument 315, 317
    Level 327
    Normal triggering 327, 329
    probes 318
    Scale 320, 348
    single trigger 329
    subtraction 344
    time per division 316
    Timebase.......... 316, 319, 359
    triggering 326
    volts per division 316, 321
    window 319, 336
    Y position 322, 347
    Y/T button 320
output ripple 426
Output variables tab............... 48
output versus input 204

P

pΩ.. 131
parallel RLC 352
Parameter 187, 283
Parameter Sweep . 281, 415, 420
    AC 281
    Analysis to sweep 284
    Description 283

frequency response 281
gain bandwidth product... 281
Group all traces on one plot
    .... 284, 418
List Sweep Variation Type
    283
    Sweep Variation Type 283
part properties................. 10, 12
parts
    ground 28
    jFET................................... 297
pass transistor 422
passive components............. 15
peak amplitude 235, 241
peak current 428
Pen Size 91, 94, 99
Period 387
PF 288
phase 233, 340, 411
    measuring 335, 337
    plots 252
    versus frequency.............. 252
phasor 233, 258
piece-wise linear................. 378
Place
    part............................... 9, 20
    part with menus 22
    Place Component.............. 23
    resistor 15
    single component................ 5
Place Led1_Red.................... 39
Place Resistor1_1.0k 38
Plot during simulation .. 48, 146, 163
PMOS enhancement device. 210
PNP bipolar junction transistor
    224
points per decade 258, 277
Postprocessor..................... 1, 47
    add trace 52
    Delete Diagram button 85
    delete graph 83
    delete page 86
    Delete Page button 86
    Delete Trace button 82
    Draw button...................... 56
    New Chart button 245
    new graph 51
    New Graph button . 62, 64, 67
    new page........................... 51
    New Page button 62, 66
power 287
power absorbed 140

power factor 287
power factor correction 290
power limitation 429
power meter.................140, 142
power supply 356, 375
power-up 474
Preferences 6
Probe 318, 454
Properties 105, 199
pulse voltage source 399, 405
Pulse Width 387, 415
Pulse_Voltage_Source 386
Pulsed Value 387, 406
pulsed voltage source ...390, 404
PWL 378

Q

Q output 466
quad OPAMP package 270

R

Range 102
rated components 427, 430
Rated Virtual Components Bar
   427
real mode............................477
real mode digital simulation.470
rectangular coordinates 287
reference ID........................274
regulated power supply 356
rename a node 32
resistance measurement........135, 292, 294
resistor................................15
    3D.....................................38
    color code 39
    current-sensing 340
    name 17
    properties........................16
resistor power limitation 429
resonant frequency 344
Restore Graph 112
Reverse Breakdown Voltage 429
right axis..............................100
Right Axis tab 101
RIGHT mouse button 21, 22
ring counter 474
ripple 358, 371, 427
ripple rejection ratio 361
Rise Time 387, 415
RLC circuit...................344, 352
RMS 12

magnitude 241
value 235
rotate component 22
Rth 157
running time average 69

S

saturation current 150
Save as Default 6
Save button 33
save graphs 118
saving a schematic 33
saving Postprocessor pages ... 76
Scale, oscilloscope 348
Schmitt Trigger... 410, 414, 464, 465
second harmonic 398
second y-axis 98
sensing terminals 345
series resistance of current
   indicator 126
series RLC circuit 344
Set button 135
Set X_Value 108
Set Y_Value 108
SEVEN_SEG_COM_K 443
SEVEN_SEG_DISPLAY 443
sharp edges 466
SHIFT-CTRL-R 22
short circuit current 159
Show all device parameters . 163
Show node names 145
Show Rated Virtual
   Components Bar 427
Show reference ID option 273
Show Simulation Error
   Log/Audit Trail 164
Show values option 273
Simulate 146
   Analyses
      AC Analysis 242, 258
      DC Operating Point 145, 162, 166
      DC Sweep 176, 186, 198
      Fourier Analysis 396
      Parameter Sweep . 281, 417
      Temperature Sweep 423
      Transient Analysis .. 47, 78, 366, 368
   Default Instrument Settings
      154
   Digital Simulation Settings
      478
   Postprocess 50, 61, 66

Simulate button 49, 129, 130, 131, 136, 141
simulation points 383
single trigger 329
single-frequency AC 234
sinusoidal steady state 233
sinusoidal voltage source ... 392, 411
slew rate 411
small-signal gain 264
small-signal model 264
solid line 27
source 2 225
source follower 298
SPICE AC Analysis 240, 257
SPICE naming convention .. 177
SPICE results graphically 47
SPICE simulation 48
SPICE Transient Analysis ... 315
square-wave voltage source 385
standard 5% resistor 19
standard components 21
standard value 21
Start frequency 242
Start menu 1
startup clear 472
Stop frequency 242
stop simulation 148
straight lines 380
subcircuit 171
Sweep Parameter 281
Sweep Type 242
Sweep Variable Type 424
Sweep Variation Type 283
switch-tail counter 469
sync circuit 468

T

temp option 154
temperature 150, 154
   dependence 426
   linear regulator 421
Temperature Sweep 421, 423
   Sweep Variable Type 424
terohm 158
text file of data points 380
THD 398
thermal voltage 150
Thevenin equivalent 156
three dimensional 38
threshold 456
threshold voltage 207
time average 69
time domain analyses 315

time per division 316
Timebase 316, 319, 359
Title 96
title block 34
   changing 36
   display 34
   modifying 36
TMAX 366
Toggle Cursors 104, 278
Toggle Legend 88
total harmonic distortion 398
trace
   color 90
   label 90
   width 90
Trace field 93
Traces tab 91
transfer curve 204, 384, 385, 389, 390, 391, 413
   NMOS inverter 207
transformer 354
Transient Analysis 315, 366
   End time 366
   Generate time steps
      automatically 366
   Maximum time step .. 366, 393
   Maximum time step settings
      366
   Minimum number of time
      points 366
   TMAX 366
   TSTART 366
   TSTOP 366
transistor 161
   operating point 161
Transistors toolbar 161
triangle wave 381, 385, 390
trigger Level 327, 467
triggering an oscilloscope 326
TS_PQ4_24 354
TSTART 366
TSTOP 366
TTL compatible 464
two y-axes 98
two-phase clock 475

U

undecoded bar graph 454, 456
unity power factor 290
upper −3 dB frequency . 268, 279
Use IC for Transient Analysis
   473
Use this option 155

V

V button 136
Value tab 158
Vcc part 477
VCE 212, 225
Vertical axis in Bode Plotter 267
Vertical scale 242, 258, 277
VGND 478
VHDL 439
View 10
   Restore Graph 112
   Show Simulation Error
    Log/Audit Trail 164
virtual 151
Voltage Amplitude 411
voltage indicator .. 123, 147, 234
   impedance 148
   Mode 127
   parallel resistance 149
   resistance 128
voltage measurement 135
Voltage Offset 411
voltage regulator 358, 365
Voltage RMS 12
voltage swing 392
voltage-controlled current
   source 150
voltage-controlled voltage
   source 147, 345
VOLTMETER_V 235
volts per division 316, 321
Vth 157
VTO 207

W

Wattmeter 140, 287, 288
waveforms versus time 315
width-to-length ratio 210
Windows Start menu 1
   auto-hide 2
wire 27
   connection 28
   correcting 32
wiring
   components 26
Word Generator 440
   Burst mode 449
   Cycle mode 449
   Pattern button 444
   programming 444
   Step button 449
   Up Counter 444
Workspace tab 34

X

x-axis
   label 95, 97
   linear 222
   logarithmic 222
   range in Bode Plotter 254
   scale 112

Y

Y position 322
Y/T button 320
y-axis
   label 95, 97
   range in Bode Plotter 254
   scale 112
   settings 96

Z

Zener
   clipping circuit 204, 385,
   464, 465
   diode 379
   voltage reference 422
zero 31
zero crossing 349, 463
zoom
   box 394
   Grapher 110, 394
   in 10
   in with Grapher 113
   out 10
   out in Grapher 112
   rectangle 111